The 3G IP Multimedia Subsystem (IMS)

The 3G IP Multimedia Subsystem (IMS)

Merging the Internet and the Cellular Worlds

Second Edition

Gonzalo Camarillo
Ericsson, Finland

Miguel A. García-Martín
Nokia Research Center, Finland

John Wiley & Sons, Ltd

Other Wiley Editorial Offices

John Wiley & Sons Inc., 111 River Street, Hoboken, NJ 07030, USA

Jossey-Bass, 989 Market Street, San Francisco, CA 94103-1741, USA

Wiley-VCH Verlag GmbH, Boschstr. 12, D-69469 Weinheim, Germany

John Wiley & Sons Australia Ltd, 42 McDougall Street, Milton, Queensland 4064, Australia

John Wiley & Sons (Asia) Pte Ltd, 2 Clementi Loop #02-01, Jin Xing Distripark, Singapore 129809

John Wiley & Sons Canada Ltd, 22 Worcester Road, Etobicoke, Ontario, Canada M9W 1L1

Wiley also publishes its books in a variety of electronic formats. Some content that appears in print may
not be available in electronic books.

Library of Congress Cataloging-in-Publication Data

Camarillo, Gonzalo.
The 3G IP multimedia subsystem (IMS) : merging the Internet and the cellular worlds / Gonzalo
Camarillo, Miguel A. García-Martín.–2nd ed.
    p. cm.
Includes bibliographical references and index.

ISBN-13: 978-0-470-01818-7 (cloth : alk. paper)
ISBN-10: 0-470-01818-6 (cloth : alk. paper)

1. Wireless communication systems. 2. Mobile communication systems. 3. Multimedia
communications. I. García-Martín, Miguel A. II. Title.
TK5103.2.C35 2006
621.384–dc22   2005026863

British Library Cataloguing in Publication Data

A catalogue record for this book is available from the British Library

ISBN-13 978-0-470-01818-7 (HB)
ISBN-10 0-470-01818-6 (HB)

Typeset by Sunrise Setting Ltd, Torquay, Devon, UK.
Printed and bound in Great Britain by Antony Rowe Ltd, Chippenham, Wiltshire.
This book is printed on acid-free paper responsibly manufactured from sustainable forestry in which at
least two trees are planted for each one used for paper production.

To my parents, Anselmo and Isabel; my brothers, Alvaro, Daniel, and Ignacio; and Viviana. They all are a source of energy and motivation in everything I do.

Gonzalo

To my daughter Maria Elizabeth, who was born at the time I started writing this book, she is the sunshine of my life; my wife Jelena, who provided me with all the support and love I needed; my parents, José and Mari-Luz, my aunt Feli, my brother Javier José who, through the distance, encouraged and supported me during this project.

Miguel Angel

Contents

Foreword by Stephen Hayes

3GPP, or the 3rd Generation Partnership Project, was formed in late 1998 to specify the evolution of GSM into a 3rd generation cellular system. Although much focus was placed on new higher bandwidth radio access methods, it was realized that the network infrastructure must also evolve in order to provide the rich services capable of taking advantage of higher bandwidths. The original GSM network infrastructure was very much circuit and voice-centric. Although data capabilities were added over time the system retained much of its circuit-switched heritage and its inherent limitations. A new approach was needed.

IMS, or the IP Multimedia Subsystem, represented that new approach. The development of IMS was very much a collaborative effort between the leading cellular standards organization (3GPP) and the leading Internet standards organization (IETF). IETF provided the base technology and protocol specifications, while 3GPP developed the architectural framework and protocol integration required to provide the capabilities expected of a world-class mobile system, such as inter-operator roaming, differentiated QoS, and robust charging.

Since the initial specification of IMS, IMS has been adopted by 3GPP2 (the other major cellular standards organization) and is the leading contender as the base of the ITU work on Next Generation Networks. In the upcoming decades an understanding of IMS will be as important a fundamental for the well-rounded telecom engineer as ISUP knowledge was in previous decades.

IMS is a system. It is designed to provide robust multimedia services across roaming boundaries and over diverse access technologies. To understand IMS, you must understand both the underlying protocols and how IMS uses them within its architectural framework. This book facilitates that understanding by explaining first the underlying protocols, such as SIP, and then explaining how IMS makes use of those protocols. This approach allows the user to easily grasp the complex relationship between the protocols and entities as developed in the IETF and their usage and extensions as defined in IMS.

The two authors are uniquely qualified to explain not just the inner workings of IMS but also the rationale and tradeoffs behind the various design choices. Miguel Angel García-Martín was and still is a key contributor within 3GPP. He was one of the principal designers of IMS and authored the initial protocol requirements draft as well as other 3GPP-specific SIP drafts and RFCs. Gonzalo Camarillo was similarly a key contributor within IETF where he is currently a SIPPING WG co-chair. He has written many RFCs that are key components of IMS. Both authors have been involved with IMS since its inception and do a good job of explaining not only what IMS is but also how it came to be.

Stephen Hayes
Chair – 3GPP Core Network

Foreword by Allison Mankin and Jon Peterson

The Session Initiation Protocol (SIP) is one of the most active initiatives underway in the Internet Engineering Task Force (IETF) today. While the IETF has standardized a number of Internet applications that have turned out to be quite successful (notably, email and the web), few efforts in the IETF have been as ambitious as SIP. Unlike previous attempts to bring telephony over the Internet, which relied extensively on the existing protocols and operational models of the Public Switched Telephone Network (PSTN), SIP elected to use the best parts of email and web technology as its building blocks, and to construct a framework for establishing real-time communication – be it voice, video, instant messaging, or what have you – that is truly native to the Internet.

SIP is a rendezvous protocol – a protocol that allows endpoints on the Internet to discover one another and negotiate the characteristics of a session they would like to share. It converges on the best way for users to communicate, given their preferences, and the capabilities of devices they have at their disposal. Even though it establishes sessions over numerous communications media, it allows policies and services to be provided at the rendezvous level, which greatly simplifies the way end-users and operators manage their needs.

This approach has garnered the attention of almost all of the major vendors and service providers interested in telephony today. But the adoption of SIP by 3GPP has been a special, definitive success for SIP in the global marketplace. 3GPP promises to place SIP firmly in the hands of millions of consumers worldwide, ushering in a whole new paradigm of Internet-based mobile multimedia communications. The IP Multimedia Subsystem (IMS) of 3GPP is the core of this strategy, and it is a SIP-based core.

The IETF has created and continues to develop SIP, and the other protocols for real-time communication and infrastructure: RTP, SDP, DNS, Diameter... As 3GPP builds its successive IMS releases, towards a SIP-based multimedia Internet, IETF and 3GPP have grown into a close, working partnership, initiated by our liaison (RFC3113). Both committed to the Internet style afforded by SIP, two worlds with very different perspectives, the 3GPP world of mobile wireless telephony, and the IETF world of the packet Internet, have learned each other's considerations. There remain some differences, in the security models, in some aspects of network control. It's a tribute to the communications, the design work, and not least, to work by the authors of the present volume, that such differences have nonetheless resulted in interoperable SIP, SIP with a coherent character.

Gonzalo Camarillo has been one of the protagonists in SIP's development. In addition to his work editing the core SIP specification (RFC3261) within the IETF, Gonzalo has chaired the SIPPING Working Group of the IETF (which studies new applications of SIP) and authored numerous documents related to interworking SIP with the traditional telephone network, ensuring that SIP is IPv6 compliant, and using SIP in a wireless context.

Miguel A. García-Martín is one of the principal designers of the IMS, and has also somehow found the time to be one of the main voices for 3GPP within the IETF SIP community. The application of SIP to the mobile handset domain gave rise to numerous new requirements for SIP functionality, many of which would not be obvious to designers unfamiliar with the intricacies of wireless roaming, bandwidth constraints, and so on. As such, Miguel provided some very valuable guidance to the IETF which ensured that SIP is well-tooled to one of its most promising applications.

This book is a milestone presenting the first in-depth coverage of the 3GPP SIP architecture. It is difficult to overestimate the importance of the 3GPP deployment, and this book will position readers to participate in the engineering of that network.

Allison Mankin
Jon Peterson
Directors of the Transport Area of the IETF

About the Authors

Gonzalo Camarillo

Gonzalo Camarillo leads the Advanced Signalling Research Laboratory of Ericsson in Helsinki, Finland. He is an active participant in the IETF, where he has authored and co-authored several specifications used in the IMS. In particular, he is a co-author of the main SIP specification, RFC 3261. In addition, he co-chairs the IETF SIPPING working group, which handles the requirements from 3GPP and 3GPP2 related to SIP, and the IETF HIP (Host Identity Protocol) working group, which deals with lower-layer mobility and security. He is the Ericsson representative in the SIP Forum and is a regular speaker at different industry conferences. During his stay as a visitor researcher at Columbia University in New York, USA, he published a book entitled "SIP Demystified". Gonzalo received an M.Sc. degree in Electrical Engineering from Universidad Politecnica de Madrid, Spain, and another M.Sc. degree (also in Electrical Engineering) from the Royal Institute of Technology in Stockholm, Sweden. He is currently continuing his studies as a Ph.D. candidate at Helsinki University of Technology, in Finland.

Miguel A. García-Martín

Miguel A. García-Martín is a Principal Research Engineer in the Networking Technologies Laboratory of the Nokia Research Center in Helsinki, Finland. Before joining Nokia Miguel was working for Ericsson in Spain, and then Ericsson in Finland. Miguel is an active participant of the IETF, and for a number of years has been a key contributor in 3GPP. Lately Miguel has also been participating in the specification of NGN in ETSI. In the IETF, he has authored and co-authored several specifications related to the IMS. In 3GPP, he has been a key contributor to the development of the IMS standard. Miguel is also a regular speaker at different industry conferences. Miguel received a B. Eng. degree in Telecommunications Engineering from Universidad de Valladolid, Spain.

Preface to the Second Edition

The pace at which new IMS-related technologies have been developed in the last year has been impressive. Based on the deployment experiences of their members and on feedback from several organizations, 3GPP and 3GPP2 have worked extensively to update the IMS architecture so that it supports a wide range of new services.

While many of these updates consist of extensions to provide more functionality, some of them consist of simplifications to the IMS architecture. These simplifications make the IMS architecture more robust and reliable, or increase the performance of services implemented on top of it.

Examples of organizations that provide feedback to 3GPP and 3GPP2 on how to evolve the IMS are the OMA (Open Mobile Alliance) and the standardization bodies involved in the developing of NGN (Next Generation Networks). These organizations use the IMS as a base to provide different types of services.

The second edition of this book, in addition to describing updates to the IMS architecture, includes extensive discussions on the NGN architecture and the services it provides, and on the OMA PoC (Push-to-talk over Cellular) service. We are confident that the reader will find the chapters on these IMS-based services useful.

From the feedback received on the first edition, it seems that many readers found the structure of the book novel and useful. Readers agreed that first describing how a technology works on the Internet before discussing how it applies to the IMS provides a wider perspective than studying the technology in the IMS context alone.

Of course, we have also updated the sections dealing with Internet technologies. These sections include some of the latest protocol extensions developed in the IETF.

Based on the feedback received during the IMS seminars we have given around the world, we have clarified those concepts which were difficult to understand in the first edition.

Finally, also new to the second edition is a companion website on which instructors and lecturers can find electronic versions of the figures. Please go to

http://www.wiley.com/go/camarillo

Preface to the First Edition

The IMS (IP Multimedia Subsystem) is the technology that will merge the Internet with the cellular world. It will make Internet technologies, such as the web, email, instant messaging, presence, and videoconferencing available nearly everywhere. We have written this book to help engineers, programmers, business managers, marketing representatives, and technically aware users understand how the IMS works and the business model behind it.

We have distributed the topics in this book into four parts: an introduction, the signaling plane in the IMS, the media plane in the IMS, and IMS service examples. All four parts follow a similar structure; they provide both Internet and IMS perspectives on each topic.

First, we describe how each technology works on the Internet. Then, we see how the same technology is adapted to work in the IMS. Following these two steps for each technology provides the reader with a wider perspective. So, this book is not a commented version of the IMS specifications. It covers a much broader field.

Reading this book will improve anyone's understanding of the Internet technologies used in the IMS. You will know how each technology is used on the Internet and which modifications are needed to make it work in the IMS. This way you will understand how the use of Internet technologies in the IMS will make it easy to take advantage of any current and future Internet service. Finally, you will appreciate how operators can reduce the operational cost of providing new services.

Engineers who are already familiar with the IMS or with any of the IMS-related Internet protocols will also benefit substantially from this book. This way, engineers from the IETF (Internet Engineering Task Force) will understand which special characteristics of the IMS makes it necessary to add or remove certain features from a few Internet protocols so that they can be used in the IMS. On the other hand, engineers from 3GPP (Third Generation Partnership Project) and 3GPP2 will gain a wider perspective on IMS technologies. In addition, any engineer who focuses on a specific technology will gain a better understanding of the system as a whole.

Readers who want to expand their knowledge of any particular topic will find multiple references to 3GPP and 3GPP2 specifications, ITU recommendations, and IETF RFCs and Internet-Drafts in the text. Moreover, Appendix B contains a list with all the 3GPP and 3GPP2 specifications that are relevant to the IMS.

Now, let us look at each part of this book. Part I provides an introduction to the IMS: its goals, its history, and its architecture. We highlight the gains the operators obtain from the IMS. Besides, we discuss what the user can expect from the IMS. In addition, we describe how existing services, such as GRPS, WAP, SMS, MMS, and video-telephony over circuits relate to the IMS.

Part II deals with the signaling plane of the IMS, which includes protocols, such as SIP (Session Initiation Protocol), SDP (Session Description Protocol), Diameter, IPsec, and

COPS (Common Open Policy Service). As we said earlier, we describe each protocol as it is used on the Internet and, then, as it is used in the IMS.

Part III describes the media plane of the IMS. We describe how to convert audio and video into a digital form and how to transport it using protocols, such as RTP (Real-Time Transport Protocol) and RTCP (RTP Control Protocol). Furthermore, we introduce Internet protocols such as DCCP (Datagram Congestion Control Protocol) and SRTP (Secure RTP) that are not currently used in the IMS, but might be in the future.

Finally, Part IV provides IMS service examples, such as presence, instant messaging, and Push-to-Talk. These examples illustrate how to build meaningful services using the technologies described in Parts II and III.

Essentially, this book is useful to a wide range of technical and business professionals because it provides a thorough overview of the IMS and its related technologies.

Acknowledgements

Without the encouragement we received from Stephen Hayes we would not have written this book. He was the first to see the need for a book on the IMS that provided the IETF perspective in addition to the 3GPP and 3GPP2 perspectives. In addition, he and Allison Mankin did an outstanding job coordinating the IMS standardization from 3GPP and from the IETF, respectively.

Once we decided, pushed by Stephen, to start writing this book our management in Ericsson Finland fully supported us in this endeavor. In particular, Stefan Von Schantz, Christian Engblom, Jussi Haapakangas, Rolf Svanback, and Markku Korpi understood from the beginning the importance of the IMS and of spreading knowledge about it.

Our technical reviewers helped us fix technical errors in early versions of the manuscript. Andrew Allen provided useful comments on the whole manuscript, and Harri Hakala, Arto Mahkonen, Miguel Angel Pallares, Janne Suotula, Vesa Torvinen, Magnus Westerlund, Brian Williams, and Oscar Novo provided suggestions on different parts of the book. Anna Reiter provided guidance on language and writing style.

Our editor at John Wiley & Sons, Ltd, Mark Hammond, believed in this book from day one and supported us at every moment.

Part I

Introduction to the IMS

Before we look at how the IMS works in Parts II and III of this book we need to provide some background information on the IMS. This part (Part I) of the book will answer questions, such as what is the IMS, why was it created, what does it provide, or which organizations are involved in its standardization? In addition, we will describe the IMS architecture and the design principles behind it.

Chapter 1

IMS Vision: Where Do We Want to Go?

Third Generation (3G) networks aim to merge two of the most successful paradigms in communications: cellular networks and the Internet. The IP (Internet Protocol) Multimedia Subsystem (IMS) is the key element in the 3G architecture that makes it possible to provide ubiquitous cellular access to all the services that the Internet provides. Picture yourself accessing your favorite web pages, reading your email, watching a movie, or taking part in a videoconference wherever you are by simply pulling a 3G hand-held device out of your pocket. This is the IMS vision.

1.1 The Internet

The Internet has experienced dramatic growth over the last few years. It has evolved from a small network linking a few research sites to a massive worldwide network. The main reason for this growth has been the ability to provide a number of extremely useful services that millions of users like. The best known examples are the World Wide Web and email, but there are many more, such as instant messaging, presence, VoIP (Voice Over IP), videoconferencing, and shared whiteboards.

The Internet is able to provide so many new services because it uses open protocols that are available on the web for any service developer. Moreover, the tools needed to create Internet services are taught at university and are described in tons of books.

Having a widespread knowledge of Internet protocols has an important implication: people that develop new services are the ones that are going to use them. Let's say that a user is interested in chess and would like to play chess over the Internet. This user will be able to program a chess application and make it work over the Internet using an existing transport protocol.

On the other hand, if the protocols were not open and there were a few individuals that had access to them, the person programming the chess application would be somebody with deep knowledge of the protocol but little of chess. It is not difficult to guess who would come up with the best chess program: the chess player that understands what to expect from a chess program or the protocol expert. In fact, this is what the Internet has achieved. The number of protocol experts is so high that there is always somebody within a given community

The 3G IP Multimedia Subsystem (IMS) Second Edition   Gonzalo Camarillo and Miguel A. García-Martín
© 2006 John Wiley & Sons, Ltd

(e.g., the chess community) that understands the requirement of the community and the protocols that need to be involved.

1.2   The Cellular World

At present, cellular telephone networks provide services to over one billion users worldwide. These services include, of course, telephone calls, but are not limited to them. Modern cellular networks provide messaging services ranging from simple text messages (e.g., SMS, Short Messaging Service) to fancy multimedia messages that include video, audio, and text (e.g., MMS, Multimedia Messaging Service). Cellular users are able to surf the Internet and read email using data connections, and some operators even offer location services which notify users when a friend or colleague is nearby.

Still, cellular networks did not become so attractive to users only for the services they offered. Their main strength is that users have coverage virtually *everywhere*. Within a country, users can use their terminals not only in cities, but also in the countryside. In addition, there exist international roaming agreements between operators that allow users to access cellular services when they are abroad.

Reduction in terminal size also helped the spread of cellular networks. Old brick-like terminals gave way to modern small terminals that work several days without having their batteries recharged. This allows people to carry their terminals everywhere with little difficulty.

1.3   Why do we need the IMS?

On the one hand, we have mentioned that the idea of the IMS is to offer Internet services everywhere and at any time using cellular technologies. On the other hand, we have also said that cellular networks already provide a wide range of services, which include some of the most successful Internet services like instant messaging. In fact, any cellular user can access the Internet using a data connection and in this way access any services the Internet may provide. So, what do we need the IMS for?

We need to further clarify what we mean by merging the Internet and the cellular worlds and what the real advantages of doing so are. To do that, we need to introduce the different domains in 3G networks, namely the circuit-switched domain and the packet-switched domain.

The circuit-switched domain is an evolution of the technology used in Second Generation (2G) networks. The circuits in this domain are optimized to transport voice and video, although they can also be used to transport instant messages.

Although circuit-switched technology has been in use since the birth of the telephone, the current trend is to substitute it with more efficient packet-switched technology. Cellular networks follow this trend and, as we said earlier, 3G networks have a packet-switched domain.

The packet-switched domain provides IP access to the Internet. While 2G terminals can act as a modem to transmit IP packets over a circuit, 3G terminals use native packet-switched technology to perform data communications. This way, data transmissions are much faster and the available bandwidth for Internet access increases dramatically. Users can surf the web, read email, download videos, and do virtually everything they can do over any other broadband Internet connection, such as ISDN (Integrated Services Digital Line) or DSL (Digital Subscriber Line). This means that any given user can install a VoIP client in their 3G terminal and establish VoIP calls over the packet-switched domain. Such a user can take advantage of all the services that service providers on the Internet offer, such as voice mail or conferencing services.

So, again the same question: why do we need the IMS, if all the power of the Internet is already available for 3G users through the packet-switched domain? The answer is threefold: QoS (Quality of Service), charging, and integration of different services.

The main issue with the packet-switched domain to provide real-time multimedia services is that it provides a best-effort service without QoS; that is, the network offers no guarantees about the amount of bandwidth a user gets for a particular connection or about the delay the packets experience. Consequently, the quality of a VoIP conversation can vary dramatically throughout its duration. At a certain point the voice of the person at the other end of the phone may sound perfectly clear and instants later it can become impossible to understand. Trying to maintain a conversation (or a videoconference) with poor QoS can soon become a nightmare.

So, one of the reasons for creating the IMS was to provide the QoS required for enjoying, rather than suffering, real-time multimedia sessions. The IMS takes care of synchronizing session establishment with QoS provision so that users have a predictable experience.

Another reason for creating the IMS was to be able to charge multimedia sessions appropriately. A user involved in a videoconference over the packet-switched domain usually transfers a large amount of information (which consists mainly of encoded audio and video). Depending on the 3G operator the transfer of such an amount of data may generate large expenses to the user, since operators typically charge based on the number of bytes transferred. The user's operator cannot follow a different business model to charge the user because the operator is not aware of the contents of those bytes: they could belong to a VoIP session, to an instant message, to a web page, or to an email.

On the other hand, if the operator is aware of the actual service that the user is using, the operator can provide an alternative charging scheme that may be more beneficial for the user. For instance, the operator might be able to charge a fixed amount for every instant message, regardless of its size. Additionally, the operator may charge for a multimedia session based on its duration, independently of the number of bytes transferred.

The IMS does not mandate any particular business model. Instead, it lets operators charge as they think more appropriate. The IMS provides information about the service being invoked by the user, and with this information the operator decides whether to use a flat rate for the service, apply traditional time-based charging, apply QoS-based, or perform any new type of charging. As a clarification, by service, in this charging context, we refer to any value offered to the user (e.g., a voice session, an audio/video session, a conference bridge, an instant message, or the provision of presence information about co-workers).

Providing integrated services to users is the third main reason for the existence of the IMS. Although large equipment vendors and operators will develop some multimedia services, operators do not want to restrict themselves to these services. Operators want to be able to use services developed by third parties, combine them, integrate them with services they already have, and provide the user with a completely new service. For example, an operator has a voicemail service able to store voice messages and a third party develops a text-to-speech conversion service. If the operator buys the text-to-speech service from the third party, it can provide voice versions of incoming text messages for blind users.

The IMS defines the standard interfaces to be used by service developers. This way, operators can take advantage of a powerful multi-vendor service creation industry, avoiding sticking to a single vendor to obtain new services.

Furthermore, the aim of the IMS is not only to provide new services but to provide all the services, current and future, that the Internet provides. In addition, users have to be able to execute all their services when roaming as well as from their home networks. To achieve these goals the IMS uses Internet technologies and Internet protocols. So, a multimedia session between two IMS users, between an IMS user and a user on the Internet, and between two users on the Internet is established using exactly the same protocol. Moreover, the interfaces for service developers we mentioned above are also based on Internet protocols. This is why the IMS truly merges the Internet with the cellular world; it uses cellular technologies to provide ubiquitous access and Internet technologies to provide appealing services.

1.4   Relation between IMS and non-IMS Services

We have just explained that the IMS is needed to provide Internet services (including real-time multimedia services) with an acceptable QoS at an acceptable price. Yet, many such services can be provided outside the IMS as well. Two users can establish a videoconference over the circuit-switched domain and send each other multimedia messages using MMS. At the same time they can surf the web and check email over the packet-switched domain (e.g., GPRS, General Packet Radio Service). They can even access a presence server on the Internet to check the availability of more people that may want to join the videoconference.

Given that all the services just described can be provided with an excellent QoS with no IMS at all, then what does the IMS really provide?

First of all, the IMS provides all the services using packet-switched technology, which is generally more efficient than circuit-switched technology. Nevertheless, the real strength of the IMS when compared with the situation above is that the IMS creates a service environment where any service can access any aspect of the session. This allows service providers to create far richer services than in an environment where all the services are independent of one another.

For example, a service could insert an announcement in a conference based on an event that happens on the Internet, like the change of the presence state of a colleague from busy to available. Another service could, for instance, display on the user's screen the web page of the person who is calling every time a call is received. Moreover, the same service could automatically set the user's presence status to busy and divert incoming calls to an email address instead of to the typical voicemail.

When services in the network can access all the aspects of a session, they can perform many operations (e.g., changing the presence status of the user) without sending any data over the air to the terminal. Spare radio capacity can be used to provide a higher QoS to existing users or to accommodate more users with the same QoS.

Another important advantage of the IMS is that it does not depend on the circuit-switched domain. This way, interworking with devices with no access to this domain, such as laptops connected to the Internet using any videoconferencing software, becomes trivial. This increments dramatically the number of people IMS users are able to communicate with using all types of media.

Chapter 2

The History of the IMS Standardization

In Chapter 1 we mentioned that the IMS (IP Multimedia Subsystem) uses Internet protocols. When the IMS needs a protocol to perform a particular task (e.g., to establish a multimedia session), the standardization bodies standardizing the IMS take the Internet protocol intended for that task and specify its use in the IMS. Still, no matter how simple this may sound the process of choosing protocols to be used in the IMS can sometimes get tricky. Sometimes, the Internet protocol that is chosen lacks some essential functionality, or does not even exist at all. When this happens the IMS standardization bodies contact the standardization body developing Internet protocols to work together on a solution. We will cover this collaboration in Section 2.5. Nevertheless, before jumping into that we will introduce in Section 2.1 all the standardization bodies involved in IMS development. We need to know who is who and which functions of the IMS each of them performs.

2.1 Relations between IMS-related Standardization Bodies

The ITU (International Telecommunication Union) IMT-2000 (International Mobile Telecommunications-2000) is the global standard for 3G networks. IMT-2000 is the result of the collaboration between different standards bodies and aims to provide access to telecommunication services using radio links, which include satellite and terrestrial networks.

We will focus on two of the standard bodies involved in IMT-2000: 3GPP (Third Generation Partnership Project) and 3GPP2 (Third Generation Partnership Project 2). Still, they are not the only ones working within IMT-2000. Other bodies, such as the ITU-R (ITU-Radiocommunication Sector), for instance, are also involved in IMT-2000 but on different areas than the IMS.

Both 3GPP and 3GPP2 have standardized their own IMS. The 3GPP IMS and the 3GPP2 IMS are fairly similar, but, nevertheless, have a few differences (Appendix A lists the most important differences).

An important similarity between the 3GPP IMS and the 3GPP2 IMS is that both use Internet protocols, which have been traditionally standardized by the IETF (Internet Engineering Task Force). Consequently, both 3GPP and 3GPP2 collaborate with the IETF in developing protocols that fulfill their requirements. The following sections introduce the IETF, 3GPP, and 3GPP2 and provide a brief history of the IETF-3GPP/3GPP2 collaboration.

The 3G IP Multimedia Subsystem (IMS) Second Edition   Gonzalo Camarillo and Miguel A. García-Martín
© 2006 John Wiley & Sons, Ltd

In addition to the standard bodies we have just mentioned, OMA (Open Mobile Alliance [169]) plays an important role in developing IMS services. While 3GPP and 3GPP2 have standardized (or are standardizing) a few IMS services, such as basic video calls or conferencing, OMA focuses on the standardization of service enablers on top of the IMS (of course, other standard bodies and third parties besides OMA may also develop services and service enablers for the IMS).

Lately, additional standardization bodies have come on the scene when IMS made its debut in the fixed broadband access arena. We are referring to Next Generation Networks (NGN) for which IMS forms a substantial part.

In 2004 the ITU-T created an NGN Focus Group (NGN-FG) that for a couple of years was studying and advancing the specification work of Next Generation Networks for fixed line accesses based on IMS. In Europe, in 2004, the European Telecommunication Standards Institute (ETSI) created the Telecoms and Internet converged Services and Protocols for Advanced Networks (TISPAN) technical committee, with the goal of standardizing a Next Generation Network for fixed network access based on IMS. In North America, the Alliance for Telecommunication Industry Solutions (ATIS), also in 2004, created the NGN Focus Group to study the applicability of NGN and IMS to North American fixed access networks. The three standardization bodies keep synchronized in the definition of NGN and the applicability of IMS to fixed access networks. Additionally, they also bring new requirements to 3GPP and 3GPP2 to support fixed broadband access to IMS.

2.2   Internet Engineering Task Force

The Internet Engineering Task Force (IETF) is a loosely self-organized collection of network designers, operators, vendors, and research institutions that work together to develop the architecture, protocols, and operation of the public Internet. The IETF is a body that is open to any interested individual. It is not a corporation and, therefore, does not have a board of directors, members, or dues.

The IETF is the standardization body that has developed most of the protocols that are currently used on the Internet. The IETF does not standardize networks, architectures combining different protocols, the internal behavior of nodes, or APIs (Application Programming Interfaces). The IETF is the protocol factory for IP-related protocols.

2.2.1   Structure of the IETF

Work in the IETF is organized in working groups. Each working group is chartered to perform specific tasks, such as the delivery of a precise set of documents. Each working group has from one to three chairs, who ensure that the working group completes its chartered tasks in time. Working groups have a temporary lifetime; so, once they have delivered their documents, either they are rechartered or they cease to exist. Figure 2.1 shows a few, but not all, of the working groups in the IETF; there are more than 100 active working groups in the IETF. The complete up-to-date list of active working groups is available at:

```
http://www.ietf.org/html.charters/wg-dir.html
```

Working groups get an acronym name that identifies the chartered task. For instance, SIPPING is the acronym of "Session Initiation Protocol Investigation", SIMPLE is the acronym of "SIP for Instant Messaging and Presence Leveraging Extensions", and AAA for "Authentication, Authorization and Accounting".

A collection of working groups form an Area Directorate. There are currently eight areas, as illustrated in Figure 2.1.

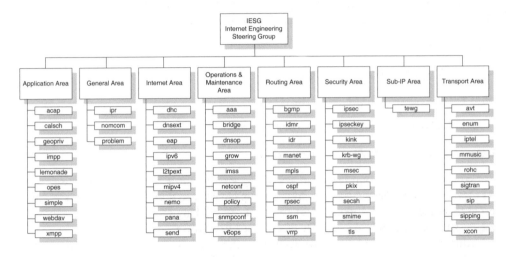

Figure 2.1: The structure of the IETF

Each area has one or two area directors who, together with the IETF chairman, form the IESG (Internet Engineering Steering Group). The IESG is the technical management team of the IETF. They decide on which areas the IETF should work on and review all the specifications that are produced.

The following web pages contain the complete list of the working groups of all areas and the charter of the SIPPING working group, respectively:

```
http://www.ietf.org/html.charters/wg-dir.html
http://www.ietf.org/html.charters/sipping-charter.html
```

The IAB (Internet Architecture Board) is the body that provides technical leadership and handles appeals. Its web page is:

```
http://www.iab.org/
```

2.2.2 Working Group Operations

The technical work in the IETF is done within the working groups. Working groups do not have any kind of membership; they are formed by a number of volunteers that work as individuals. That is, they do not represent their companies when working for the IETF.

Most of the technical discussions within a working group take place in its mailing list. Even the decisions made at face-to-face meetings (held three times a year) have to be confirmed in the mailing list.

The technical documents used within the working groups are called Internet-Drafts. There are two types of them: individual submissions and working group items. Individual submissions are technical proposals submitted by an individual or individuals. If the working group decides that an individual submission is a good starting point to work on a particular topic, it becomes a working group item.

Individual submissions and working group items can be distinguished by the name of the file where they are stored. Individual submissions start with:

```
draft-author's_name
```

while working group items start with:

```
draft-ietf-name_of_the_working_group
```

A list of all the Internet-Drafts can be found at:

```
http://www.ietf.org/internet-drafts/
```

When a working group feels that a working group item is ready for publication as an RFC (Request for Comments) the working group chairs send it to the IESG. The IESG may provide feedback to the working group (e.g., ask the working group to change something in the draft) and, eventually, decides whether or not a new RFC is to be published.

Although most of the Internet-Drafts that the IESG receives come from working groups, an individual may also submit an Internet-Draft to the IESG. This usually happens with topics that are not large enough to grant the creation of a working group, but which, nevertheless, are of interest to the Internet community.

It is important to note that Internet-Drafts, even if they are working group items, represent work in progress and should only be referenced as such. Internet-Drafts are temporary documents that expire and cease to exist six months after they are issued. They can change at any time without taking into consideration backward compatibility issues with existing implementations. Only when a particular Internet-Draft becomes an RFC can it be considered a stable specification.

2.2.3   Types of RFCs

The technical documents produced by the IETF are called RFCs. According to the contents of the document there are three main types of RFCs.

- Standards-track RFCs.

- Non-standards-track RFCs.

- BCP (Best Current Practise) RFCs.

Standards-track RFCs typically define protocols and extensions to protocols. According to the maturity of the protocol there are three levels of standards-track RFCs: proposed standard, draft standard, and Internet standard. Standards-track specifications are supposed to advance from proposed to draft and, finally, to Internet standard as they get more and more mature. An important requirement in this process is that a particular specification is implemented by several people to show that different independently built implementations that follow the same specification can successfully interoperate.

Nevertheless, in practice only a few RFCs reach the draft standard level and even fewer become Internet standards. At present, the specifications of many protocols that are used massively on the Internet are proposed standards.

There are three types of non-standards-track RFCs: experimental, informational, and historical (which are called *historic* RFCs). Experimental RFCs specify protocols with a very limited use, while informational RFCs provide information for the Internet community about some topic, such as a requirements document or a process. When a standards-track RFC becomes obsolete, it becomes a historic RFC.

BCP RFCs record the best current practice known to the community to perform a particular task. They may deal with protocol issues or with administrative issues.

Figure 2.2 shows the relations between all the RFC types. A list of all the RFCs published so far and their status can be fetched from:

```
http://www.ietf.org/iesg/1rfc_index.txt
```

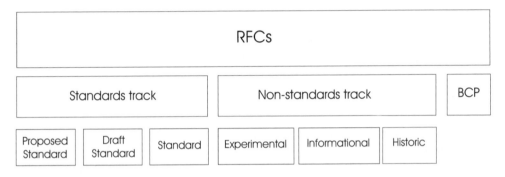

Figure 2.2: RFC types

RFCs can be downloaded from the following web page by just introducing the RFC number:

```
http://www.ietf.org/rfc.html
```

Additionally, the RFC Editor offers a web page that allows us to search for RFCs by title, number, author, and keywords:

```
http://www.rfc-editor.org/rfcsearch.html
```

2.3 Third Generation Partnership Project

The Third Generation Partnership Project (3GPP) was born in 1998 as a collaboration agreement between a number of regional telecommunication standard bodies, known as *organizational partners*. The current 3GPP organizational partners are:

1. ARIB (Association of Radio Industries and Business) in Japan,

   ```
   http://www.arib.or.jp/english/
   ```

2. CCSA (China Communications Standards Associations) in China,

   ```
   http://www.ccsa.org.cn/english/
   ```

3. ETSI (European Telecommunications Standards Institute) in Europe,

   `http://www.etsi.org/`

4. Committee T1 in the United States of America,

   `http://www.t1.org/`

5. TTA (Telecommunications Technology Association) of Korea

   `http://www.tta.or.kr/English/`

6. TTC (Telecommunication Technology Committee) in Japan,

   `http://www.ttc.or.jp/e/`

3GPP was originally chartered to develop globally applicable Technical Specifications and Technical Reports for a third-generation mobile system based on GSM (Global System for Mobile communication). The scope has been reinforced to include maintenance and development of GSM specifications including the supported and evolved radio networks, technologies, and packet access technologies.

Besides the organizational partners, *market representation partners* provide the partnership with market requirements. Market representation partners include, among others, the UMTS Forum, 3G Americas, the GSM Association, the Global mobile Suppliers Association, the TD-SCDMA Forum, and the IPv6 Forum.

3GPP maintains an up-to-date web site at:

`http://www.3gpp.org/`

2.3.1   3GPP Structure

3GPP is organized in a Project Co-ordination Group (PCG) and Technical Specification Groups (TSGs), as illustrated in Figure 2.3. The PCG is responsible for the overall management of 3GPP, time plans, allocation of work, etc. The technical work is produced in the TSGs. At the moment there are four TSGs, responsible for the Core Network and Terminals (CT), System and Services Aspects (SA), GSM EDGE Radio Access Network (GERAN), and Radio Access Network (RAN). Each of the TSGs is further divided into Working Groups. Each of the Working Groups is allocated particular tasks. For instance, CT WG1 is responsible for all the detailed design of the usage of SIP and SDP in the IMS, CT WG3 for interworking aspects, and CT WG4 for all the detailed design of the usage of Diameter. SA WG1 is responsible for the requirements, SA WG2 for the architecture, SA WG3 for the security aspects, SA WG4 for the codecs, and SA WG5 for the operation and maintenance of the network.

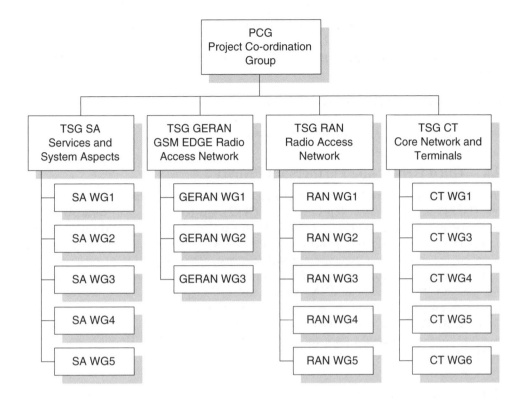

Figure 2.3: The structure of 3GPP

2.3.2 3GPP Deliverables

3GPP working groups do not produce standards. Instead, they produce Technical Specifications (TS) and Technical Reports (TR) that are approved by the TSGs. Once approved they are submitted to the organizational partners to be submitted to their respective standardization processes. The final part of the process is in the organizational partners' hands when they approve the TSs or TRs as part of their standards procedures. As a result, there is a set of globally developed standards that are ready to be used in a particular region.

3GPP TSs and TRs are numbered according to a sequence of four or five digits that follow the pattern "xx.yyy". The first two digits "xx" identify the series number, and the last two or three digits "yy" or "yyy" identify a particular specification within a series. For instance, 3GPP TS 23.228 [23] describes the architectural aspects of the IMS.

3GPP groups its specifications in what is called a *Release*. 3GPP Release 5 contains the first version of the IMS. 3GPP Release 6 contains enhancements to the IMS. The reader must note that the IMS is just a fraction of the 3GPP deliverables in a particular Release, as there are other non-IMS specifications included in a 3GPP Release. 3GPP TSs and TRs include a version number that follows the pattern "x.y.z", where "x" represents the 3GPP Release where the specification is published, "y" is the version number, and "z" is a sub-version number. So, 3GPP TS 23.228 version 5.8.0 means version 8.0 of the Release 5 version of TS 23.228.

3GPP TSs and TRs are publicly available at the 3GPP web site at either of the following URIs:

```
http://www.3gpp.org/specs/specs.htm
```

```
http://www.3gpp.org/ftp/Specs/archive/
```

2.4   Third Generation Partnership Project 2

If 3GPP was born to evolve GSM specifications into a third-generation cellular system, the Third Generation Partnership Project 2 (3GPP2) was born to evolve North American and Asian cellular networks based on ANSI/TIA/EIA-41 standards and CDMA2000® radio access into a third-generation system. 3GPP2 like 3GPP is a partnership project whose members are also known as *organizational partners*. The current list of organizational partners include ARIB (Japan), CCSA (China), TIA (Telecommunications Industry Association) (North America), TTA (Korea), and TTC (Japan). Probably, the reader has noticed that most of them are also organizational partners of 3GPP.

Like 3GPP, 3GPP2 gets market requirements and advice from *market representation partners*. At the moment the list includes the IPv6 Forum, the CDMA Development Group, and the International 450 Association.

2.4.1   3GPP2 Structure

The 3GPP2 structure mimics the structure of 3GPP, as illustrated in Figure 2.4. A Steering Committee (SC) is responsible for the overall standardization process and the planning. The technical work is done in Technical Specification Groups (TSGs). TSG-A is focused on the Access Network Interface, TSG-C on CDMA2000® technology, TSG-S on Service and System Aspects, and TSG-X in Intersystems Operations. TSG-X was born as a merger between the former TSG-N (Core Networks) and TSG-P (Packet Data) TSGs. The structure of TSG-X is not yet completely defined.

2.4.2   3GPP2 Deliverables

Like 3GPP, 3GPP2 does not produce standards but, instead, Technical Specifications and Technical Reports. The documents are created by the TSGs and approved by the SC. Then, they are submitted to the organizational partners to be submitted to their respective standardization processes.

3GPP2 TSs and TRs are numbered with a sequence of letters and digits that follows the scheme "A.Bxxxx-yyy-R" where "A" is a letter that represents the name of the TSG that delivers the document, "B" can be an "R" letter to indicate a TR or a requirements document, but it can also be an "S" letter to indicate a TS. "xxxx" is a sequential number allocated to the document. An optional "yyy" sequence of digits can further identify the chapter within a specification series. The "R" letter identifies the revision. The version number follows the specification and indicates a major and minor version. For instance, the specification X.S0013-002-A v1.0 represents the IP Multimedia Subsystem (IMS) chapter 2, revision A, version 1.0.

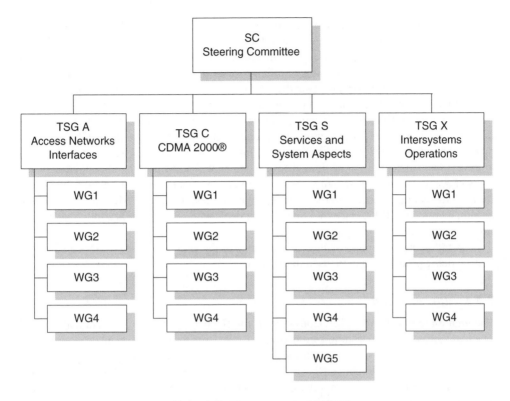

Figure 2.4: The structure of 3GPP2

3GPP2 TSs and TRs are publicly available at the 3GPP2 web site at:

`http://www.3gpp2.org/Public_html/specs/`

Since 3GPP2 IMS specifications are based on corresponding 3GPP IMS ones, we focus on the IMS defined by 3GPP. Sometimes, we will highlight differences between both networks, when those differences are relevant for the discussion. Appendix A provides an overview of the 3GPP2 IMS architecture and the main differences with respect to the IMS defined by 3GPP.

2.5 IETF-3GPP/3GPP2 Collaboration

As we mentioned in Chapter 1 the IMS aims to use Internet protocols. However, some of the protocols chosen for use in the IMS architecture were not completely suitable for the IMS environment. There were even cases where the IETF did not have any solution at all to address some of the issues the IMS was facing.

One possibility would have been to take whatever IETF protocols that were already there and modify them to meet the requirements of the IMS. However, the goal of 3GPP and 3GPP2 was clear. They wanted the IMS to use Internet technologies. This way they could take advantage of any future service created for the Internet. Modifying Internet protocols on their own was not an option. Instead, they established a collaboration with the IETF to make sure that the protocols developed there met their requirements.

The terms of this collaboration were documented in RFC 3113 [161] (3GPP-IETF) and in RFC 3131 [58] (3GPP2-IETF). Both 3GPP and 3GPP2 nominated liaisons with the IETF (Ileana Leuca and later Stephen Hayes from 3GPP, and Tom Hiller and later AC Mahendran from 3GPP2), and the IETF nominated a liaison with them (Thomas Narten). In any case these collaborations took part mostly at the working group level, without involving the official liaisons most of the time: for example, groups of engineers discussing technical issues in mailing lists, IETF face-to-face meetings, and special workshops. 3G engineers collaborated in providing wireless expertise and requirements from the operators while IETF engineers provided protocol knowledge. The goal was to find solutions that addressed the requirements of the IMS and that, at the same time, were general enough to be used in other environments.

So far, several protocol specifications and protocol extensions have been published in the form of RFCs and Internet-Drafts as the fruit of this collaboration. Most of them do not need to mention the IMS, since they specify protocols with general applicability that are not IMS-specific at all.

The following sections provide a brief history of the areas where the IETF collaborated in developing protocols that are used in the IMS.

2.5.1 Internet Area

The area director driving the collaboration in the IETF Internet area was Thomas Narten. The main areas of collaboration were IPv6 and DNS (Domain Name System).

The IPv6 working group produced a specification (RFC 3316 [50]) that provides guidelines on how to implement IPv6 in cellular hosts. When such a host detects that it is using a GPRS access, it follows the guidelines provided in that specification. On the other hand, if the same host is using a different access (e.g., WLAN), it behaves as a regular Internet host. So, terminals behave differently depending on the type of access they are using, not on the type of terminals they are.

In the DNS area there were discussions on how to perform DNS server discovery in the IMS. It was decided not to use DHCP (Dynamic Host Configuration Protocol), but to use GPRS-specific mechanisms instead. At that point there was no working group agreement on stateless DNS server discovery procedures that could be used in the IMS.

2.5.2 Operations and Management Area

The main protocols in the IETF operations and management area where there was collaboration between 3GPP and the IETF were COPS (Common Open Policy Service) and Diameter. Both area directors, Bert Wijnen and Randy Bush, were involved in the discussions; Bert Wijnen in COPS-related discussions and Randy Bush in Diameter-related discussions. Bert Wijnen even participated in 3GPP CN3 meetings as part of this collaboration.

In the COPS area the IMS had decided to use COPS-PR in the *Go* interface and, so, 3GPP needed to standardize the *Go* Policy Information Base (PIB). However, in the IETF it was not clear whether using COPS-PR for 3GPP's purposes was a good idea. After a lot of discussions the *Go* PIB was finally created (the IETF produced RFC 3317 [77] and RFC 3318 [219]).

In the Diameter area the IMS needed to define three Diameter applications, to support the *Cx*, *Sh*, and *Ro* interfaces, respectively. Nevertheless, although new Diameter codes could only be defined in RFCs, there was not enough time to produce an RFC describing these two Diameter applications and the new command codes that were needed. At last, the IETF agreed to provide 3GPP with a number of command codes (allocated in RFC 3589 [156]) to be used in 3GPP Release 5 with one condition: 3GPP needed to collaborate with the IETF on improving those Diameter applications until they became general enough to be documented in the Internet-Drafts "Diameter SIP Application" [106] and "Diameter Credit Control Application" [162]. The IMS is supposed to migrate to these new Diameter applications in feature releases.

2.5.3   Transport Area

Collaboration in the transport area was mainly driven by 3GPP (not much from 3GPP2). Two people were essential to the collaboration in this area: Stephen Hayes, 3GPP's liaison with the IETF and chairman of CN, and Allison Mankin, transport area director in the IETF. They ensured that all the issues related to signaling got the appropriate attention in both organizations.

Everything began when 3GPP decided that SIP was going to be the session control protocol in the IMS. At that point SIP was still an immature protocol that did not meet most of the requirements 3GPP had. SIP was defined in RFC 2543 [116]. At that time there was an Internet-Draft, commonly known as 2543bis, that had fixed some of the issues present in RFC 2543 and was supposed to become the next revision of the protocol specification. However, 2543bis only had two active editors (namely Henning Schulzrinne and Jonathan Rosenberg) and the 3GPP deadlines were extremely tough. A larger team was needed if the SIP working group, where SIP was being developed, wanted to meet those deadlines. That is how Gonzalo Camarillo, Alan Johnston, Jon Peterson, and Robert Sparks were recruited to edit the SIP specification.

After tons of emails, conference calls, and face-to-face meetings the main outcome of the team was RFC 3261 [215]. However, it was soon clear that 3GPP's requirements were not going to be met with a single protocol. Many extensions were needed to fulfill them all. In fact, there were so many requirements and so many extensions needed that the SIP working group was overloaded (other working groups, like MMUSIC, SIMPLE, or ROHC, were also involved, but the main body of the work was tackled by SIP). A new process was needed to handle all of this new work.

The IETF decided to create a new working group to assist SIP in deciding how to best use its resources. The new working group was called SIPPING, and its main function was to gather requirements for SIP, prioritize them and send them to the SIP working group, which was in charge of doing the actual protocol work. This new process was documented in RFC 3427 [160].

At present, most of the protocol extensions related to session establishment needed by 3GPP are finished or quite advanced. As a consequence the 3GPP focus is moving towards the SIMPLE working group which develops SIP extensions for presence and instant messaging.

2.6   Open Mobile Alliance

In June 2002, the Open Mobile Alliance (OMA) was created to provide interoperable mobile data services. A number of existing forums at that time, such as the WAP Forum and Wireless Village, were integrated into OMA. Nowadays, OMA includes companies representing most segments of industry. Vendors, service providers, and content providers are all represented in OMA. The OMA web site can be found at the following link:

```
http://www.openmobilealliance.org/
```

OMA pays special attention to usability. That is, OMA services need to be easy to use. In OMA, spending time thinking about how users will interact with a particular service is routine.

Figure 2.5 shows the structure of OMA. The technical plenary is responsible for the approval and maintenance of the OMA specifications. It consists of a number of Technical Working Groups and, at the time of writing, two Committees.

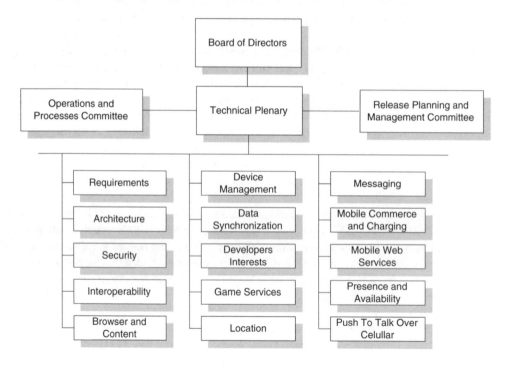

Figure 2.5: OMA structure

The Operations and Processes Committee defines and supports the operational processes of the Technical Plenary. The Release Planning and Management Committee plans and manages the different OMA releases, which are based on the specifications developed by the Technical Working Groups.

2.6.1   OMA Releases and Specifications

OMA produces Release Packages. Each of these packages consists of a set of OMA specifications, which are the documents produced by the OMA Technical Working Groups.

For example, the Enabler Release Package for PoC Version 1.0 [175] includes an Enabler Release Definition document [172] that provides a high-level definition of the PoC (Push-to-talk over Cellular) service and lists the specifications contained in the

Enabler Release Package. Additionally, the Enabler Release Package includes the following specifications:

- Architecture [176]

- Requirements [178]

- Control Plane Specification [177]

- User Plane Specification [179]

- XDM (XML Document Management) Specification [180] (which defines data formats and XCAP (XML Configuration Access Protocol) application usages for PoC).

OMA defines different maturity levels for its releases. The maturity levels are called *phases* in OMA terminology. Each OMA Release Package can be in one of the following phases.

- Phase 1: Candidate Enabler Release – initial state of the release.

- Phase 2: Approved Enabler Release – the release has successfully passed interoperability tests.

- Phase 3: OMA Interoperability Release – the release has successfully passed exhaustive interoperability tests that may involve interoperability with other OMA service enablers.

The Enabler Release Package for PoC Version 1.0, which we discussed in the earlier example, is still in Phase 1. Therefore, it is a Candidate Enabler Release Package.

As the definitions of the different release phases clearly state, interoperability tests play a key role in OMA. The OMA interoperability tests are referred to as Test Fests and are organized by the Interoperability (IOP) Technical Working Group, which specifies the processes and programs for the Test Fests.

2.6.2 Relationship between OMA and 3GPP/3GPP2

A number of OMA Technical Working Groups use the IMS in some manner. As a consequence, we need to look at the relationship between OMA and some of its Technical Working Groups with 3GPP and 3GPP2 with respect to the IMS.

Some of the OMA work uses the IMS as a base. Due to this, there are situations where an OMA Technical Working Group comes up with new requirements on the IMS that need to be met in order to implement a new service.

In general, the agreement between 3GPP, 3GPP2, and OMA is that OMA generates requirements on the IMS, and 3GPP and 3GPP2 extend the IMS to meet these new requirements. This agreement tries to avoid having different versions of the IMS: the 3GPP IMS and the IMS as extended by OMA. Having a single organization managing and maintaining the specifications of the IMS ensures interoperability between the IMS implementations of different vendors.

Still, there is no clear cut distinction between the IMS and the services on top of it. A multimedia session between two participants may be considered a service by some, but it is part of the IMS, as discussed in Chapter 5. Conferencing can also be considered to be a service, but it is specified by 3GPP as part of the IMS. Presence is an interesting area as well, because both 3GPP and OMA have ongoing work related to presence.

Yet, even when both 3GPP and OMA work on similar issues, such as presence, they aim to have compatible specifications. For example, the OMA specifications developed by the Presence and Availability Technical Working Group focus on different aspects of presence than the 3GPP presence-related specifications. Nevertheless, all these specifications are compatible.

Another example of an area where both OMA and 3GPP perform activities is messaging. While 3GPP focuses on specifying instant messaging services for the IMS using the work of the IETF SIMPLE WG as a base (e.g., the SIP MESSAGE method and MSRP), OMA focuses on the interworking between SIMPLE-based and Wireless Village-based instant messaging and on the evolution of MMS (Multimedia Messaging Service).

In order to ensure that every OMA service uses the IMS (as specified by 3GPP and 3GPP2) in a consistent and interoperable way, OMA has produced the IMSinOMA Enabler Release Package [173]. This release package includes an Enabler Release Definition document [171], a Requirements document [182] and an Architecture document [181]. This release package also describes how non-IMS-based OMA services can interoperate with IMS-based OMA services.

2.6.3   Relationship between OMA and the IETF

In the same way as the 3GPP and 3GPP2 IMS specifications refer to IETF protocols and extensions, OMA specifications also include references to IETF documents. The standardization collaboration between OMA and the IETF (documented in RFC 3975 [125]) consists mainly of working group level communications. A set of engineers collaborate with both OMA and the IETF. They bring OMA requirements to the relevant IETF working groups, which analyze them and develop appropriate solutions.

However, sometimes communications at the working group level are insufficient. To handle these cases, both OMA and the IETF have appointed a liaison to each other. At the time of writing, the OMA liaison to the IETF is Ileana Leuca and the IETF liaison to OMA is Dean Willis.

OMA maintains a web page at the following link that allows both organizations to track the status of the IETF Internet-Drafts that the OMA Technical Working Groups need:

`http://www.openmobilealliance.org/collaborating/ietf.html`

Chapter 3

General Principles of the IMS Architecture

In Chapter 1 we introduced the circuit-switched and the packet-switched domains and described why we need the IMS to provide rich Internet services. Chapter 2 introduced the players standardizing the IMS and defining its architecture. In this chapter we will describe the history of the circuit-switched and the packet-switched domains. In addition, we will introduce the design principles that lay behind the IMS architecture and its protocols. We will also tackle in this chapter the IMS network nodes and the different ways in which users are identified in the IMS.

3.1 From Circuit-switched to Packet-switched

Let us look at how cellular networks have evolved from circuit-switched networks to packet-switched networks and how the IMS is the next step in this evolution. We will start with a brief introduction to the history of the 3G circuit-switched and packet-switched domains.

The Third Generation Partnership Project (3GPP) is chartered to develop specifications for the evolution of GSM. That is, 3GPP uses the GSM specifications as a design base for a third-generation mobile system.

GSM has two different modes of operation: circuit-switched and packet-switched. The 3G circuit-switched and packet-switched domains are based on these GSM modes of operation.

3.1.1 GSM Circuit-switched

Not surprisingly, GSM circuit-switched uses circuit-switched technologies, which are also used in the PSTN (Public Switched Telephone Network). Circuit-switched networks have two different planes: the signaling plane and the media plane.

The signaling plane includes the protocols used to establish a circuit-switched path between terminals. In addition, service invocation also occurs in the signaling plane.

The media plane includes the data transmitted over the circuit-switched path between the terminals. The encoded voice exchanged between users belongs to the media plane.

Signaling and media planes followed the same path in early circuit-switched networks. Nevertheless, at a certain point in time the PSTN started to differentiate the paths the signaling plane and the media plane follow. This differentiation was triggered by the introduction of services based on IN (Intelligent Network). Calls to toll-free numbers are an example of an IN service. The GSM version of IN services is known as CAMEL services (Customized Applications for Mobile networks using Enhanced Logic).

In both IN and CAMEL the signaling plane follows the media plane until a point where the call is temporarily suspended. At that point the signaling plane performs a database query (e.g., a query for a routing number for an 800 number) and receives a response. When the signaling plane receives the response to the query the call setup is resumed and both the signaling plane and the media plane follow the same path until they reach the destination.

3GPP has gone a step further in the separation of signaling and media planes with the introduction of the split architecture for the MSC (Mobile Switching Center). The MSC is split into an MSC server and a media gateway. The MSC server handles the signaling plane and the media gateway handles the media plane. The split architecture was introduced in Release 4 of the 3GPP specifications.

We will see that the IMS also keeps signaling and media paths separate, but goes even further in this separation. The only nodes that need to handle both signaling and media are the IMS terminals; no network node needs to handle both.

3.1.2   GSM Packet-switched

The GSM packet-switched network, also known as GPRS (General Packet Radio Service, specified in 3GPP TS 23.060 [15]) was the base for the 3GPP Release 4 packet-switched domain. This domain allows users to connect to the Internet using native packet-switched technologies.

Initially, there were three applications designed to boost the usage of the packet-switched domain.

1. The Wireless Application Protocol (WAP) [233].

2. Access to corporate networks.

3. Access to the public Internet.

Nevertheless, none of these applications was attracting enough customers to justify the enormous cost to deploy packet-switched mobile networks.

3.2   IMS Requirements

The situation operators were facing right before the conception of the IMS was not encouraging at all. The circuit-switched voice market had become a commodity, and operators found it difficult to make a profit by only providing and charging for voice calls. On the other hand, packet-switched services had not taken off yet, so, operators were not making much money from them either.

Thus, operators needed a way to provide more attractive packet-switched services to attract users to the packet-switched domain. That is, the mobile Internet needed to become more attractive to its users. In this way the IMS (IP Multimedia Subsystem) was born. With the vision described in Chapter 1 in mind, equipment vendors and operators started designing the IMS.

So, the IMS aims to:

1. combine the latest trends in technology;

2. make the *mobile Internet* paradigm come true;

3. create a common platform to develop diverse multimedia services;

4. create a mechanism to boost margins due to extra usage of mobile packet-switched networks.

Let us look at the requirements that led to the design of the 3GPP IMS (captured in 3GPP TS 22.228 [34] Release 5). In these requirements the IMS is defined as an architectural framework created for the purpose of delivering IP multimedia services to end-users. This framework needs to meet the following requirements.

1. Support for establishing IP Multimedia Sessions.

2. Support for a mechanism to negotiate Quality of Service (QoS).

3. Support for interworking with the Internet and circuit-switched networks.

4. Support for roaming.

5. Support for strong control imposed by the operator with respect to the services delivered to the end-user.

6. Support for rapid service creation without requiring standardization.

The release 6 version of 3GPP TS 22.228 [34] added a new requirement to support access from networks other than GPRS. This is the so-called *access independence* of the IMS, since the IMS provides support for different access networks.

3.2.1 IP Multimedia Sessions

The IMS can deliver a broad range of services. Still, there is one service of special importance for users: audio and video communications. This requirement stresses the need to support the main service to be delivered by the IMS: multimedia sessions over packet-switched networks. Multimedia refers to the simultaneous existence of several media types. The media types in this case are audio and video.

Multimedia communications were already standardized in previous 3GPP releases, but those multimedia communications take place over the circuit-switched network rather than the packet-switched network.

3.2.2 QoS

Continuing with the analysis of the requirements we find the requirement to negotiate a certain QoS (Quality of Service). This is a key component of the IMS.

The QoS for a particular session is determined by a number of factors, such as the maximum bandwidth that can be allocated to the user based on the user's subscription or the current state of the network. The IMS allows operators to control the QoS a user gets, so that operators can differentiate certain groups of customers from others.

3.2.3   Interworking

Support for interworking with the Internet is an obvious requirement, given that the Internet offers millions of potential destinations for multimedia sessions initiated in the IMS. By requiring interworking with the Internet the number of potential sources and destinations for multimedia sessions is dramatically expanded.

The IMS is also required to interwork with circuit-switched networks, such as the PSTN (Public Switched Telephone Network), or existing cellular networks. The first audio/video IMS terminals that will reach the market will be able to connect to both circuit-switched and packet-switched networks. So, when a user wants to call a phone in the PSTN or a cellular phone the IMS terminal chooses to use the circuit-switched domain.

So, interworking with circuit-switched networks is not strictly required although, effectively, most of the IMS terminals will also support the circuit-switched domain.[1] The requirement to support interworking with circuit-switched networks can be considered a long-term requirement. This requirement will be implemented when it is possible to build IMS terminals with packet-switched support only.

3.2.4   Roaming

Roaming support has been a general requirement since the second generation of cellular networks; users have to be able to roam to different networks (e.g., a user visiting a foreign country). Obviously the IMS inherits this requirement, so it should be possible for users to roam to different countries (subject to the existence of a roaming agreement signed between the home and the visited network).

3.2.5   Service Control

Operators typically want to impose policies on the services delivered to the user. We can divide these policies into two categories.

- General policies applicable to all the users in the network.

- Individual policies that apply to a particular user.

The first type of policy comprises a set of restrictions that apply to all users in the network. For instance, operators may want to restrict the usage of high-bandwidth audio codecs, such as G.711 (ITU-T Recommendation G.711 [131]), in their networks. Instead, they may want to promote lower bandwidth codecs like AMR (Adaptive Multi Rate, specified in 3GPP TS 26.071 [5]).

The second type of policy includes a set of policies which are tailored to each user. For instance, a user may have some subscription to use IMS services that do not include the use of video. The IMS terminal will most likely support video capabilities, but in case the user attempts to initiate a multimedia session that includes video the operator will prevent that session being set up. This policy is modeled on a user-by-user basis, as they are dependent on the terms of usage in the user's subscription.

---

[1] IMS terminals supporting audio capabilities are required to support the circuit-switched domain due to the inability of the IMS (at least in the first phases) to provide support for emergency calls. So, emergency calls are placed over the circuit-switched domain. Emergency calls are further analyzed in Section 5.10.

3.2.6 Rapid Service Creation

The requirement about service creation had a strong impact on the design of IMS architecture. This requirement states that IMS services do not need to be standardized.

This requirement represents a milestone in cellular design, because in the past, every single service was either standardized or had a proprietary implementation. Even when services were standardized there was no guarantee that the service would work when roaming to another network. The reader may already have experienced the lack of support for call diversion to voicemail in GSM networks when the user is visiting another country.

The IMS aims to reduce the time it takes to introduce a new service. In the past the standardization of the service and interoperability tests caused a significant delay. The IMS reduces this delay by standardizing *service capabilities* instead of *services*.

3.2.7 Multiple Access

The multiple access requirement introduces other means of access than GPRS. The IMS is just an IP network and, like any other IP network, it is lower layer and access-independent. Any access network can in principle provide access to the IMS. For instance, the IMS can be accessed using a WLAN (Wireless Local Access Network), an ADSL (Asymmetric Digital Subscriber Line), an HFC (Hybrid Fiber Coax), or a Cable Modem.

Still, 3GPP, as a project committed to developing solutions for the evolution of GSM, has focused on GPRS access (both in GSM and UMTS, Universal Mobil Telecommunication System) for the first release of the IMS (i.e., Release 5). Future releases will study other accesses, such as WLAN.

3.3 Overview of Protocols used in the IMS

When the European Telecommunications Standard Institute (ETSI) developed the GSM standard, most of its protocols were specially designed for GSM (especially those dealing with the radio interface and with mobility management). ETSI reused only a few protocols developed by the International Telecommunication Union-Telecommunications (ITU-T). Most of the protocols were developed from scratch because there were no existing protocols to take as a base.

A few years later, 3GPP began developing the IMS, a system based on IP protocols, which had been traditionally developed by the IETF (Internet Engineering Task Force). 3GPP analyzed the work done in the past by ETSI in developing its own protocols and decided to reuse protocols which had already been developed (or were under development at that time) in other Standards Development Organizations (SDOs) such as the IETF or ITU-T. This way, 3GPP takes advantage of the experience of the IETF and the ITU-T in designing robust protocols, reducing at the same time standardization and development costs.

3.3.1 Session Control Protocol

The protocols that control the calls play a key role in any telephony system. In circuit-switched networks the most common call control protocols are TUP (Telephony User Part, ITU-T Recommendation Q.721 [130]), ISUP (ISDN User Part, ITU-T Recommendation Q.761 [139]), and the more modern BICC (Bearer Independent Call Control, ITU-T Recommendation Q.1901 [140]). The protocols considered to be used as the session control protocol for the IMS were obviously all based on IP. The candidates were as follows.

Bearer Independent Call Control (BICC): BICC (specified in ITU-T Recommendation Q.1901 [140]) is an evolution of ISUP. Unlike ISUP, BICC separates the signaling plane from the media plane, so that signaling can traverse a separate set of nodes than the media plane. Additionally, BICC supports and can run over a different set of technologies, such as IP, SS7 (Signaling System No. 7, ITU-T Recommendation Q.700 [134]), or ATM (Asynchronous Transfer Mode).

H.323: like BICC, H.323 (ITU-T Recommendation H.323 [145]) is an ITU-T protocol. H.323 defines a new protocol to establish multimedia sessions. Unlike BICC, H.323 was designed from scratch to support IP technologies. In H.323, signaling and the media do not need to traverse the same set of hosts.

SIP (Session Initiation Protocol, RFC 3261 [215]): specified by the IETF as a protocol to establish and manage multimedia sessions over IP networks, SIP was gaining momentum at the time 3GPP was choosing its session control protocol. SIP follows the well-known client–server model, so much used by many protocols developed by the IETF. SIP designers borrowed design principles from SMTP (Simple Mail Transfer Protocol, RFC 2821 [151]) and especially, from HTTP (Hypertext Transfer Protocol, RFC 2616 [101]). SIP inherits most of its characteristics from these two protocols. This is an important strength of SIP, because HTTP and SMTP are the most successful protocols on the Internet. SIP, unlike BICC and H.323, does not differentiate the User-to-Network Interface (UNI) from a Network-to-Network Interface (NNI). In SIP there is just a single protocol which works end to end. Unlike BICC and H.323, SIP is a text-based protocol. This means that it is easier to extend, debug, and use to build services.

SIP was chosen as the session control protocol for the IMS. The fact that SIP makes it easy to create new services carried great weight in this decision. Since SIP is based on HTTP, SIP service developers can use all the service frameworks developed for HTTP, such as CGI (Common Gateway Interface) and Java servlets.

3.3.2   The AAA Protocol

In addition to the session control protocol there are a number of other protocols that play important roles in the IMS. Diameter (whose base protocol is specified in RFC 3588 [60]) was chosen to be the AAA (Authentication, Authorization, and Accounting) protocol in the IMS.

Diameter is an evolution of RADIUS (specified in RFC 2865 [195]), which is a protocol that is widely used on the Internet to perform AAA. For instance, when a user dials up to an Internet Service Provider (ISP) the network access server uses RADIUS to authenticate and authorize the user accessing the network.

Diameter consists of a base protocol that is complemented with so-called *Diameter Applications*. Diameter applications are customizations or extensions to Diameter to suit a particular application in a given environment.

The IMS uses Diameter in a number of interfaces, although not all the interfaces use the same Diameter application. For instance, the IMS defines a Diameter application to interact with SIP during session setup and another one to perform credit control accounting.

3.3.3   Other Protocols

In addition to SIP and Diameter there are other protocols that are used in the IMS. The COPS (Common Open Policy Service) protocol (specified in RFC 2748 [85]) is used to transfer policies between PDPs (Policy Decision Points) and PEPs (Policy Enforcement Points).

H.248 (ITU-T Recommendation H.248 [143]) and its packages are used by signaling nodes to control nodes in the media plane (e.g., a media gateway controller controlling a media gateway). H.248 was jointly developed by ITU-T and IETF and is also referred to as the MEGACO (MEdia GAteway COntrol) protocol.

RTP (Real-Time Transport Protocol, defined in RFC 3550 [225]) and RTCP (RTP Control Protocol, defined in RFC 3550 [225] as well) are used to transport real-time media, such as video and audio.

We have mentioned a few application-layer protocols used in the IMS. We will describe these in Parts II and III of this book, along with other application-layer Internet protocols that may be used in the IMS in the future, and other protocols that belong to other layers.

3.4   Overview of IMS Architecture

Before exploring the general architecture in the IMS we should keep in mind that 3GPP does not standardize *nodes*, but *functions*. This means that the IMS architecture is a collection of functions linked by standardized interfaces. Implementers are free to combine two functions into a single node (e.g., into a single physical box). Similarly, implementers can split a single function into two or more nodes.

In general, most vendors follow the IMS architecture closely and implement each function into a single node. Still, it is possible to find nodes implementing more than one function and functions distributed over more than one node.

Figure 3.1 depicts an overview of the IMS architecture as standardized by 3GPP. The figure shows most of the signaling interfaces in the IMS, typically referred to by a two- or three-letter code. We do not include all the interfaces defined in the IMS, but only the most relevant ones. The reader can refer to 3GPP TS 23.002 [28] to find a complete list of all the interfaces.

On the left side of Figure 3.1 we can see the IMS mobile terminal, typically referred to as the User Equipment (UE). The IMS terminal attaches to a packet network, such as the GPRS network, through a radio link.

Note that, although the figure shows an IMS terminal attaching to the network using a radio link, the IMS supports other types of devices and accesses. PDAs (Personal Digital Assistants) and computers are examples of devices that can connect to the IMS. Examples of alternative accesses are WLAN or ADSL.

The remainder of Figure 3.1 shows the nodes included in the so-called IP Multimedia Core Network Subsystem. These nodes are:

- one or more user databases, called HSSs (Home Subscriber Servers) and SLFs (Subscriber Location Functions);

- one or more SIP servers, collectively known as CSCFs (Call/Session Control Functions);

- one or more ASs (Application Servers);

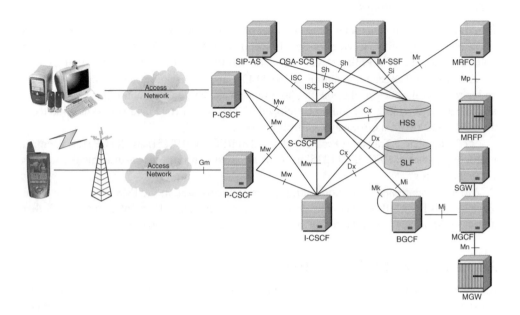

Figure 3.1: 3GPP IMS architecture overview

- one or more MRFs (Media Resource Functions), each one further divided into MRFC (Media Resource Function Controllers) and MRFP (Media Resource Function Processors);

- one or more BGCFs (Breakout Gateway Control Functions);

- one or more PSTN gateways, each one decomposed into an SGW (Signaling Gateway), an MGCF (Media Gateway Controller Function), and an MGW (Media Gateway).

Note that Figure 3.1 does not contain a reference to charging collector functions. These are described in Section 7.4.

3.4.1   The Databases: the HSS and the SLF

The Home Subscriber Server (HSS) is the central repository for user-related information. Technically, the HSS is an evolution of the HLR (Home Location Register), which is a GSM node. The HSS contains all the user-related subscription data required to handle multimedia sessions. These data include, among other items, location information, security information (including both authentication and authorization information), user profile information (including the services that the user is subscribed to), and the S-CSCF (Serving-CSCF) allocated to the user.

A network may contain more than one HSS, in case the number of subscribers is too high to be handled by a single HSS. In any case, all the data related to a particular user are stored in a single HSS.[2]

---

[2]The HSS is typically implemented using a redundant configuration, to avoid a single point of failure. Nevertheless, we consider a redundant configuration of HSSs as a single logical node.

Networks with a single HSS do not need a Subscription Locator Function (SLF). On the other hand, networks with more than one HSS do require an SLF.

The SLF is a simple database that maps users' addresses to HSSs. A node that queries the SLF, with a user's address as the input, obtains the HSS that contains all the information related to that user as the output.

Both the HSS and the SLF implement the Diameter protocol (RFC 3588 [60]) with an IMS-specific Diameter application.

3.4.2   The CSCF

The CSCF (Call/Session Control Function), which is a SIP server, is an essential node in the IMS. The CSCF processes SIP signaling in the IMS. There are three types of CSCFs, depending on the functionality they provide. All of them are collectively known as CSCFs, but any CSCF belongs to one of the following three categories.

- P-CSCF (Proxy-CSCF).

- I-CSCF (Interrogating-CSCF).

- S-CSCF (Serving-CSCF).

3.4.2.1   The P-CSCF

The P-CSCF is the first point of contact (in the signaling plane) between the IMS terminal and the IMS network. From the SIP point of view the P-CSCF is acting as an outbound/inbound SIP proxy server. This means that all the requests initiated by the IMS terminal or destined for the IMS terminal traverse the P-CSCF. The P-CSCF forwards SIP requests and responses in the appropriate direction (i.e., toward the IMS terminal or toward the IMS network).

The P-CSCF is allocated to the IMS terminal during IMS registration and does not change for the duration of the registration (i.e., the IMS terminal communicates with a single P-CSCF during the registration).

The P-CSCF includes several functions, some of which are related to security. First, it establishes a number of IPsec security associations toward the IMS terminal. These IPsec security associations offer integrity protection (i.e., the ability to detect whether the contents of the message have changed since its creation).

Once the P-CSCF authenticates the user (as part of security association establishment) the P-CSCF asserts the identity of the user to the rest of the nodes in the network. This way, other nodes do not need to further authenticate the user, because they trust the P-CSCF. The rest of the nodes in the network user's identity (asserted by the P-CSCF) have a number of purposes, such as providing personalized services and generating account records.

Additionally, the P-CSCF verifies the correctness of SIP requests sent by the IMS terminal. This verification keeps IMS terminals from creating SIP requests that are not built according to SIP rules.

The P-CSCF also includes a compressor and a decompressor of SIP messages (IMS terminals include both as well). SIP messages can be large, given that SIP is a text-based protocol. While a SIP message can be transmitted over a broadband connection in a fairly short time, transmitting large SIP messages over a narrowband channel, such as some radio links, may take a few seconds. The mechanism used to reduce the time to transmit a SIP message is to compress the message, send it over the air interface, and decompress it at the other end.[3]

The P-CSCF may include a PDF (Policy Decision Function). The PDF may be integrated with the P-CSCF or be implemented as a stand-alone unit. The PDF authorizes media plane resources and manages Quality of Service over the media plane.

The P-CSCF also generates charging information toward a charging collection node.

An IMS network usually includes a number of P-CSCFs for the sake of scalability and redundancy. Each P-CSCF serves a number of IMS terminals, depending on the capacity of the node.

3.4.2.2   P-CSCF Location

The P-CSCF may be located either in the visited network or in the home network. In case the underlying packet network is based on GPRS, the P-CSCF is always located in the same network where the GGSN (Gateway GPRS Support Node) is located. So both P-CSCF and GGSN are either located in the visited network or in the home network. Due to current deployments of GPRS, it is expected that the first IMS networks will inherit this mode and will be configured with the GGSN and P-CSCF in the home network. It is also expected that once IMS reaches the mass market, operators will migrate the configuration and will locate the P-CSCF and the GGSN in the visited network.

3.4.2.3   The I-CSCF

The I-CSCF is a SIP proxy located at the edge of an administrative domain. The address of the I-CSCF is listed in the DNS (Domain Name System) records of the domain. When a SIP server follows SIP procedures (described in RFC 3263 [214]) to find the next SIP hop for a particular message the SIP server obtains the address of an I-CSCF of the destination domain.

Besides the SIP proxy server functionality the I-CSCF has an interface to the SLF and the HSS. This interface is based on the Diameter protocol (RFC 3588 [60]). The I-CSCF retrieves user location information and routes the SIP request to the appropriate destination (typically an S-CSCF).

Additionally, the I-CSCF may optionally encrypt the parts of the SIP messages that contain sensitive information about the domain, such as the number of servers in the domain, their DNS names, or their capacity. This functionality is referred to as THIG (Topology Hiding Inter-network Gateway). THIG functionality is optional and is not likely to be deployed by most networks.

A network will include typically a number of I-CSCFs for the sake of scalability and redundancy.

3.4.2.4   I-CSCF Location

The I-CSCF is usually located in the home network, although in some especial cases, such as an I-CSCF(THIG), it may be located in a visited network as well.

---

[3]There is a misconception that compression between the IMS terminal and the P-CSCF is enabled just to save a few bytes over the air interface. This is not the motivation lying behind compression. In particular, it is not worth saving a few bytes of signaling when the IMS terminal will be establishing a multimedia session (e.g., audio, video) that will use much more bandwidth than the signaling. The main motivation for compression is to reduce the time to transmit SIP messages over the air interface.

3.4.2.5 The S-CSCF

The S-CSCF is the central node of the signaling plane. The S-CSCF is essentially a SIP server, but it performs session control as well. In addition to SIP server functionality the S-CSCF also acts as a SIP registrar. This means that it maintains a binding between the user location (e.g., the IP address of the terminal the user is logged on) and the user's SIP address of record (also known as a Public User Identity).

Like the I-CSCF the S-CSCF also implements a Diameter (RFC 3588 [60]) interface to the HSS. The main reasons to interface the HSS are as follows.

- To download the authentication vectors of the user who is trying to access the IMS from the HSS. The S-CSCF uses these vectors to authenticate the user.

- To download the user profile from the HSS. The user profile includes the service profile, which is a set of triggers that may cause a SIP message to be routed through one or more application servers.

- To inform the HSS that this is the S-CSCF allocated to the user for the duration of the registration.

All the SIP signaling the IMS terminals sends, and all the SIP signaling the IMS terminal receives, traverses the allocated S-CSCF. The S-CSCF inspects every SIP message and determines whether the SIP signaling should visit one or more application servers en route toward the final destination. Those application servers would potentially provide a service to the user.

One of the main functions of the S-CSCF is to provide SIP routing services. If the user dials a telephone number instead of a SIP URI (Uniform Resource Identifier) the S-CSCF provides translation services, typically based on DNS E.164 Number Translation (as described in RFC 2916 [100]).

The S-CSCF also enforces the policy of the network operator. For example, a user may not be authorized to establish certain types of sessions. The S-CSCF keeps users from performing unauthorized operations.

A network usually includes a number of S-CSCFs for the sake of scalability and redundancy. Each S-CSCF serves a number of IMS terminals, depending on the capacity of the node.

3.4.2.6 S-CSCF Location

The S-CSCF is always located in the home network.

3.4.3 The AS

The AS (Application Server) is a SIP entity that hosts and executes services. Depending on the actual service the AS can operate in SIP proxy mode, SIP UA (User Agent) mode (i.e., endpoint), or SIP B2BUA (Back-to-Back User Agent) mode (i.e., a concatenation of two SIP User Agents). The AS interfaces the S-CSCF using SIP.

Figure 3.2 depicts three different types of Application Servers.

SIP AS (Application Server): this is the native Application Server that hosts and executes IP Multimedia Services based on SIP. It is expected that new IMS-specific services will likely be developed in SIP Application Servers.

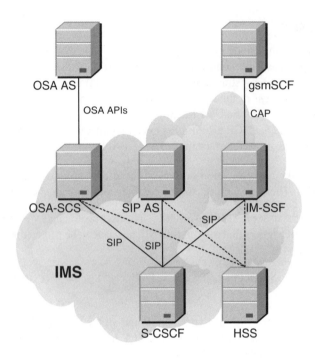

Figure 3.2: Three types of Application Servers

OSA-SCS (Open Service Access–Service Capability Server): this   application   server
   provides an interface to the OSA framework Application Server. It inherits all the
   OSA capabilities, especially the capability to access the IMS securely from external
   networks.   This node acts as an Application Server on one side (interfacing the
   S-CSCF with SIP) and as an interface between the OSA Application Server and the
   OSA Application Programming Interface (API, described in 3GPP TS 29.198 [29]).

IM-SSF (IP Multimedia Service Switching Function): this specialized application server
   allows us to reuse CAMEL (Customized Applications for Mobile network Enhanced
   Logic) services that were developed for GSM in the IMS. The IM-SSF allows a
   gsmSCF (GSM Service Control Function) to control an IMS session. The IM-SSF
   acts as an Application Server on one side (interfacing the S-CSCF with SIP). On the
   other side, it acts as an SSF (Service Switching Function), interfacing the gsmSCF with
   a protocol based on CAP (CAMEL Application Part, defined in 3GPP TS 29.278 [10]).

All three types of application servers behave as SIP application servers toward the IMS
network (i.e., they act as either a SIP proxy server, a SIP User Agent, a SIP redirect server,
or a SIP Back-to-back User Agent). The IM-SSF AS and the OSA-SCS AS have other roles
when interfacing CAMEL or OSA, respectively.

   In addition to the SIP interface the AS may optionally provide an interface to the HSS.
The SIP-AS and OSA-SCS interfaces toward the HSS are based on the Diameter protocol
(RFC 3588 [60]) and are used to download or upload data related to a user stored in the HSS.
The IM-SSF interface toward the HSS is based on MAP (Mobile Application Part, defined in
3GPP TS 29.002 [26]).

3.4.3.1 AS Location

The AS can be located either in the home network or in an external third-party network to which the home operator maintains a service agreement. In any case, if the AS is located outside the home network, it does not interface the HSS.

3.4.4 The MRF

The MRF (Media Resource Function) provides a source of media in the home network. The MRF provides the home network with the ability to play announcements, mix media streams (e.g., in a centralized conference bridge), transcode between different codecs, obtain statistics, and do any sort of media analysis.

The MRF is further divided into a signaling plane node called the MRFC (Media Resource Function Controller) and a media plane node called the MRFP (Media Resource Function Processor). The MRFC acts as a SIP User Agent and contains a SIP interface towards the S-CSCF. The MRFC controls the resources in the MRFP via an H.248 interface.

The MRFP implements all the media-related functions, such as playing and mixing media.

3.4.4.1 MRF Location

The MRF is always located in the home network.

3.4.5 The BGCF

The BGCF is essentially a SIP server that includes routing functionality based on telephone numbers. The BGCF is only used in sessions that are initiated by an IMS terminal and addressed to a user in a circuit-switched network, such as the PSTN or the PLMN. The main functionality of the BGCF is:

(a) to select an appropriate network where interworking with the circuit-switched domain is to occur;

(b) or, to select an appropriate PSTN/CS gateway, if interworking is to occur in the same network where the BGCF is located.

3.4.6 The IMS-ALG and the TrGW

As we describe later in Section 5.2, IMS supports two IP versions, namely IP version 4 (IPv4, specified in RFC 791, [189]) and IP version 6 (IPv6, specified in RFC 2460 [81]). At some point in an IP multimedia session or communication, interworking between the two versions may occur. In order to facilitate interworking between IPv4 and IPv6 without requiring terminal support, the IMS adds two new functional entities that provides translation between both protocols. These new entities are the IMS Application Layer Gateway (IMS-ALG) and the Transition Gateway (TrGW). The former processes control plane signaling (e.g., SIP and SDP messages); the latter processes user plane traffic (e.g., RTP, RTCP).

Figure 3.3 shows the relation of the IMS-ALG with the TrGW and the rest of the IMS nodes. The IMS-ALG acts as a SIP B2BUA by maintaining two independent signaling legs: one towards the internal IMS network and the other towards the other network. Each of these legs are running over a different IP version. Additionally, the IMS-ALG rewrites the SDP

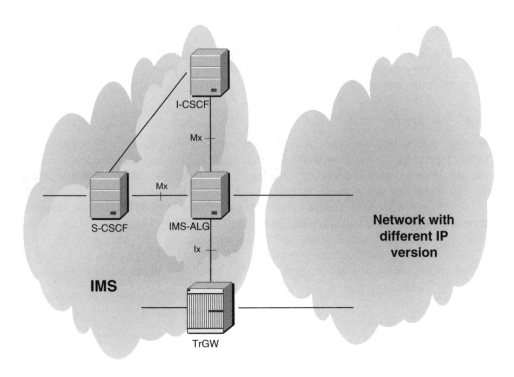

Figure 3.3: The IMS-ALG and the TrGW

by changing the IP addresses and port numbers created by the terminal with one or more IP addresses and port numbers allocated to the TrGW. This allows the user plane traffic to be routed to the TrGW.

The IMS-ALG interfaces the I-CSCF for incoming traffic and the S-CSCF for outgoing traffic through the *Mx* interface. The *Mx* interface is not standardized by 3GPP Release 6, although it is believed to be SIP as in any of the other CSCF interfaces.

The TrGW is effectively a NAT-PT/NAPT-PT (Network Address Port Translator–Protocol Translator). The TrGW is configured with a pool of IPv4 addresses that are dynamically allocated for a given session. The TrGW does the translation of IPv4 and IPv6 at the media level (e.g., RTP, RTCP).

3GPP standardizes the details of the IPv4/IPv6 interworking of the IMS-ALG and TrGW in 3GPP TS 29.162 [18].

The IMS-ALG interfaces the TrGW through the *Ix* interface, which is not standardized in 3GPP Release 6.

3.4.7   The PSTN/CS Gateway

The PSTN gateway provides an interface toward a circuit-switched network, allowing IMS terminals to make and receive calls to and from the PSTN (or any other circuit-switched network).

Figure 3.4 shows a BGCF and a decomposed PSTN gateway that interfaces the PSTN. The PSTN gateway is decomposed into the following functions.

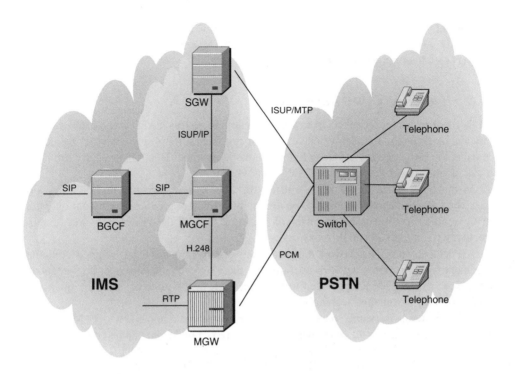

Figure 3.4: The PSTN/CS gateway interfacing a CS network

SGW (Signaling Gateway): the Signaling Gateway interfaces the signaling plane of the CS network (e.g., the PSTN). The SGW performs lower layer protocol conversion. For instance, an SGW is responsible for replacing the lower MTP (ITU-T Recommendation Q.701 [133]) transport with SCTP (Stream Control Transmission Protocol, defined in RFC 2960 [230]) over IP. So, the SGW transforms ISUP (ITU-T Recommendation Q.761 [139]) or BICC (ITU-T Recommendation Q.1901 [140]) over MTP into ISUP or BICC over SCTP/IP.

MGCF (Media Gateway Control Function): the MGCF is the central node of the PSTN/CS gateway. It implements a state machine that does protocol conversion and maps SIP (the call control protocol on the IMS side) to either ISUP over IP or BICC over IP (both BICC and ISUP are call control protocols in circuit-switched networks). In addition to the call control protocol conversion the MGCF controls the resources in an MGW (Media Gateway). The protocol used between the MGCF and the MGW is H.248 (ITU-T Recommendation H.248 [143]).

MGW (Media Gateway): the Media Gateway interfaces the media plane of the PSTN or CS network. On one side the MGW is able to send and receive IMS media over the Real-Time Protocol (RTP) (RFC 3550 [225]). On the other side the MGW uses one or more PCM (Pulse Code Modulation) time slots to connect to the CS network. Additionally, the MGW performs transcoding when the IMS terminal does not support the codec used by the CS side. A common scenario occurs when the IMS terminal is using the AMR (3GPP TS 26.071 [5]) codec and the PSTN terminal is using the G.711 codec (ITU-T Recommendation G.711 [131]).

3.4.8   Home and Visited Networks

The IMS borrows a few concepts from GSM and GPRS, such as having a home and a visited network. In the cellular model, when we use our cellphones in the area where we reside, we are using the infrastructure provided by our network operator. This infrastructure forms the so-called home network.

On the other hand, if we roam outside the area of coverage of our home network (e.g., when we visit another country), we use an infrastructure provided not by our operator, but by another operator. This infrastructure is what we call the visited network, because effectively we are a *visitor* in this network.

In order to use a visited network the visited network operator has to have signed a roaming agreement with our home network operator. In these agreements both operators negotiate some aspects of the service provided to the user, such as the price of calls, the quality of service, or how to exchange accounting records.

The IMS reuses the same concept of having a visited and a home network. Most of the IMS nodes are located in the home network, but there is a node that can be either located in the home or the visited network. That node is the P-CSCF (Proxy-CSCF). The IMS allows two different configurations, depending on whether the P-CSCF is located in the home or the visited network.

Additionally, when the IP-CAN (IP Connectivity Access Network) is GPRS the location of the P-CSCF is subordinated to the location of the GGSN. In roaming scenarios, GPRS allows location of the GGSN either in the home or in the visited network (the SGSN is always located in the visited network).

In the IMS, both the GGSN and the P-CSCF share the same network. This allows the P-CSCF to control the GGSN over the so-called *Go* interface. As both the P-CSCF and the GGSN are located in the same network the *Go* interface is always an intra-operator interface, which makes its operation simpler.

Figure 3.5 shows a configuration where the P-CSCF (and the GGSN) is located in the visited network. This configuration represents a longer-term vision of the IMS, because it requires IMS support from the visited network (i.e., the GGSN has to be upgraded to be 3GPP Release 5-compliant).

It is not expected that all networks in the world will deploy IMS simultaneously. Consequently, it is not expected that all roaming partners will upgrade their GGSNs to a Release 5 GGSN at the same time the home network operator starts to provide the IMS service. So, we expect that early IMS deployments will locate the P-CSCF in the home network, as shown in Figure 3.6.

Figure 3.6 shows a near-term configuration where both the P-CSCF and the GGSN are located in the home network. This configuration does not require any IMS support from the visited network. Particularly, the visited network does not need to have a 3GPP Release 5-compliant GGSN. The visited network only provides the radio bearers and the SGSN. So, this configuration can be deployed from the very first day of the IMS. As a consequence, it is expected that this will be the most common configuration in the early years of IMS deployments.

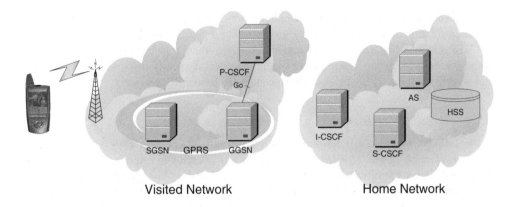

Figure 3.5: The P-CSCF located in the visited network

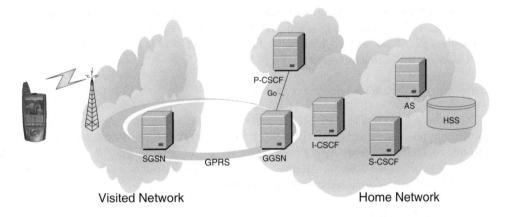

Figure 3.6: The P-CSCF located in the home network

Even so, this configuration has a severe disadvantage with respect to the configuration where the P-CSCF and GGSN are located in the visited network. Since the media plane traverses the GGSN and the GGSN is located in the home network the media are first routed to the home network and then to their destination. This creates an undesired trombone effect that causes delays in the media plane.

3.5 Identification in the IMS

In a network of any kind, it must be possible to uniquely identify users. This is the property that allows a particular phone to ring (as opposed to a different telephone) when we dial a sequence of digits in the PSTN (Public Switched Telephone Network).

Central to any network is the ability of the operator to identify users, so that calls can be directed to the appropriate user. In the PSTN, users are identified by a telephone number (i.e., a collection of ordered digits that identify the telephone subscriber). The telephone number that identifies a subscriber may be represented in different formats: a local short number, a long-distance number, or an international number. In essence, these are just different representations of the same telephone subscriber. The length of the digits depends on the destination of the call (e.g., same area, another region, or another country).

Additionally, when a service is provided, sometimes there is a need to identify the service. In the PSTN services are identified by special numbers, typically through a special prefix, such as 800 numbers. IMS also provides mechanisms to identify services.

3.5.1   Public User Identities

In the IMS there is also a deterministic way to identify users. An IMS user is allocated with one or more *Public User Identities*. The home operator is responsible for allocating these Public User Identities to each IMS subscriber. A Public User Identity is either a SIP URI (as defined in RFC 3261 [215]) or a TEL URI (as defined in RFC 3966 [220]). Public User Identities are used as contact information on business cards. In the IMS, Public User Identities are used to route SIP signaling. If we compare the IMS with GSM, a Public User Identity is to the IMS what an MSISDN (Mobile Subscriber ISDN Number) is to GSM.

When the Public User Identity contains a SIP URI, it typically takes the form of `sip:first.last@operator.com`, although IMS operators are able to change this scheme and address their own needs. Additionally, it is possible to include a telephone number in a SIP URI using the following format:

`sip:+1-212-555-0293@operator.com;user=phone`

This format is needed because SIP requires that the URI under registration be a SIP URI. So, it is not possible to register a TEL URI in SIP, although it is possible to register a SIP URI that contains a telephone number.

The TEL URI is the other format that a Public User Identity can take. The following is a TEL URI representing a phone number in international format:

`tel:+1-212-555-0293`

TEL URIs are needed to make a call from an IMS terminal to a PSTN phone, because PSTN numbers are represented only by digits. On the other hand, TEL URIs are also needed if a PSTN subscriber wants to make a call to an IMS user, because a PSTN user can only dial digits.

We envision that operators will allocate at least one SIP URI and one TEL URI per user. There are reasons for allocating more than one Public User Identity to a user, such as having the ability to differentiate personal (e.g., private) identities, that are known to friends and family from business Public User Identities (that are known to colleagues), or for triggering a different set of services.

The IMS brings an interesting concept: *a set of implicitly registered public user identities*. In regular SIP operation, each identity that needs to be registered requires a SIP REGISTER request. In the IMS, it is possible to register several Public User Identities in one message, saving time and bandwidth (the complete mechanism is described in Section 5.6).

3.5.2   Private User Identities

Each IMS subscriber is assigned a *Private User Identity*. Unlike Public User Identities, Private User Identities are not SIP URIs or TEL URIs; instead, they take the format of a NAI (Network Access Identifier, specified in RFC 2486 [45]). The format of a NAI is `username@operator.com`.

Unlike Public User Identities, Private User Identities are not used for routing SIP requests; instead, they are exclusively used for subscription identification and authentication purposes. A Private User Identity performs a similar function in the IMS as an IMSI (International Mobile Subscriber Identifier) does in GSM. A Private User Identity need not be known by the user, because it might be stored in a smart card, in the same way that an IMSI is stored in a SIM (Subscriber Identity Module).

3.5.3 The Relation between Public and Private User Identities

Operators assign one or more Public User Identities and a Private User Identity to each user. In the case of GSM/UMTS (Universal Mobile Telecommunication System) the smart card stores the Private User Identity and at least one Public User Identity. The HSS, as a general database for all the data related to a subscriber, stores the Private User Identity and the collection of Public User Identities allocated to the user. The HSS and the S-CSCF also correlate the Public and Private User Identities.

The relation between an IMS subscriber, the Private User Identity and the Public User Identities is shown in Figure 3.7. An IMS subscriber is assigned one Private User Identity and a number of Public User Identities. This is the case of the IMS as standardized in 3GPP Release 5.

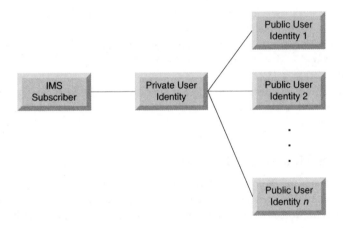

Figure 3.7: Relation of Private and Public User Identities in 3GPP R5

3GPP Release 6 has extended the relationship of Private and Public User Identities, as shown in Figure 3.8. An IMS subscriber is allocated not with one, but with a number of Private User Identities. In the case of UMTS, only one Private User Identity is stored in the smart card, but users may have different smart cards that they insert in different IMS terminals. It might be possible that some of those Public User Identities are used in combination with more than a single Private User Identity. This is the case of Public User Identity #2 in Figure 3.8, because it is assigned to both Private User Identity #1 and #2. This allows Public User Identity #2 to be used simultaneously from two IMS terminals, each one assigned with a different Private User Identity (e.g., different smart cards are inserted in different terminals).

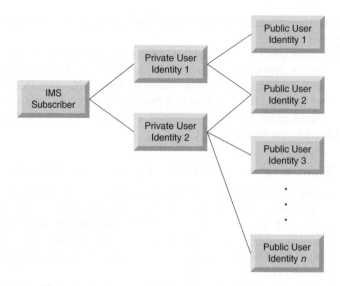

Figure 3.8: Relation of Private and Public User Identities in 3GPP R6

3.5.4 Public Service Identities

The concept of Public Service Identities (PSIs) is introduced in Release 6 of the 3GPP specifications. Unlike Public User Identities, which are allocated to users, a PSI is an identity allocated to a service hosted in an Application Server. For instance, an Application Server hosting a chat room is identified by a PSI. Like Public User Identities, PSIs may take the format of a SIP URI or a TEL URI.

Unlike Public User Identities, PSIs do not have an associated Private User Identity. This is because the Private User Identity is used for user authentication purposes. PSIs are not applicable to users.

3.6 SIM, USIM, and ISIM in 3GPP

Central to the design of 3GPP terminals is the presence of a UICC (Universal Integrated Circuit Card). The UICC is a removable smart card that contains a limited storage of data. The UICC is used to store, among other things, subscription information, authentication keys, a phonebook, and messages.

GSM and 3GPP specifications rely on the presence of a UICC in the terminal for its operation. Without a UICC present in the terminal the user can only make emergency calls.

The UICC allows users to easily move their user subscriptions (including the phonebook) from one terminal to another. The user simply removes the smart card from a terminal and inserts it into another terminal.

UICC is a generic term that defines the physical characteristics of the smart card (like the number and disposition of pins, voltage values, etc.). The interface between the UICC and the terminal is standardized.

A UICC may contain several logical applications, such as a SIM (Subscriber Identity Module), a USIM (Universal Subscriber Identity Module), and an ISIM (IP multimedia Services Identity Module). Additionally, a UICC can contain other applications, such as a telephone book. Figure 3.9 represents a UICC that contains several applications.

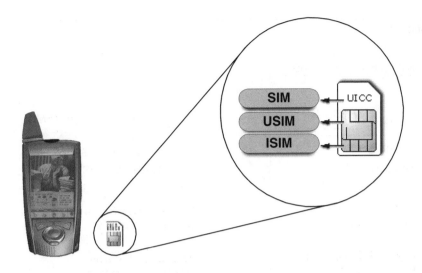

Figure 3.9: SIM, USIM, and ISIM in the UICC of 3GPP IMS terminals

3.6.1 SIM

SIM provides storage for a collection of parameters (e.g., user subscription information, user preferences, authentication keys, and storage of messages) that are essential for the operation of terminals in GSM networks. Although the terms UICC and SIM are often interchanged, UICC refers to the physical card, whereas SIM refers to a single application residing in the UICC that collects GSM user subscription information. SIM is widely used in 2G (Second Generation) networks, such as GSM networks.

The SIM application was standardized in the early stages of GSM. 3GPP inherited the specifications (currently SIM is specified in 3GPP TS 11.11 [37] and 3GPP TS 51.011 [36]).

3.6.2 USIM

USIM (standardized in 3GPP TS 31.102 [8]) is another example of an application that resides in third-generation UICCs. USIM provides another set of parameters (similar in nature, but different from those provided by SIM) which include user subscriber information, authentication information, payment methods, and storage for messages. USIM is used to access UMTS networks, the third-generation evolution of GSM.

A USIM is required if a circuit-switched or packet-switched terminal needs to operate in a 3G (Third Generation) network. Obviously, both SIM and USIM can co-exist in the same UICC, so that if the terminal is capable, it can use both GSM and UMTS networks.

Figure 3.10 shows a simplified version of the structure of USIM. USIM stores, among others, the following parameters.

IMSI (International Mobile Subscriber Identity): IMSI is an identity which is assigned to each user. This identity is not visible to the users themselves, but only to the network. IMSI is used as the user identification for authentication purposes. The Private User Identity is the equivalent of the IMSI in IMS.

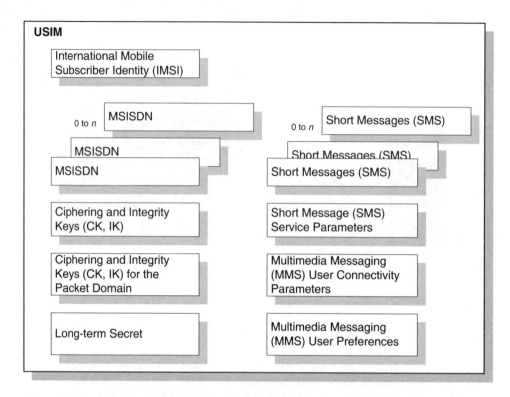

Figure 3.10: Simplified representation of the structure of the USIM application

MSISDN (Mobile Subscriber ISDN Number): this field stores one or more telephone numbers allocated to the user. A Public User Identity is the equivalent of the MSISDN in the IMS.

CK (Ciphering Key) and IK (Integrity Key): these are the keys used for ciphering and integrity protection of data over the air interface. USIM separately stores the keys used in circuit-switched and packet-switched networks.

Long-term secret: USIM stores a long-term secret that is used for authentication purposes and for calculating the integrity and cipher keys used between the terminal and the network.

SMS (Short Messages Service): USIM provides a storage for short messages and their associated data (e.g., sender, receiver, and status).

SMS parameters: this field in the USIM stores configuration data related to the SMS service, such as the address of the SMS center or the protocols that are supported.

MMS (Multimedia Messaging Service) user connectivity parameters: this field stores configuration data related to the MMS service, such as the address of the MMS server and the address of the MMS gateway.

MMS user preferences: this field stores the user preferences related to the MMS service, such as the delivery report flag, read-reply preference, priority, and time of expiration.

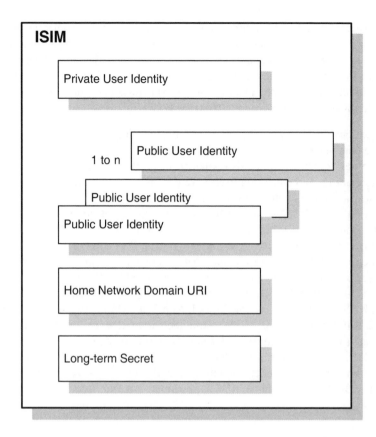

Figure 3.11: Structure of an ISIM application

3.6.3 ISIM

A third application that may be present in the UICC is ISIM (standardized in 3GPP TS 31.103 [7]). ISIM is of especial importance for the IMS, because it contains the collection of parameters that are used for user identification, user authentication, and terminal configuration when the terminal operates in the IMS. ISIM can co-exist with a SIM, a USIM, or both applications in the same UICC.

Figure 3.11 depicts the structure of the ISIM application. The relevant parameters stored in ISIM are as follows.

Private User Identity: ISIM stores the Private User Identity allocated to the user. There can only be one Private User Identity stored in ISIM.

Public User Identity: ISIM stores one or more SIP URIs of Public User Identities allocated to the user.

Home Network Domain URI: ISIM stores the SIP URI that contains the home network domain name. This is used to find the address of the home network during the registration procedure. There can only be one home network domain name URI stored in ISIM.

Long-term secret: ISIM stores a long-term secret that is used for authentication purposes and for calculating the integrity and cipher keys used between the terminal and the network. The IMS terminal uses the integrity key to integrity-protect the SIP signaling that the IMS terminal sends to or receives from the P-CSCF. If the signaling is ciphered, the IMS terminal uses the cipher key to encrypt and decrypt the SIP signaling that the IMS terminal sends to or receives from the P-CSCF.

All of the above-mentioned fields are read-only, meaning that the user cannot modify the values of the parameters.

From the description of the fields contained in ISIM the reader probably realized that ISIM is important for authenticating users. We describe in detail the access to the IMS and the authentication of users with an ISIM in Section 9.1.2.

Access to a 3GPP IMS network relies on the presence of either an ISIM or a USIM application in the UICC. ISIM is preferred because it is tailored to the IMS, although access with USIM is also possible. This allows operation in an IMS network of users who have not upgraded their UICCs to IMS-specific ones that contain an ISIM application. We describe in detail the access to the IMS and authentication with a USIM in Section 9.1.3.

Due to the lower degree of security contained in a SIM application, access to a 3GPP IMS network with a SIM application is not allowed. Non-3GPP IMS networks that do not support UICC in the IMS terminals (e.g., 3GPP2) store the parameters contained in the ISIM as part of the terminal's configuration or in the terminal's built-in memory.

3GPP2 IMS networks also allow the above-mentioned parameters to be stored in an R-UIM (Removable User Identity Module). The R-UIM is a smart card secure storage, equivalent to a 3GPP UICC with an ISIM application.

Part II

The Signaling Plane in the IMS

As we saw in Chapter 3 the IMS has a signaling plane and a media plane that traverse different paths. This part of the book (Part II) tackles the signaling plane in the IMS while Part III tackles the media plane.

We explained in Chapter 2 that one of the goals of the IMS is to use Internet technologies and protocols as much as possible. One of the main advantages of Internet protocols is that they are designed to be general enough to work within a wide range of architectures. While we are mainly interested in showing how these protocols are used in the IMS, it is essential to understand how they are used in other environments as well.

So, we will divide the explanation of every protocol into two chapters. First, we describe how each protocol works in the Internet environment and how it can be extended for use in architectures with different requirements than the public Internet (e.g., the IMS). Once we introduce a particular protocol and its design principles in a chapter, we can study how that protocol is used in the IMS in the following chapter, which protocol extensions are used, and how that protocol fits into the IMS architecture. Following these steps in the explanation of every protocol gives the reader a broad perspective of the Internet technologies used in the IMS. Understanding the similarities and differences in the operation of a particular protocol on the Internet and in the IMS will help the reader to understand how the IMS provides Internet services to its users.

Chapter 4

Session Control on the Internet

Many think that the most important component of the signaling plane is the protocol that performs session control. The protocol chosen to perform this task in the IMS is the Session Initiation Protocol (SIP) (defined in RFC 3261 [215]).

SIP was originally developed within the SIP working group in the IETF. Even though SIP was initially designed to invite users to existing multimedia conferences, today it is mainly used to create, modify and terminate multimedia sessions. In addition, there exist SIP extensions to deliver instant messages and to handle subscriptions to events. We will first look at the core protocol (used to manage multimedia sessions), and then we will deal with the most important extensions.

4.1 SIP Functionality

Protocols developed by the IETF have a well-defined scope. The functionality to be provided by a particular protocol is carefully defined in advance before any working group starts working on it. In our case the main goal of SIP is to deliver a session description to a user at their current location. Once the user has been located and the initial session description delivered, SIP can deliver new session descriptions to modify the characteristics of the ongoing sessions and terminate the session whenever the user wants.

4.1.1 Session Descriptions and SDP

A session description is, as its name indicates, a description of the session to be established. It contains enough information for the remote user to join the session. In multimedia sessions over the Internet this information includes the IP address and port number where the media needs to be sent and the codecs used to encode the voice and the images of the participants.

Session descriptions are created using standard formats. The most common format for describing multimedia sessions is the Session Description Protocol (SDP), defined in RFC 2327 [115]. Note that although the "P" in SDP stands for "Protocol", SDP is simply a textual format to describe multimedia sessions. Figure 4.1 shows an example of an SDP session description that Alice sent to Bob. It contains, among other things, the subject of the conversation (Swimming techniques), Alice's IP address (192.0.0.1), the port number where Alice wants to receive audio (20000), the port number where Alice wants to receive video

```
v=0
o=Alice 2790844676 2867892807 IN IP4 192.0.0.1
s=Let's talk about swimming techniques
c=IN IP4 192.0.0.1
t=0 0
m=audio 20000 RTP/AVP 0
a=sendrecv
m=video 20002 RTP/AVP 31
a=sendrecv
```

Figure 4.1: Example of an SDP session description

(20002), and the audio and video codecs that Alice supports (0 corresponds to the audio codec G.711 μ-law and 31 corresponds to the video codec H.261).

As we can see in Figure 4.1 an SDP description consists of two parts: session-level information and media-level information. The session-level information applies to the whole session and comes before the m= lines. In our example, the first five lines correspond to session-level information. They provide version and user identifiers (v= and o= lines), the subject of the session (s= line), Alice's IP address (c= line), and the time of the session (t= line). Note that this session is supposed to take place at the moment this session description is received. That is why the t= line is t=0 0.

The media-level information is media stream-specific and consists of an m= line and a number of optional a= lines that provide further information about the media stream. Our example has two media streams and, thus, has two m= lines. The a= lines indicate that the streams are bidirectional (i.e., users send and receive media).

As Figure 4.1 illustrates, the format of all the SDP lines consists of *type=value*, where type is always one character long. Table 4.1 shows all the types defined by SDP.

Even if SDP is the most common format to describe multimedia sessions, SIP does not depend on it. SIP is session description format-independent. That is, SIP can deliver a description of a session written in SDP or in any other format. For example, after the video conversation above about swimming techniques, Alice feels like inviting Bob to a real training session this evening in the swimming pool next to her place. She uses a session description format for swimming sessions to create a session description and uses SIP to send it to Bob. Alice's session description looks something like Figure 4.2.

This example intends to stress that SIP is completely independent of the format of the objects it transports. Those objects may be session descriptions written in different formats or any other piece of information. We will see in subsequent sections that SIP is also used to deliver instant messages, which of course are written using a different format than SDP and than our description format for swimming sessions.

4.1.2   The Offer/Answer Model

In the SDP example in Figure 4.1, Alice sent a session description to Bob that contained Alice's transport addresses (IP address plus port numbers). Obviously, this is not enough to establish a session between them. Alice needs to know Bob's transport addresses as well. SIP provides a two-way session description exchange called the offer/answer model (which is described in RFC 3264 [212]). One of the users (the offerer) generates a session description

Table 4.1: SDP types

Type	Meaning
v	Protocol version
b	Bandwidth information
o	Owner of the session and session identifier
z	Time zone adjustments
s	Name of the session
k	Encryption key
i	Information about the session
a	Attribute lines
u	URL containing a description of the session
t	Time when the session is active
e	Email address to obtain information about the session
t	Times when the session will be repeated
p	Phone number to obtain information about the session
m	Media line
c	Connection information
i	Information about the media line

```
Subject: Swimming Training Session
Time: Today from 20:00 to 21:00
Place: Lane number 4 of the swimming-pool near my place
```

Figure 4.2: Example of a session description without using SDP

(the offer) and sends it to the remote user (the answerer) who then generates a new session description (the answer) and sends it to the offerer.

RFC 3264 [212] provides the rules for offer and answer generation. After the offer/answer exchange, both users have a common view of the session to be established. They know, at least, the formats they can use (i.e., formats that the remote end understands) and the transport addresses for the session. The offer/answer exchange can also provide extra information, such as cryptographic keys to encrypt traffic.

Figure 4.3 shows the answer that Bob sent to Alice after having received Alice's offer in Figure 4.1. Bob's IP address is 192.0.0.2, the port number where Bob will receive audio is 30000, the port number where Bob will receive video is 30002, and, fortunately, Bob supports the same audio and video codecs as Alice (G.711 μ-law and H.261). After this offer/answer exchange, all they have left to do is to have a nice video conversation.

4.1.3 SIP and SIPS URIs

SIP identifies users using SIP URIs, which are similar to email addresses; they consist of a user name and a domain name. Additionally, SIP URIs can contain a number of parameters

```
v=0
o=Bob 234562566 236376607 IN IP4 192.0.0.2
s=Let's talk about swimming techniques
c=IN IP4 192.0.0.2
t=0 0
m=audio 30000 RTP/AVP 0
a=sendrecv
m=video 30002 RTP/AVP 31
a=sendrecv
```

Figure 4.3: Bob's SDP session description

(e.g., transport), which are encoded using semi-colons. The following are examples of SIP URIs:

```
sip:Alice.Smith@domain.com
sip:Bob.Brown@example.com
sip:carol@ws1234.domain2.com;transport=tcp
```

Additionally, users can be identified using SIPS URIs. Entities contacting a SIPS URI use TLS (Transport Layer Security, see Section 8.3) to secure their messages. The following are examples of SIP URIs:

```
sips:Alice.Smith@domain.com
sips:Bob.Brown@example.com
```

4.1.4   User Location

We said earlier that the main purpose of SIP is to deliver a session description to a user at their current location, and we have already seen what a session description looks like. Now, let us look at how SIP tracks the location of a given user.

SIP provides personal mobility. That is, users can be reached using the same identifier no matter where they are. For example, Alice can be reached at

```
sip:Alice.Smith@domain.com
```

regardless of her current location. This is her public URI, also known as her AoR (Address of Record).

Nevertheless, when Alice is logged in at work her SIP URI is

```
sip:asmith@ws1234.company.com
```

and when she is working at her computer at the university her SIP URI is

```
sip:alice@pc12.university.edu
```

Therefore, we need a way to map Alice's public URI

```
sip:Alice.Smith@domain.com
```

to her current URI (at work or at the university) at any given moment.

To do this, SIP introduces a network element called the registrar of a particular domain. A registrar handles requests addressed to its domain. Thus, SIP requests sent to

`sip:Alice.Smith@domain.com`

will be handled by the SIP registrar at domain.com.

Every time Alice logs into a new location, she registers her new location with the registrar at domain.com, as shown in Figure 4.4. This way the registrar at domain.com can always forward incoming requests to Alice wherever she is.

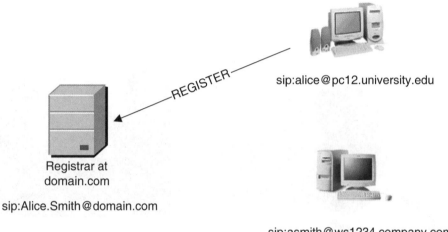

sip:alice@pc12.university.edu

REGISTER

Registrar at
domain.com

sip:Alice.Smith@domain.com

sip:asmith@ws1234.company.com

Figure 4.4: Alice registers her location with the domain.com registrar

On reception of the registration the registrar at domain.com can store the mapping between Alice's public URI and her current location in two ways: it can use a local database or it can upload this mapping into a location server. If the registrar uses a location server, it will need to consult it when it receives a request for Alice. Note that the interface between the registrar and the location server is not based on SIP, but on other protocols.

4.2 SIP Entities

Besides the registrars, which were introduced in the previous section, SIP defines user agents, proxy servers, and redirect servers. UAs (User Agents) are SIP endpoints that are usually handled by a user. In any case, user agents can also establish sessions automatically with no user intervention (e.g., a SIP voicemail). Sessions are typically established between user agents.

User agents come in all types of flavors. Some are software running on a computer, others, like the commercial SIP phones shown in Figure 4.5, look like desktop phones, and others still are embedded in mobile devices like laptops, PDAs, or mobile phones. Some of them are not even used for telephony and do not have speakers or microphones.

Proxy servers, typically referred to as proxies, are SIP routers. A proxy receives a SIP message from a user agent or from another proxy and routes it toward its destination.

Figure 4.5: Three examples of commercial SIP phones

Routing the request involves relaying the message to the destination user agent or to another proxy in the path.

It is important to understand fully how SIP routing works, because it is one of the key components of the protocol. A given user can be available at several user agents at the same time. For instance, Alice can be reachable on her computer at the university

`sip:alice@pc12.university.edu`

and on her PDA with a wireless connection

`sip:alice@pda.com`

She has registered both locations with the registrar at domain.com. If the registrar receives a SIP message addressed to Alice's public URI

`sip:Alice.Smith@domain.com`

it has to decide whether to route it to Alice's computer or to Alice's PDA. In this case, Alice has programmed the registrar to route SIP messages to her computer between 8:00 and 13:00

and to her PDA from 13:00 to 14:00. The registrar simply checks the current time and routes the SIP message accordingly.

Being able to route SIP messages on the basis of any criteria is a very powerful tool for building services that are specially tailored to the needs of each user. Users typically choose to route SIP messages based on the sender, the time of the day, whether the subject is business-related or personal, the type of session (e.g., route video calls to the computer with the big screen), etc, the combinations are infinite.

In the previous example we saw that the registrar routed the SIP message to Alice's user agent. Yet, the entities handling routing of messages are called proxies. Proxies and registrars are only logical roles. In our example the same physical box acts as a registrar when Alice registered her current location and as a proxy when it was routing SIP messages toward Alice's user agent. This configuration is shown in Figure 4.6.

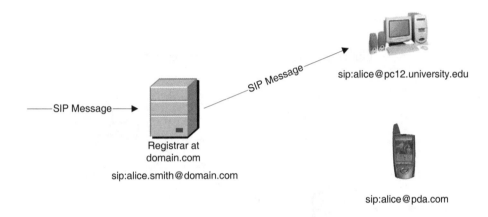

Figure 4.6: Proxy co-located with the registrar of the domain

A different configuration could consist of using a separate physical box for each role, as shown in Figure 4.7. Here, the proxy needs to access the information about Alice's location that the registrar got in the first place. This is resolved by adding a location server. The registrar uploads Alice's location to the location server, and the proxy consults the location server in order to route incoming messages.

4.2.1 Forking Proxies

In the previous examples the proxy chose a single user agent as the destination of the SIP message. Still, sometimes it is useful to receive calls on several user agents at the same time. For instance, in a house all the telephones ring at once, giving us the chance to pick up the call in the kitchen or in the living room. SIP proxy servers that route messages to more than one destination are called forking proxies, as shown in Figure 4.8.

A forking proxy can route messages in parallel or in sequence. An example of parallel forking is the simultaneous ringing of all the telephones in a house. Sequential forking consists of the proxy trying the different locations one after the other. A proxy can, for example, let a user agent ring for a certain period of time and, if the user does not pick up, try a new user agent.

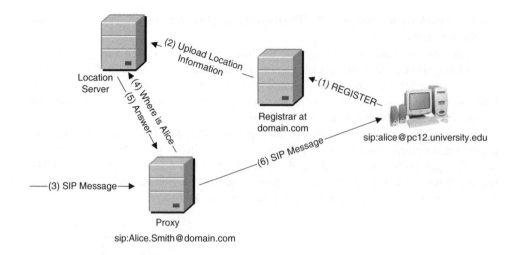

Figure 4.7: Proxy and registrar kept separate

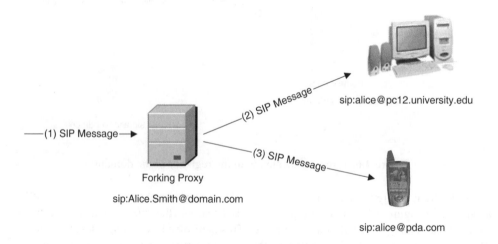

Figure 4.8: Forking proxy operation

4.2.2 Redirect Servers

Redirect servers are also used to route SIP messages, but they do not relay the message to its destination as proxies do. Redirect servers instruct the entity that sent the message (a user agent or a proxy) to try a new location instead. Figure 4.9 shows how redirect servers work. A user agent sends a SIP message to

`sip:Alice.Smith@domain.com`

and the redirect server tells it to try the alternative address

`sip:alice@pda.com.`

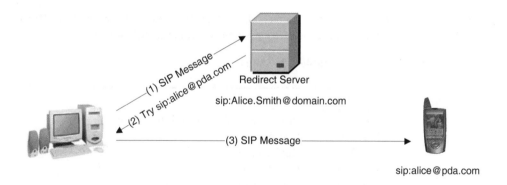

Figure 4.9: Redirect server operation

4.3   Message Format

SIP is based on HTTP [101] and, so, is a textual request-response protocol. Clients send requests, and servers answer with responses. A SIP transaction consists of a request from a client, zero or more provisional responses, and a final response from a server. We will introduce the format of SIP requests and responses before explaining, in Section 4.8, the types of transactions that SIP defines.

Figure 4.10 shows the format of SIP messages. They start with the *start line*, which is called the *request line* in requests and the *status line* in responses. The *start line* is followed by a number of header fields that follow the format *name:value* and an empty line that separates the header fields from the optional message body.

```
Start line
A number of header fields
Empty line
Optional message body
```

Figure 4.10: SIP message format

4.4   The Start Line in SIP Responses: the Status Line

As we said earlier the start line of a response is referred to as the status line. The status line contains the protocol version (SIP/2.0) and the status of the transaction, which is given in numerical (status code) and in human-readable (reason phrase) formats. The following is an example of a status line:

```
SIP/2.0 180 Ringing
```

The protocol version is always set to SIP/2.0 (a history of previous versions of the protocol is given in *SIP Demystified* [61]). We will see in Section 4.11 how SIP is extended without needing to increase its protocol version.

The status code 180 indicates that the remote user is being alerted. Ringing is the reason phrase and is intended to be read by a human (e.g., displayed to the user). Since it is intended for human consumption the reason phrase can be written in any language.

Responses are classified by their status codes, which are integers that range from 199 to 699. Table 4.2 shows how status codes are classified according to their values.

Table 4.2: Status code ranges

Status code range	Meaning
100–199	Provisional (also called informational)
200–299	Success
300–399	Redirection
400–499	Client error
500–599	Server error
600–699	Global failure

Apart from the start line (status line in responses and request line in requests) the format of requests and responses is identical, as shown in Figure 4.10. So, let's now tackle the format of the request line and then the format of the rest of the message.

4.5   The Start Line in SIP Requests: the Request Line

The start line in requests is referred to as the request line. It consists of a *method name*, the *Request-URI*, and the protocol version SIP/2.0. The method name indicates the purpose of the request and the *Request-URI* contains the destination of the request. Below, you have an example of a request line:

```
INVITE sip:Alice.Smith@domain.com SIP/2.0
```

The method name in this example is INVITE. It indicates that the purpose of this request is to invite a user to a session. The Request-URI shows that this request is intended for Alice.

Table 4.3 shows the methods that are currently defined in SIP and their meaning.

Figure 4.11 shows a SIP transaction. The User Agent Client (UAC) sends a BYE request, and the User Agent Server (UAS) sends back a 200 (OK) response. Note that, usually, SIP message flows only show the method name of the request and the status code and the reason phrase of the response. These pieces of information are usually enough to understand any message flow.

Before explaining the types of SIP transactions and how to use them, we will study the formats of SIP header fields and bodies. After that, we will provide the reader with some message flows that will help her understand how to perform useful tasks, such as establishing a session using SIP.

4.6   Header Fields

Right after the start line, SIP messages (both requests and responses) contain a set of header fields (see Figure 4.10). There are mandatory header fields that appear in every message and optional header fields that only appear when needed. A header field consists of the header field's name, a colon, and the header field's value, as shown in the example below:

```
To: Alice Smith <sip:Alice.Smith@domain.com>;tag=1234
```

Table 4.3: SIP methods

Method name	Meaning
ACK	Acknowledges the establishment of a session
BYE	Terminates a session
CANCEL	Cancels a pending request
INFO	Transports PSTN telephony signaling
INVITE	Establishes a session
NOTIFY	Notifies the user agent about a particular event
OPTIONS	Queries a server about its capabilities
PRACK	Acknowledges the reception of a provisional response
PUBLISH	Uploads information to a server
REGISTER	Maps a public URI with the current location of the user
SUBSCRIBE	Requests to be notified about a particular event
UPDATE	Modifies some characteristics of a session
MESSAGE	Carries an instant message
REFER	Instructs a server to send a request

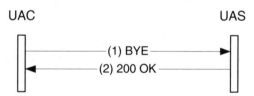

Figure 4.11: SIP transaction

As we can see, the value of a header field can consist of multiple items. The To header field above contains a display name (Alice Smith), a URI

```
sip:Alice.Smith@domain.com
```

and a tag parameter.

Some header fields can have more than one entry in the same message, as shown in the example below:

```
Route: <sip:p1.domain1.com>
Route: <sip:p34.domain2.com>
```

Multi-entry header fields can appear in a single-value-per-line form, as shown above, or in a comma-separated value form, as shown below. Both formats are equivalent:

```
Route: <sip:p1.domain1.com>, <sip:p34.domain2.com>
```

Note that in all the examples so far there is a space between the colon and the value of the header field. In the example above, we can also see a space after the comma separating both Route entries. SIP parsers ignore these spaces, but they are typically included in the messages to improve their readability by humans.

Let us have a look at the most important SIP header fields: the six mandatory header fields that appear in every SIP message. They are To, From, Cseq, Call-ID, Max-Forwards, and Via.

To: the To header field contains the URI of the destination of the request. However, this URI is not used to route the request. It is intended for human consumption and for filtering purposes. For example, a user can have a private URI and a professional URI and filter requests depending on which URI appears in the To field. The tag parameter is used to distinguish, in the presence of forking proxies, different user agents that are identified with the same URI.

From: the From header field contains the URI of the originator of the request. Like the To header field, it is mainly used for human consumption and for filtering purposes.

Cseq: the Cseq header field contains a sequence number and a method name. They are used to match requests and responses.

Call-ID: the Call-ID provides a unique identifier for a SIP message exchange.

Max-Forwards: the Max-Forwards header field is used to avoid routing loops. Every proxy that handles a request decrements its value by one, and if it reaches zero, the request it discarded.

Via: the Via header field keeps track of all the proxies a request has traversed. The response uses these Via entries so that it traverses the same proxies as the request did in the opposite direction.

4.7 Message Body

As Figure 4.10 shows, the message body is separated from the header fields by an empty line. SIP messages can carry any type of body and even multipart bodies using MIME (Multipurpose Internet Mail Extensions) encoding.

RFC 2045 [103] defines the MIME format which allows us to send emails with multiple attachments in different formats. For example, a given email message can carry a JPEG picture and an MPEG video as attachments.

SIP uses MIME to encode its message bodies. Consequently, SIP bodies are described in the same way as attachments to an email message. A set of header fields provide information about the body: its length, its format, and how it should be handled. For example, the header fields below describe the SDP session description of Figure 4.1:

```
Content-Disposition: session
Content-Type: application/sdp
Content-Length: 193
```

The Content-Disposition indicates that the body is a session description, the Content-Type indicates that the session description uses the SDP format, and the Content-Length contains the length of the body in bytes.

Figure 4.12 shows an example of a multipart body encoded using MIME. The first body part is an SDP session description and the second body part consists of the text "This is the second body part". Note that the Content-Type for the whole body is multipart/mixed

```
Content-Type: multipart/mixed; boundary="--0806040504000805090"
Content-Length: 384

--0806040504000805090
Content-Type: application/sdp
Content-Disposition: session

v=0
o=Alice 2790844676 2867892807 IN IP4 192.0.0.1
s=Let's talk about swimming techniques
c=IN IP4 192.0.0.1
t=0 0
m=audio 20000 RTP/AVP 0
a=sendrecv
m=video 20002 RTP/AVP 31
a=sendrecv
--0806040504000805090
Content-Type: text/plain;

This is the second body part
--0806040504000805090
```

Figure 4.12: MIME encoding of a multipart body

and that each body part has its own Content-Type: namely application/sdp and text/plain.

An important property of bodies is that they are transmitted end to end. That is, proxies do not need to parse the message body in order to route the message. In fact, the user agents may choose to encrypt the contents of the message body end-to-end. In this case, proxies would not even be able to tell which type of session is being established between both user agents.

4.8 SIP Transactions

Now that we know all the elements in a SIP network and the elements of SIP messages, we can study the three types of transactions that SIP defines: regular transactions, INVITE–ACK transactions, and CANCEL transactions. The type of a particular transaction depends on the request initiating it.

Regular transactions are initiated by any request but INVITE, ACK, or CANCEL. Figure 4.13 shows a regular BYE transaction. In a regular transaction, the user agent server receives a request and generates a final response that terminates the transaction. In theory, it would be possible for the user agent server to generate one or more provisional responses before generating the final response, although in practice, provisional responses are seldom sent within a regular transaction.

An INVITE–ACK transaction involves two transactions; an INVITE transaction and an ACK transaction, as shown in Figure 4.14. The user agent server receives an INVITE request and generates zero or more provisional responses and a final response. When the user agent

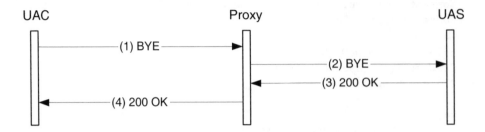

Figure 4.13: Regular transaction

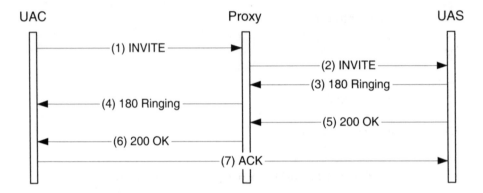

Figure 4.14: INVITE–ACK transaction

client receives the final response, it generates an ACK request, which does not have any response associated with it.

CANCEL transactions are initiated by a CANCEL request and are always connected to a previous transaction (i.e., the transaction to be cancelled). CANCEL transactions are similar to regular transactions, with the difference that the final response is generated by the next SIP hop (typically a proxy) instead of by the user agent server. Figure 4.15 shows a CANCEL transaction cancelling an INVITE transaction. Note that the INVITE transaction, once it is cancelled, terminates as usual (i.e., final response plus ACK).

4.9   Message Flow for Session Establishment

Now that we have introduced the different types of SIP transactions, let's see how we can use SIP to establish a multimedia session. First of all, Alice registers her current location

```
sip:alice@pda.com
```

with the registrar at domain.com, as shown in Figure 4.16. To do this, Alice sends a REGISTER request (Figure 4.17) indicating that requests addressed to the URI in the To header field

```
sip:Alice.Smith@domain.com
```

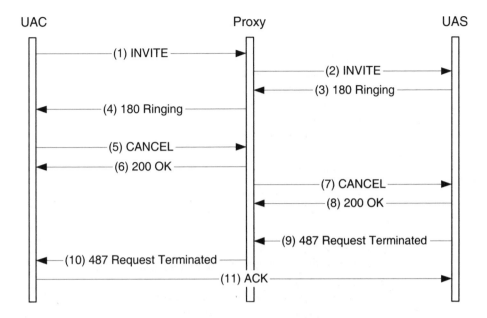

Figure 4.15: CANCEL transaction

Figure 4.16: Alice registers her location

should be relayed to the URI in the `Contact` header field

`sip:alice@pda.com`

The *Request-URI* of the REGISTER request contains the domain of the registrar (domain.com). The registrar responds with a 200 (OK) response (Figure 4.18) indicating that the transaction was successfully completed.

At a later time, Bob invites Alice to an audio session. Figure 4.19 shows the establishment of the audio session between Bob and Alice through the proxy server at domain.com.

Bob sends an INVITE request (Figure 4.20) using Alice's public URI

`sip:Alice.Smith@domain.com`

as the *Request-URI*. The proxy at domain.com relays the INVITE request (Figure 4.21) to Alice at her current location (her PDA). Alice accepts the invitation sending a 200 (OK) response (Figure 4.22), which is relayed by the proxy to Bob (Figure 4.23).

```
REGISTER sip:domain.com SIP/2.0
Via: SIP/2.0/UDP 192.0.0.1:5060;branch=z9hG4bKna43f
Max-Forwards: 70
To: <sip:Alice.Smith@domain.com>
From: <sip:Alice@pda.com>;tag=453448
Call-ID: 843528637684230@998sdasdsfgt
Cseq: 1 REGISTER
Contact: <sip:alice@pda.com>
Expires: 7200
Content-Length: 0
```

Figure 4.17: (1) REGISTER

```
SIP/2.0 200 OK
Via: SIP/2.0/UDP 192.0.0.1:5060;branch=z9hG4bKna43f
    ;received=192.0.0.1
To: <sip:Alice.Smith@domain.com>;tag=54262
From: <sip:Alice@pda.com>;tag=453448
Call-ID: 843528637684230@998sdasdsfgt
Cseq: 1 REGISTER
Contact: <sip:alice@pda.com>
Expires: 7200
Content-Length: 0
```

Figure 4.18: (2) 200 OK

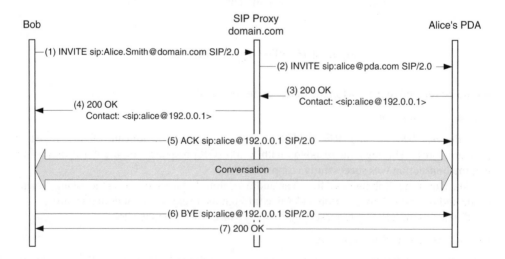

Figure 4.19: Session establishment through a proxy

```
INVITE sip:Alice.Smith@domain.com SIP/2.0
Via: SIP/2.0/UDP ws1.domain2.com:5060;branch=z9hG4bK74gh5
Max-Forwards: 70
From: Bob <sip:Bob.Brown@domain2.com>;tag=9hx34576sl
To: Alice <sip:Alice.Smith@domain.com>
Call-ID: 6328776298220188511@192.0.100.2
Cseq: 1 INVITE
Contact: <sip:bob@192.0.100.2>
Content-Type: application/sdp
Content-Length: 151

v=0
o=bob 2890844526 2890844526 IN IP4 ws1.domain2.com
s=-
c=IN IP4 192.0.100.2
t=0 0
m=audio 20000 RTP/AVP 0
a=rtpmap:0 PCMU/8000
```

Figure 4.20: (1) INVITE

Note that Alice has included a `Contact` header field in her 200 (OK) response. This header field is used by Bob to send subsequent messages to Alice. This way, once the proxy at domain.com has helped Bob locate Alice, Bob and Alice can exchange messages directly between them.

Bob uses the URI in the `Contact` header field of the 200 (OK) response to send his ACK (Figure 4.24). Now that the session (i.e., an audio stream) is established, Bob and Alice can talk about whatever they want. If, in the middle of the session, they wanted to make any changes to the session (e.g., add video), all they would need to do would be to issue another INVITE request with an updated session description. INVITE requests sent within an ongoing session are usually referred to as re-INVITEs. (UPDATE requests can also be used to modify ongoing sessions. In any case, UPDATEs are used when no interactions with the callee are expected. In this case, we use re-INVITE because the callee is typically prompted before adding video to a session.)

When Bob and Alice finish their conversation, Bob sends a BYE request to Alice (Figure 4.25). Note that, as with the ACK, this request is sent directly to Alice, without the intervention of the proxy. Alice responses with a 200 (OK) response to the BYE request (Figure 4.26).

4.10 SIP Dialogs

In Figure 4.19, Bob and Alice exchange a number of SIP messages in order to establish (and terminate) a session. The exchange of a set of SIP messages between two user agents is referred to as a SIP dialog. In our example the SIP dialog is established by the "INVITE–200 OK" transaction and is terminated by the "BYE–200 OK" transaction. Note, however, that, in addition to INVITE, there are other methods that can create dialogs as well (e.g., SUBSCRIBE). We will study them in later sections.

```
INVITE sip:Alice.Smith@domain.com SIP/2.0
Via: SIP/2.0/UDP p1.domain.com:5060;branch=z9hG4bK543fg
Via: SIP/2.0/UDP ws1.domain2.com:5060;branch=z9hG4bK74gh5
   ;received=192.0.100.2
Max-Forwards: 69
From: Bob <sip:Bob.Brown@domain2.com>;tag=9hx34576sl
To: Alice <sip:Alice.Smith@domain.com>
Call-ID: 6328776298220188511@192.0.100.2
Cseq: 1 INVITE
Contact: <sip:bob@192.0.100.2>
Content-Type: application/sdp
Content-Length: 151


v=0
o=bob 2890844526 2890844526 IN IP4 ws1.domain2.com
s=-
c=IN IP4 192.0.100.2
t=0 0
m=audio 20000 RTP/AVP 0
a=rtpmap:0 PCMU/8000
```

Figure 4.21: (2) INVITE


```
SIP/2.0 200 OK
Via: SIP/2.0/UDP p1.domain.com:5060;branch=z9hG4bK543fg
   ;received=192.1.0.1
Via: SIP/2.0/UDP ws1.domain2.com:5060;branch=z9hG4bK74gh5
   ;received=192.0.100.2
From: Bob <sip:Bob.Brown@domain2.com>;tag=9hx34576sl
To: Alice <sip:Alice.Smith@domain.com>;tag=1df345fkj
Call-ID: 6328776298220188511@192.0.100.2
Cseq: 1 INVITE
Contact: <sip:alice@192.0.0.1>
Content-Type: application/sdp
Content-Length: 151


v=0
o=alice 2890844545 2890844545 IN IP4 192.0.0.1
s=-
c=IN IP4 192.0.0.1
t=0 0
m=audio 30000 RTP/AVP 0
a=rtpmap:0 PCMU/8000
```

Figure 4.22: (3) 200 OK

```
SIP/2.0 200 OK
Via: SIP/2.0/UDP ws1.domain2.com:5060;branch=z9hG4bK74gh5
   ;received=192.0.100.2
From: Bob <sip:Bob.Brown@domain2.com>;tag=9hx34576sl
To: Alice <sip:Alice.Smith@domain.com>;tag=1df345fkj
Call-ID: 6328776298220188511@192.0.100.2
Cseq: 1 INVITE
Contact: <sip:alice@192.0.0.1>
Content-Type: application/sdp
Content-Length: 151

v=0
o=alice 2890844545 2890844545 IN IP4 192.0.0.1
s=-
c=IN IP4 192.0.0.1
t=0 0
m=audio 30000 RTP/AVP 0
a=rtpmap:0 PCMU/8000
```

Figure 4.23: (4) 200 OK

```
ACK sip:alice@192.0.0.1 SIP/2.0
Via: SIP/2.0/UDP ws1.domain2.com:5060;branch=z9hG4bK74765
Max-Forwards: 70
From: Bob <sip:Bob.Brown@domain2.com>;tag=9hx34576sl
To: Alice <sip:Alice.Smith@domain.com>;tag=1df345fkj
Call-ID: 6328776298220188511@192.0.100.2
Cseq: 1 ACK
Contact: <sip:bob@192.0.100.2>
Content-Length: 0
```

Figure 4.24: (5) ACK

```
BYE sip:alice@192.0.0.1 SIP/2.0
Via: SIP/2.0/UDP ws1.domain2.com:5060;branch=z9hG4bK745gh
Max-Forwards: 70
From: Bob <sip:Bob.Brown@domain2.com>;tag=9hx34576sl
To: Alice <sip:Alice.Smith@domain.com>;tag=1df345fkj
Call-ID: 6328776298220188511@192.0.100.2
Cseq: 2 BYE
Contact: <sip:bob@192.0.100.2>
Content-Length: 0
```

Figure 4.25: (6) BYE

```
SIP/2.0 200 OK
Via: SIP/2.0/UDP ws1.domain2.com:5060;branch=z9hG4bK745gh
    ;received=192.0.100.2
From: Bob <sip:Bob.Brown@domain2.com>;tag=9hx34576sl
To: Alice <sip:Alice.Smith@domain.com>;tag=1df345fkj
Call-ID: 63287762982201885110192.0.100.2
Cseq: 2 BYE
Contact: <sip:alice@192.0.0.1>
Content-Length: 0
```

Figure 4.26: (7) 200 OK

When a SIP dialog is established (e.g., with an INVITE transaction), all the subsequent requests within that dialog follow the same path. In our example, all the requests after the INVITE (the ACK (5) and the BYE (6)) are sent end to end between the user agents. However, some proxies choose to remain in the signaling path for subsequent requests within a dialog instead of routing the first INVITE request and stepping down after the 200 (OK) response. Let us study the mechanism used by proxies to stay in the path after the first INVITE request. It consists of three header fields: Record-Route, Route, and Contact.

4.10.1   Record-Route, Route, *and* Contact *Header Fields*

Figure 4.27 shows a message flow where the proxy at domain.com remains in the path for all the requests sent within the dialog. The proxy requests to remain in the path by adding a Record-Route header field to the INVITE request (2). The lr parameter that appears at the end of the URI indicates that this proxy is RFC 3261-compliant (older proxies used a different routing mechanism).

Alice obtains the Record-Route header field with the proxy's URI in the INVITE request (2), and Bob obtains it in the 200 (OK) response (4). From that point on, both Bob and Alice insert a Route header field in their requests, indicating that the proxy at domain.com needs to be visited. The ACK (5 and 6) is an example of a request with a Route header field sent from Bob to Alice. The BYE (7 and 8) shows that requests in the opposite direction (i.e., from Alice to Bob) use the same Route mechanism.

4.11   Extending SIP

So far, we have focused on describing the core SIP protocol, as defined in RFC 3261 [215]. Now that the main SIP concepts (such as registrars, proxies, redirect servers, forking, SIP encoding, and SIP routing) are clear, it is time to study how SIP is extended.

SIP's extension negotiation mechanism uses three header fields: Supported, Require, and Unsupported. When a SIP dialog is being established the user agent client lists all the names of the extensions it wants to use for that dialog in a Require header field, and all the names of the extensions it supports not listed previously in a Supported header field. The names of the extensions are referred to as *option tags*.

The user agent server inspects the Require header field and, if it does not support any of the extensions listed there, it sends back an error response indicating that the dialog could not be established. This error response contains an Unsupported header field listing the extensions the user agent server did not support.

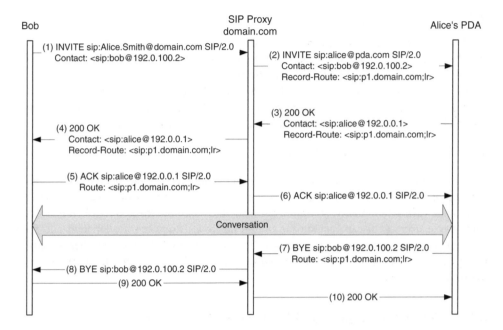

Figure 4.27: Usage of Record-Route, Route, and Contact

If the user agent server supports (and is willing to use) all the required extensions, it should decide whether or not it wants to use any extra extension for this dialog and, if so, it includes the option tag for the extension in the Require header field of its response. If this option tag was included in the Supported header field of the client, the dialog will be established. Otherwise, the client does not support the extension (or is not willing to use it). In this case the user agent server includes the extension which is required by the server in a Require header field of an error response. Such an error response terminates the establishment of the dialog.

Figure 4.28 shows a successful extension negotiation between Bob and Alice. They end up using the extensions whose option tags are foo1, foo2, and foo4.

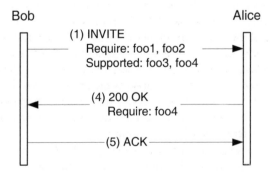

Figure 4.28: Extension negotiation in SIP

4.11.1   New Methods

In addition to option tags, SIP can be extended by defining new methods. We saw in Table 4.3 that there are many SIP methods, but that the core protocol only uses a subset of them. The rest of the methods are defined in SIP extensions.

In a SIP dialog the user agents need to know which methods the other end understands. For this purpose, each of the user agents include an `Allow` header field in their messages listing all the methods it supports. An example of an `Allow` header field is:

`Allow: INVITE, ACK, CANCEL, OPTIONS, BYE`

As we can see the `Allow` header field lets user agents advertise the methods they support, but cannot be used to express that a particular method is required for a particular dialog. To provide such functionality, an option tag associated with the method required is defined. This way a user agent can include the option tag in its `Require` header field and force the remote end to apply the extension and, so, to understand the method. The extension for reliable provisional responses described in Section 4.13 is an example of an option tag associated with a method.

4.12   Caller Preferences and User Agent Capabilities

We saw in Section 4.9 how Alice's user agents can register their location in a registrar using a REGISTER request. When a proxy server in the same domain as the registrar receives a request for Alice, it relays the request to all the locations registered by Alice's user agents. Still, Alice might not want to receive personal calls on her office phone or business calls on her home phone. Moreover, the person calling Alice may not want to talk to her, but only leave a message on her voicemail. The following two SIP extensions make it possible to do what we have just described.

The user agent capabilities extension (defined in RFC 3840 [217]) allows user agents to provide more information about themselves when they register. A user agent can indicate, among other things: the SIP methods it supports; whether or not it supports video, audio, and text communications; whether it is used for business or for personal communications; and whether it is handled by a human or by an automaton (e.g., voicemail). Figure 4.29 shows a REGISTER that carries user agent capabilities in its `Contact` header field. In this case the user agent registering supports both audio and video, is fixed (as opposed to mobile), and implements the following SIP methods: INVITE, BYE, OPTIONS, ACK, and CANCEL.

The user agent capabilities defined originally by the IETF consisted of simple properties, such as support for audio or video or being a mobile or a fixed device. By contrast, the current trend in the industry is to define whole services as single capabilities. For instance, a user agent can inform the registrar that it supports the conferencing service provided by the operator. Supporting such a conferencing service may include supporting a particular floor control protocol and stereophonic audio, but both capabilities are contained in a single user agent capability: the conferencing service capability.

The caller preferences extension (defined in RFC 3841 [216]) allows callers to indicate the type of user agent they want to reach. For instance, a caller may only want to speak to a human (no voicemails) or may want to reach a user agent with video capabilities. The caller preferences are carried in the `Accept-Contact`, `Reject-Contact`, and `Request-Disposition` header fields.

```
REGISTER sip:domain.com SIP/2.0
Via: SIP/2.0/UDP 192.0.0.1:5060;branch=z9hG4bKna43f
Max-Forwards: 70
To: <sip:Alice.Smith@domain.com>
From: <sip:Alice@pda.com>;tag=453448
Call-ID: 843528637684230@998sdasdsfgt
Cseq: 1 REGISTER
Contact: <sip:alice@pda.com;audio;video>
        ;mobility="fixed"
        ;methods="INVITE,BYE,OPTIONS,ACK,CANCEL"
Expires: 7200
Content-Length: 0
```

Figure 4.29: REGISTER carrying UA capabilities

The `Accept-Contact` header field contains a description of the destination user agents where it is OK to send the request. On the other hand, the `Reject-Contact` header field contains a description of the destination user agents where it is *not* OK to send the request. The `Request-Disposition` header field indicates how servers dealing with the request should handle it: whether they should proxy or redirect and whether they should perform sequential or parallel searches for the user.

Figure 4.30 shows an INVITE request that carries caller preferences. The caller that sent this request wants to reach a mobile user agent that implements the INVITE, OPTIONS, BYE, CANCEL, ACK, and MESSAGE methods and that does not support video. In addition, the caller wants proxies to perform parallel searches for the callee.

4.13 Reliability of Provisional Responses

When only the core protocol is used, SIP provisional responses are not transmitted reliably. Only requests and final responses are considered important and, thus, transmitted reliably. However, some applications need to ensure that provisional responses are delivered to the user agent client. For example, a telephony application may find it important to let the caller know whether or not the callee is being alerted. Since SIP transmits this information in a 180 (Ringing) provisional response this telephony application needs to use an extension for reliable provisional responses.

However, before describing such an extension, let's study how requests and final responses are transmitted. SIP is transport protocol-agnostic and, thus, can run over reliable transport protocols, such as TCP (Transport Control Protocol), and over unreliable transport protocols, such as UDP (User Datagram Protocol). The reader can find in the IEEE Network article [74] an evaluation of transport protocols for SIP that analyzes the pros and cons of UDP, TCP, and SCTP [230] to be used underneath SIP.

Regardless of the transport protocol, SIP provides an application-layer acknowledgement message that confirms the reception of the original message by the other end. When unreliable transport protocols are used, messages are retransmitted at the application layer until the acknowledge message arrives.

Some SIP messages are transmitted hop by hop (e.g., INVITE requests), while others are transmitted end to end (e.g., 200 (OK) responses for an INVITE request). Figure 4.31 shows

```
INVITE sip:Alice.Smith@domain.com SIP/2.0
Via: SIP/2.0/UDP ws1.domain2.com:5060;branch=z9hG4bK74gh5
Max-Forwards: 70
From: Bob <sip:Bob.Brown@domain2.com>;tag=9hx34576sl
To: Alice <sip:Alice.Smith@domain.com>
Call-ID: 6328776298220188511@192.0.100.2
Cseq: 1 INVITE
Request-Disposition: proxy, parallel
Accept-Contact: *;mobility="mobile"
      ;methods="INVITE,OPTIONS,BYE,CANCEL,ACK,MESSAGE"
Reject-Contact: *;video
Contact: <sip:bob@192.0.100.2>
Content-Type: application/sdp
Content-Length: 151

v=0
o=bob 2890844526 2890844526 IN IP4 ws1.domain2.com
s=-
c=IN IP4 192.0.100.2
t=0 0
m=audio 20000 RTP/AVP 0
a=rtpmap:0 PCMU/8000
```

Figure 4.30: INVITE carrying caller preferences

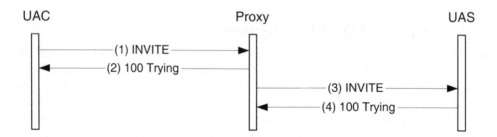

Figure 4.31: Hop-by-hop transmission in SIP

a hop-by-hop message. Upon reception of the 100 Trying response the user agent client knows that the next hop (i.e., the proxy) has received the request.

Figure 4.32 shows the previous hop-by-hop message followed by an end-to-end message. In the end-to-end message the user agent server, upon reception of the ACK request, knows that the remote end (i.e., the user agent client, as opposed to the proxy) has received the request.

Coming back to the provisional responses (other than 100 Trying), there is no application-layer acknowledgement message for them in core SIP. Therefore, there is an extension, defined in RFC 3262 [213], whose option tag is 100rel that creates such a message: a PRACK request. Figure 4.33 shows how this works.

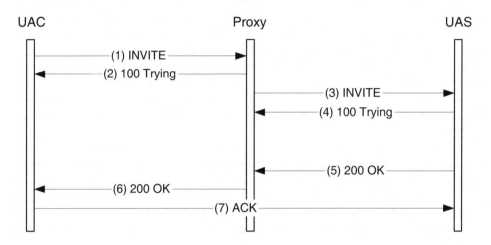

Figure 4.32: End-to-end transmission in SIP

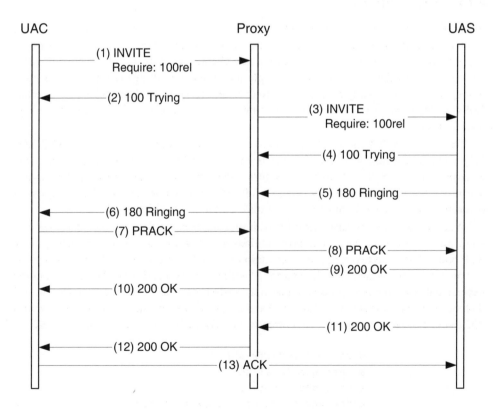

Figure 4.33: Reliable provisional responses and PRACK

The INVITE request in Figure 4.33 contains the 100rel option tag, which requests the use of reliable provisional responses. When the user agent server sends a 180 (Ringing) provisional response (5), it will retransmit it until the PRACK request arrives. At this point the user agent server knows that the provisional response was received by the user agent client (i.e., end-to-end transmission). It is worth mentioning that the 200 (OK) response (9) for the PRACK request would not be needed to provide reliability. However, in order not to make a special case for this method, it was designed to work like any other method. The only SIP method that does not have a response associated with it is ACK, for historical reasons.

As a last note about transport protocols for SIP the reader must bear in mind that, since SIP can run over different transports, a single SIP dialog may involve different transport protocols. It is common that, within the same dialog, the transport protocol between a user agent and a proxy is not the same as the one between that proxy and the other user agent or the next proxy. This leads to situations where user agents retransmit messages over a reliable transport protocol to cope with possible losses in the leg that uses an unreliable transport protocol such as UDP.

4.14   Preconditions

In all the examples we have seen so far the only conditions for a session to be established were that the callee accepted the invitation and that the user agent server supported the extensions required by the user agent client. However, some clients require that some preconditions are met before establishing a session. For example, a user may be willing to speak to another user as long as the voice quality is acceptable. In this case, if the network cannot ensure a certain QoS (Quality of Service) for the duration of the session, the caller prefers not to establish the session at all.

The extension that allows user agents to express preconditions is defined in RFC 3312 [67]. This extension, whose option tag is precondition, is a mixture between a SIP extension and an SDP extension; it defines a SIP option tag and new SDP attributes. Therefore, this extension can only be used with sessions described using SDP. Other session description formats may define their own extension for preconditions in the future.

When a user agent receives an offer with preconditions, it does not alert the user until those preconditions are met. The preconditions are encoded, as mentioned earlier, in the SDP body. There are two types of preconditions: namely, access preconditions and end-to-end preconditions. Figure 4.34 shows an SDP with access preconditions. The user agent that generated this session description is requesting (a=des:qos means desired QoS) QoS in both directions (sendrecv) in both accesses; its local access (local) and the remote access (remote). The a=curr:qos lines indicate that, currently, there is no QoS in either of the accesses.

```
m=audio 20000 RTP/AVP 0
a=curr:qos local none
a=curr:qos remote none
a=des:qos mandatory local sendrecv
a=des:qos mandatory remote sendrecv
```

Figure 4.34: Access preconditions

Figure 4.35 shows an SDP with end-to-end preconditions. The user agent that generated this session description is requesting optional end-to-end (e2e) QoS in both directions (sendrecv).

```
m=audio 20000 RTP/AVP 0
a=curr:qos e2e none
a=des:qos optional e2e sendrecv
```

Figure 4.35: End-to-end preconditions

When mandatory preconditions appear in a session description the callee is alerted only when the current QoS conditions are equal to or better than the desired conditions. Therefore, the INVITE request is typically answered with a 183 (Session Progress) provisional response that does not imply either alerting (180 (Ringing) does) or acceptance of the session (200 (OK) does).

In order to know when all the preconditions are met both user agents need to exchange session descriptions. For example, when the user agent that generated the session description in Figure 4.34 obtains QoS in its access network, it will update the a=curr:qos line from none to sendrecv and send a session description as shown in Figure 4.36. This session description is sent using the UPDATE method, defined in RFC 3311 [199], as shown in Figure 4.37. Using this method, both user agents keep each other up to date on the status of the preconditions.

```
m=audio 20000 RTP/AVP 0
a=curr:qos local sendrecv
a=curr:qos remote none
a=des:qos mandatory local sendrecv
a=des:qos mandatory remote sendrecv
```

Figure 4.36: Updated current QoS conditions

In Figure 4.37, once the first session descriptions are exchanged, both user agents perform QoS reservations in their respective accesses. When the user agent client finishes its reservation, it sends an UPDATE request (5) informing the user agent server (a=curr:qos local sendrecv). When the user agent server finishes its own QoS reservations, all the preconditions are met and the callee is alerted (7).

4.15 Event Notification

So far, we have seen how to use SIP to establish sessions. Now we will see how to use SIP to obtain the status of a given resource and track changes in that status. For example, at a given moment the state of Alice's presence is "online". When she logs off from her computer her presence status changes to "offline". In this example the resource is Alice and the status information is her presence information.

RFC 3265 [198] defines a framework for event notification in SIP. It uses the SUB-SCRIBE and NOTIFY methods. The entity interested in the status information of a resource *subscribes* to that information. The entity that keeps track of the resource state will send a NOTIFY request with the current status information of the resource and a new NOTIFY

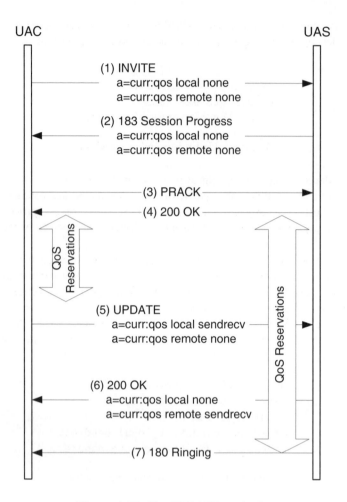

Figure 4.37: The UPDATE method

request every time the status changes. The type of status information is defined by an Event header field. Specifications defining new values for the Event header field are called *event packages*.

The event notification framework defines two new roles in SIP: namely, the *subscriber* and the *notifier*. A subscriber is a SIP UA that sends a SUBSCRIBE request for a particular event. A subscriber gets NOTIFY requests containing state information related to the subscribed event. A notifier is a SIP UA that receives SUBSCRIBE requests for a particular event and generates a NOTIFY request containing the state information related to the subscribed event. A notifier keeps a subscription state for each of the subscribers.

Figure 4.38 shows an example where Alice, acting in the role of a subscriber, subscribes to her voicemail. This application is described RFC 3842 [158]. The voicemail server is acting as a notifier. In this case, the status information she is interested in is the number of messages that have arrived to the voicemail. This corresponds to an Event header field of value message-summary.

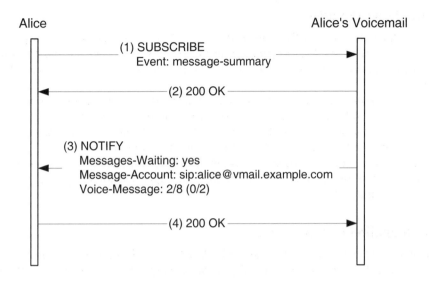

Figure 4.38: Voicemail status information

Note that the 200 (OK) response (2) to the SUBSCRIBE request only indicates that the SUBSCRIBE transaction has been successful. That is, the subscription has been accepted by the voicemail. The information about the resource always arrives in a NOTIFY transaction. The NOTIFY request (3) in Figure 4.38 shows the body of the NOTIFY. It indicates that Alice's voicemail has two new messages and five old messages, of which one new and two old messages are urgent (figures enclosed in parentheses).

If, before the subscription expires, Alice's voicemail receives a new message, it will send a new NOTIFY request to Alice informing her about the new arrival.

4.15.1 High Notification Rates

The event notification framework offers a powerful tool that allows a subscriber to be informed about changes in the state of a resource. Still, in some situations the amount of information that the subscriber has to process might be large. Imagine, for instance, a subscriber to the presence information of a user who is driving on a highway.

Our subscriber gets frequent updates, because the user's geographical position changes rapidly. In any case, our subscriber might not be interested in receiving accurate real-time information. The consequence of having very detailed and accurate information is that the bandwidth needed to transport all of this information increases and so does the power needed to process it.

The event framework provides a mechanism, called event throttling, to limit the rate at which notifications are sent to a subscriber. This mechanism is helpful for devices with low processing power capabilities, limited battery life, or low-bandwidth accesses. Event throttling is used to build such services as presence (we describe the presence service in Chapter 16).

4.15.1.1 Event Throttling

The event framework allows the notifier of the actual event package to set up a policy for the notification rate. While most packages use a default policy, there are situations where subscribers want to communicate to the notifier the amount of information they are willing to receive.

The event-throttling mechanism allows a subscriber to an event package to indicate the minimum period of time between two consecutive notifications. So, if the state information changes rapidly, the notifier holds those notifications until the throttling timer has expired, at which point the notifier sends a single notification to the subscriber.

With this mechanism the watcher does not have a real-time view of the subscribe-state information, but it has approximate information. This approximation is good enough for a number of applications.

4.16 Signaling Compression

As we explained in Section 4.15.1, users with low-bandwidth access need to minimize the amount of data they send and receive if they want an acceptable user experience. Otherwise, performing a simple operation would take so long that users would lose their patience and stop using the service. However, SIP is not an efficient protocol regarding message size. Its textual encoding makes SIP messages grow dramatically as soon as several extensions are used at the same time. Fortunately, we have signaling compression to help us.

RFC 3486 [62] describes how to signal that a SIP message needs to be compressed. It defines the comp=sigcomp parameter to be used in URIs (if a request is to be compressed) or in Via entries (if a response is to be compressed). When this parameter appears in a message the message is compressed using the mechanism described in RFC 3320 [191], which is known as SigComp (Signaling Compression). Figure 4.39 shows how to use the comp=sigcomp parameter to signal SIP traffic compression between a user agent and a proxy. The Route header field indicates that the requests (INVITE and ACK) need to be compressed, and the Via header field indicates that the response (200 (OK)) needs to be compressed.

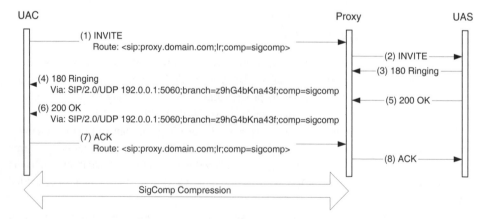

Figure 4.39: SigComp compression between a UA and a proxy

The SigComp specification, RFC 3320 [191], defines a Universal Decompressor Virtual Machine (UDVM) intended to run decompression algorithms. The entity compressing the message provides the entity receiving it with the algorithm to be used to decompress the message. This way the sender is free to choose any compression algorithm to compress its messages.

Compression algorithms substitute long expressions that are used often in the message with shorter pointers to those expressions. The compressor builds a dictionary that maps the long expressions to short pointers and sends this dictionary to the decompressor, as shown in Figure 4.40.

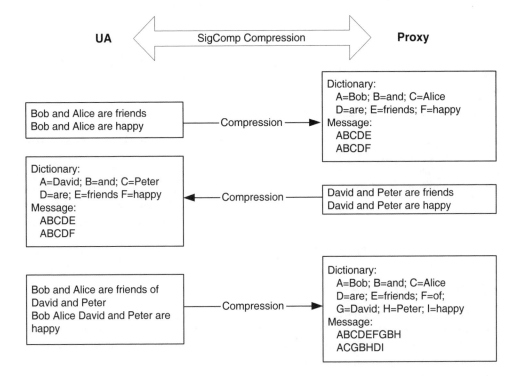

Figure 4.40: Regular SigComp

Figure 4.40 shows how a very simple compression algorithm works. This algorithm substitutes words with letters (note that advanced compression algorithms can substitute much longer expressions by pointers). We can see that every message that is compressed is sent together with a dictionary.

4.16.1 SigComp Extended Operations

However, we can see that dictionaries corresponding to different messages have many words in common. If we could use a single dictionary this redundancy would disappear. RFC 3321 [117], SigComp extended operations, defines a way to use a shared dictionary. Figure 4.41 shows how SigComp extended operations work. The first message is compressed in the same way as in Figure 4.40, but subsequent messages reference previous terms in the shared dictionary. We can see that the compression ratio achieved using extended operations is better than that achieved with regular SigComp.

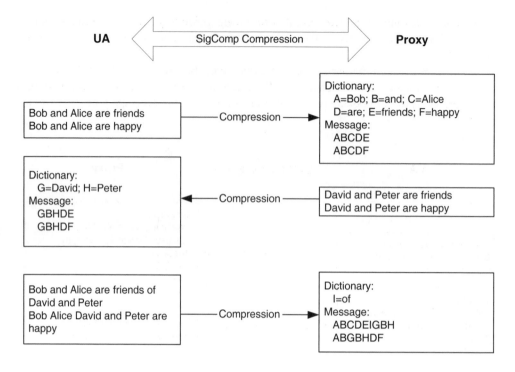

Figure 4.41: SigComp extended operations

4.16.2 Static SIP/SDP Dictionary

As we have seen in Figure 4.41 the dictionary is built up as the compressors get more input: the longer the input the better the dictionary and the more efficient the compression. However, this way of building up the decompression dictionary does not suit SIP. The first message a user agent sends to establish a session is typically an INVITE request. At this point the user is waiting for the other party to answer in a reasonable time. If sending the INVITE request takes too long the user will have a poor user experience. Unfortunately, this INVITE request cannot take advantage of any previously built dictionary and, so, it will not be compressed as much as subsequent messages will (e.g., BYE). Therefore, this way of building up the dictionary gives us slow INVITE and fast BYE requests. This means that the user has a bad user experience.

To improve the compression efficiency for the first message, SIP provides a static SIP/SDP dictionary (defined in RFC 3485 [107]) which is supported to be present in every implementation that supports SigComp. This dictionary contains the SIP and SDP terms that are used most often. Figure 4.42 shows how the dictionary is used. The compression efficiency improves significantly compared with Figure 4.41.

4.17 Content Indirection

Compressing SIP using the techniques described in Section 4.16 helps to reduce the size of the messages. Nevertheless, sometimes SIP message bodies are just too large, even after compression.

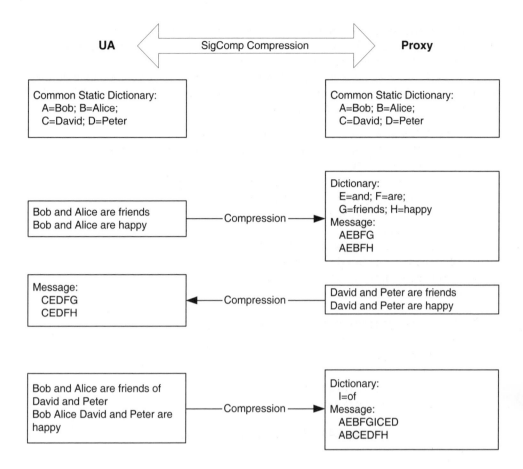

Figure 4.42: SIP/SDP static dictionary

Large messages have two important disadvantages: service behavior is too slow on low-bandwidth links and, more importantly, messages get fragmented when transported over UDP. We will describe the problems associated with IP fragmentation and then look at a SIP extension that resolves this issue.

While TCP provides transport-layer fragmentation, UDP does not. This means that if an application-layer message (e.g., a SIP message) is larger than the maximum size that link layers in the path can handle the message will be fragmented at the IP layer.

If one of the fragments of this message gets lost the sender needs to retransmit the whole message, which is clearly quite an inefficient way to perform packetloss recovery. Moreover, some port-based firewalls and NATs (Network Address Translators) cannot handle fragments. This is because only the first fragment carries the port numbers of the datagram carrying the message. When a firewall or a NAT receives a fragment which is not the first one, it cannot find the port number of the datagram and simply discards the packet.

So, in some situations (e.g., a UA behind a NAT that cannot handle fragments), it might be impossible to transmit large SIP messages. In this situation a SIP extension called *content indirection* (defined in the Internet-Draft "A Mechanism for Content Indirection in SIP Messages" [59]) is our friend.

Content indirection allows us to replace a MIME body part with an external reference, which is typically an HTTP URI. The destination UA fetches the contents of that MIME body part using the references contained in the SIP message.

Figures 4.43 and 4.44 show an INVITE request carrying a session description in the body and as an external reference, respectively. In the second case, Bob will fetch Alice's SDP using the HTTP URI provided in the INVITE request.

```
INVITE sip:Alice.Smith@domain.com SIP/2.0
Via: SIP/2.0/UDP ws1.domain2.com:5060;branch=z9hG4bK74gh5
Max-Forwards: 70
From: Bob <sip:Bob.Brown@domain2.com>;tag=9hx34576sl
To: Alice <sip:Alice.Smith@domain.com>
Call-ID: 6328776298220188511@192.0.100.2
Cseq: 1 INVITE
Contact: <sip:bob@192.0.100.2>
Content-Type: application/sdp
Content-Length: 151


v=0
o=bob 2890844526 2890844526 IN IP4 ws1.domain2.com
s=-
c=IN IP4 192.0.100.2
t=0 0
m=audio 20000 RTP/AVP 0
a=rtpmap:0 PCMU/8000
```

Figure 4.43: SDP carried in the message body

```
INVITE sip:Alice.Smith@domain.com SIP/2.0
Via: SIP/2.0/UDP ws1.domain2.com:5060;branch=z9hG4bK74gh5
Max-Forwards: 70
From: Bob <sip:Bob.Brown@domain2.com>;tag=9hx34576sl
To: Alice <sip:Alice.Smith@domain.com>
Call-ID: 6328776298220188511@192.0.100.2
Cseq: 1 INVITE
Contact: <sip:bob@192.0.100.2>
Content-Type: message/external-body;
             ACCESS-TYPE=URL;
             URL="http://www.domain.com/mysdp";
             size=151
Content-Length: 30

Content-Type: application/sdp
```

Figure 4.44: SDP provided as a external reference

Content indirection is especially useful for optional body parts. For example, if Alice uses content indirection to include her photo in her INVITEs the callees can choose whether or

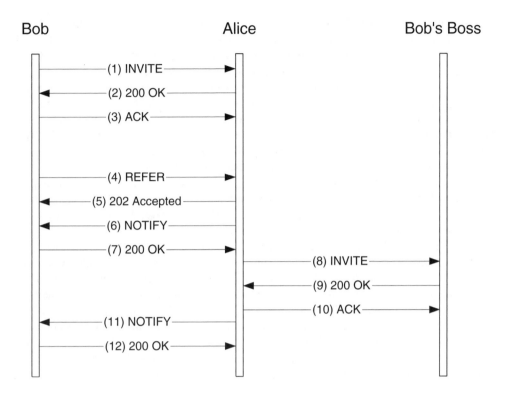

Figure 4.45: The REFER method

not they want to fetch it. A callee on a low-bandwidth link can probably live without seeing Alice's photo, while another callee using a high-speed access will most likely enjoy seeing it. In addition, SIP proxies in the path will not need to route large messages, which is important for proxies handling a large number of users.

4.18   The REFER Method

The REFER method (specified in RFC 3515 [229]) is used to request a user agent to contact a resource. This resource is identified by a URI, which may or may not be a SIP URI.

   The most common usage of REFER is call transfer. For example, Bob is in a call (i.e., an audio session) with Alice. Bob is using his mobile SIP phone because he is walking on the street. When Bob arrives at his office, he wants to transfer the session with Alice to his fixed SIP phone. Bob sends a REFER request to Alice's user agent requesting it to contact Bob's fixed phone's SIP URI. Once the transfer is finished, Bob continue with this call on this fixed phone.

   Bob would use the same REFER mechanism to transfer his call with Alice to another person (Alice may want to talk to Bob's boss after her conversation with Bob). In this case, Bob would like to make sure that Alice has been able to contact Bob's boss before hanging up. The REFER mechanism includes an implicit subscription to the result of the operation initiated by the recipient of the REFER. That is, Alice will inform Bob (with a NOTIFY request) whether or not she managed to contact Bob's boss successfully.

Bob can also use REFER to request Alice's user agent to contact a non-SIP URI. For example, if Bob wants Alice to have a look at his new personal web page, he sends a REFER request to Alice's user agent requesting it to contact his personal web page's HTTP URI. In this case, Alice will inform Bob when she downloads the web page.

Figure 4.45 shows the message flow involved in the example described earlier where Alice is requested to contact Bob's boss. The message flow shows how the REFER request creates an implicit subscription to the result of the operation (in this case, contacting Bob's boss). This subscription has the same properties as subscriptions created with a SUBSCRIBE request (see Section 4.15). That is, Bob receives NOTIFY requests carrying information about the result of the transaction initiated by Alice toward Bob's boss.

On receiving a REFER request, a user agent always generates an NOTIFY request immediately (6). At a later point, when the operation requested in the REFER request has been attempted, further NOTIFY requests are sent. The initial NOTIFY request is needed to handle forking scenarios.

Even if a REFER request forks and arrives at several user agent servers, the user agent client that sent the REFER request only receives one final response (routing rules for proxies ensure this property for non-INVITE transactions). Such a user agent client discovers all the user agent servers that received the REFER request when it receives the NOTIFY requests.

Although the implicit subscription associated with REFER is useful in many situations, some applications using the REFER method do not need the implicit subscription. It is possible to suppress the implicit subscription of a particular REFER request by using a REFER extension (specified in the Internet-Draft "Suppression of SIP REFER Method Implicit Subscription" [154]).

Chapter 5

Session Control in the IMS

We saw in Section 4.1 how SIP is used in a public Internet environment. We have also explored the core SIP functionality and a few important extensions that SIP User Agents may support. Each implementation of SIP is free to implement the options or SIP extensions that the particular application requires.

3GPP is one of those particular applications of SIP where SIP is used in a wireless environment. In the case of 3GPP, SIP is used over an underlying packet network that defines a number of constraints. The result of the evaluation of SIP in wireless environments led to the definition of a set of requirements that accommodates SIP in 3GPP networks. The implementation of solutions to these wireless requirements led 3GPP to mandate the use of a number of options and extensions to SIP and other protocols. We can consider 3GPP's function as creating a profile of utilization of SIP and other protocols in the IP Multimedia Subsystem. The 3GPP SIP profile utilization for IMS is specified in 3GPP TS 24.229 [16]. We call it a *profile* because there are no differences with respect to the usage of SIP on the public Internet. However, 3GPP has mandated the implementation of a number of extensions and options in both the IMS network nodes and IMS terminals. This section focuses on describing how SIP is used in the IMS as well as highlighting differences in the utilization of SIP with respect to the public Internet.

When 3GPP began the work on session control for the IMS, SIP was chosen as the protocol to control sessions. At that time the IETF (Internet Engineering Task Force) was working on a revision of SIP that led to the migration and extension of the protocol from RFC 2543 [116] to RFC 3261 [215] and other RFCs. Previously, the performance of SIP in wireless environments had never been evaluated.

Wireless environments have a number of strict requirements for session control protocols like SIP. These requirements range from extra security requirements to the capability of providing the same services no matter whether the mobile station is located in the home network or roaming to a visited network. The IETF analyzed these requirements and took most of them into consideration. This led to the design of a number of SIP solutions that were either included in the core SIP specification in RFC 3261 [215] or documented as separate extensions to SIP. We will analyze these extensions when we delve deeper into session control in the IMS.

5.1   Prerequisites for Operation in the IMS

Before an IMS terminal starts any IMS-related operation there are a number of prerequisites that have to be met. Figure 5.1 shows a high-level view of the required prerequisites.

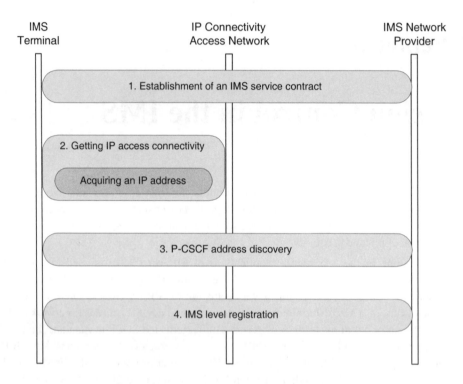

Figure 5.1: Prerequisites to get the IMS service

First, the IMS service provider has to authorize the end-user to use the IMS service. This typically requires a subscription or contract signed between the IMS network operator and the user. This contract is similar to the subscription that authorizes an end-user to receive and establish telephone calls over a wireless network.

Second, the IMS terminal needs to get access to an IP-CAN (IP Connectivity Access Network) such as GPRS (in GSM/UMTS networks), ADSL (Asymmetric Digital Subscriber Line), or WLAN (Wireless Local Access Network). IP-CAN provides access to the IMS home network or to an IMS visited network. As part of this prerequisite the IMS terminal needs to acquire an IP address (the procedures for GPRS access are described in 3GPP TS 23.060 [15]). This IP address is typically dynamically allocated by the IP-CAN operator for a determined period of time.

When these two prerequisites are fulfilled the IMS terminal needs to discover the IP address of the P-CSCF that will be acting as an outbound/inbound SIP proxy server. All the SIP signaling sent by the IMS terminal traverses this P-CSCF. When the P-CSCF discovery procedure is completed the IMS terminal is able to send and receive SIP signaling to or from the P-CSCF. The P-CSCF is allocated permanently for the duration of IMS registration, a procedure that is typically triggered when the IMS terminal is switched on or off.

Depending on the IP Connectivity Access Network in use the P-CSCF discovery procedure may take place as part of the process to obtain IP-CAN connectivity or as a separate procedure. A separate procedure is achieved by means of DHCP (Dynamic Host Configuration Protocol, specified in RFC 2131 [83]) or DHCPv6 (DHCP for IPv6, specified in RFC 3315 [84]).

When the previous prerequisites are fulfilled the IMS terminal registers at the SIP application level to the IMS network. This is accomplished by regular SIP registration. IMS terminals need to register with the IMS before initiating or receiving any other SIP signaling. As the IMS is modeled in different layers the IP-CAN layer is independent of the IMS application (SIP) layer. Therefore, registration at the IMS level is independent of registration with IP-CAN (e.g., attachment to a GPRS network). The IMS registration procedure allows the IMS network to locate the user (i.e., the IMS obtains the terminal's IP address). It also allows the IMS network to authenticate the user, establish security associations, and authorize the establishment of sessions. We describe the security functions of the IMS in Chapter 9.

5.2   IPv4 and IPv6 in the IMS

When 3GPP was designing the IMS, version 6 of the Internet Protocol (also known as IPv6) was being standardized in the IETF. 3GPP did an analysis of the applicability of IPv6 for IMS and concluded that, by the time the first IMS implementations would go into operation, IPv6 would most likely be the common IP version in the Internet. Any large scale deployment of IPv4 required the allocation of private IP addresses and the presence of some type of Network Address Translation (NAT) in the path of the communication.

SIP and its associated protocols (SDP, RTP, RTCP, etc.) were known examples of protocols that suffered problems when traversing NATs. Allowing IPv4 for IMS would have required a major analysis of NAT traversal techniques.

For all these reasons 3GPP decided to select IPv6 as the only allowed version of IP for IMS connectivity.

Unfortunately, when the first IMS products reached the market, the situation was quite different from the one foreseen a few years earlier: IPv6 had not taken off; IPv4 and NATs were becoming ubiquitous; work had been done in the field of helping SIP and associated protocols to easily traverse NATs.

In June 2004 3GPP re-examined the IPv4/IPv6 dilemma once more. Market indications at that time revealed that IPv6 had not yet gone into the mainstream. Most of the Internet was still running IPv4, and only a few mobile networks were ready to start a big IPv6 deployment, like the one IMS would require. Furthermore, the work on NAT traversal for SIP had progressed substantially, and SIP was a friendly NAT-traversal protocol. IPv4 was already implemented in most of the early IMS products since it did not require an extra effort.

Based on all these indications 3GPP decided to allow early deployments of IPv4 for IMS, starting even from the very first IMS release, 3GPP Release 5. The work describing the support for IPv4 in IMS was collected in 3GPP TR 23.981 [17]. The main 3GPP architecture document, 3GPP TS 23.221 [2], was amended to refer to 3GPP TR 23.981 [17] for those early IMS implementations that supported IPv4.

Dual stack implementations (IPv4 and IPv6) in both IMS terminals and network nodes are now allowed, as well as single IP version implementations. Because of this, two new nodes, the IMS-ALG (Application Layer Gateway) and the Transition Gateway (TrGW) were added to the IMS architecture. The former deals with SIP interworking and the latter with RTP interworking, both between IPv4 and IPv6 (and vice versa).

The consequence of adding IPv4 to the IMS is a delay in deploying IPv6 for the public Internet. Having IMS as the main driver of IPv6 would have accelerated the deployment of IPv6 in the Internet. While mostly everyone agrees that IPv6 will, one day, be the common version of IP in the Internet, it has not happened yet, and IMS has to go along with that fact.

The rest of this book, while considering both IPv4 and IPv6 IMS, gives precedence in examples to IPv6. We believe IPv6 is a future-proof protocol, and we believe most of our readers will appreciate our effort to devoting a more careful analysis to IPv6 IMS.

5.3   IP Connectivity Access Network

There are multiple types of IP Connectivity Access Networks. Examples of IP Connectivity Access Networks in fixed environments are Digital Subscriber Lines (DSL), Dial-up Lines, enterprise Local Access Networks, etc. In wireless environments we have Packet Data Access networks, such as GPRS or Wireless Local Access Networks. The procedures to register and acquire an IP address are different for different IP Connectivity Access Networks.

For instance, in GPRS, the IMS terminal first undertakes a set of procedures, globally known as *GPRS attach procedures*. These procedures involve several nodes, ranging from the SGSN to the HLR and the GGSN. The procedures are illustrated in Figure 5.2. Once these procedures are complete the terminal sends an *Activate PDP Context Request* message to the SGSN requesting connection to an either IPv4 or IPv6 network. The message includes a request for connectivity to a particular APN (Access Point Name) and packet connection type. The APN identifies the network to connect to and the address space where the IP address belongs. In the case of an IMS terminal the APN indicates a desired connection to the IMS network and the connectivity type indicates either IPv4 or IPv6. The SGSN, depending on the APN and the type of network connection, chooses an appropriate GGSN. The SGSN sends a *Create PDP Context Request* message to the GGSN. The GGSN is responsible for allocating IP addresses.

In case the terminal requested an IPv6 connection, the GGSN does not provide the terminal with a full IPv6 address belonging to the IMS address space. Instead, the GGSN provides the terminal with a 64-bit IPv6 prefix and includes it in a *Create PDP Context Response* message. The SGSN transparently forwards this IPv6 prefix in an *Activate PDP Context Accept*. When the procedure is completed the IMS terminal has got a 64-bit IPv6 prefix. The terminal is able to choose any 64-bit IPv6 suffix. Together they form a 128-bit IPv6 address (i.e., the IPv6 address that the terminal will use for its IMS traffic).

If the terminal requested an IPv4 connection, the GGSN provides the terminal with its IPv4 address.

When the IP Connectivity Access Network is not GPRS the protocol used to configure the IMS terminal will most likely be DHCP (specified in RFC 2131 [83]) or DHCP for IPv6 (DHCPv6, specified in RFC 3315 [84]). DHCP is used to send configuration parameters to a terminal. Its main purpose is to provide the terminal with an IP address, although the DHCP server can also send, if requested by the terminal, other types of configuration data such as the address of an outbound SIP proxy server or the address of an HTTP proxy server.

Sometimes, the procedure to get an IP Connectivity Access Network requires a registration and a payment of some form. It has become very popular to provide Wireless LAN access at hot spots, such as airports and hotels. Getting access to these networks typically requires some form of subscription to the service, and some form of payment. It may be required that the user log into a web page and introduce a user name and password, or some

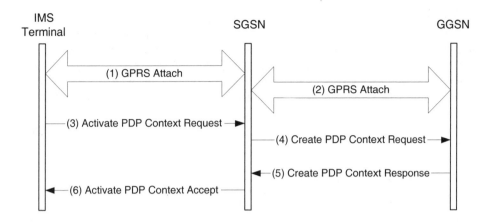

Figure 5.2: Getting IP connectivity in GPRS

credit card data for payment service usage. In other cases, a 3GPP Wireless LAN access network can be used.

5.4 P-CSCF Discovery

P-CSCF discovery is the procedure by which an IMS terminal obtains the IP address of a P-CSCF. This is the P-CSCF that acts as an outbound/inbound SIP proxy server toward the IMS terminal (i.e., all the SIP signaling sent by or destined for the IMS terminal traverses the P-CSCF).

P-CSCF discovery may take place in different ways:

(a) integrated into the procedure that gives access to the IP-CAN;

(b) as a stand-alone procedure.

The integrated version of P-CSCF discovery depends on the type of IP Connectivity Access Network. If IP-CAN is a GPRS network, once the *GPRS attach procedures* are complete the terminal is authorized to use the GPRS network. Then, the IMS terminal does a so-called *Activate PDP Context Procedure*. The main goal of the procedure is to configure the IMS terminal with an IP address,[4] but in this case the IMS terminal also discovers the IP address of the P-CSCF to which to send SIP requests.

The stand-alone version of the P-CSCF discovery procedure is based on the use of DHCP (specified in RFC 2131 [83]), or DHCPv6, for IPv6 (specified in RFC 3315 [84]) and DNS (Domain Name System, specified in RFC 1034 [166]).

In DHCPv6 the terminal does not need to know the address of the DHCP server, because it can send its DHCP messages to a reserved multicast address. If DHCP is used (for IPv4), the terminal broadcasts a discover message on its local physical subnet. In some configurations a DHCP relay may be required to relay DHCP messages to an appropriate network, although the presence of the DHCP relay is transparent to the terminal.

---

[4]If IPv6 is used, in fact, the IMS terminal is equipped with an IPv6 prefix of 64 bits. The IMS terminal is free to select any 64-bit suffix, completing the 128 bits in an IPv6 address.

The procedure for DHCPv6 is illustrated in steps 1 and 2 of Figure 5.3. Once the IMS terminal has got connectivity to the IP-CAN the IMS terminal sends a DHCPv6 Information-Request (1) where it requests the DHCPv6 Options for SIP servers (specified in RFC 3319 [228]). In the case of the IMS the P-CSCF performs the role of an outbound/inbound SIP proxy server, so the DHCP server returns a DCHP Reply message (2) that contains one or more domain names and/or IP addresses of one or more P-CSCFs.

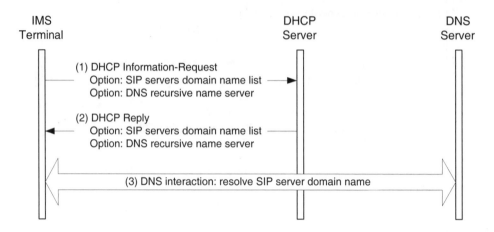

Figure 5.3: P-CSCF discovery procedure based on DHCP and DNS

At the discretion of the IMS terminal implementation, there are two possible ways in which the IMS terminal can specify the request for the DHCPv6 Option for SIP servers.

(a) The IMS terminal requests the *SIP Servers domain name list* Option in the DHCPv6 Information-Request message.[5] The DHCPv6 Reply message contains a list of the domain names of potential P-CSCFs. The IMS terminal needs to resolve at least one of these domain names into an IPv6 address. A query–response dialog with DNS resolves the P-CSCF domain name, but prior to any DNS interaction the IMS terminal also needs to get the address of one or more DNS servers to send its DNS messages. To solve this problem the DHCP Information-Request message, (1) in Figure 5.3, not only contains a request for the Option for SIP servers but also includes a request for the *DNS recursive name server Option*. The DHCPv6 Reply message (2) contains a list of IPv6 addresses of DNS servers, in addition to the domain name of the P-CSCF. Then, the IMS terminal queries the just learnt DNS server in order to resolve the P-CSCF domain name into one or more IPv6 addresses. The procedures to resolve a SIP server into one or more IP addresses are standardized in RFC 3263 [214].

(b) The alternative consists of the IMS terminal requesting the *SIP Servers IPv6 address list* Option in the DHCPv6 Information-Request message. The DHCP server answers in a DHCP Reply message that contains a list of IPv6 addresses of the P-CSCF

---

[5]The DHCPv6 Option for SIP servers differs from a similar Option in DHCPv4. In DHCPv6 there are two different Option Codes to request: either the domain name or the IP address of the SIP server. In DHCPv4 there is a single Option Code, with two possible answers: domain names or IPv4 addresses. It seems that the maximum number of DHCPv4 options is limited to 256, whereas in DHCPv6 the maximum number of options is 65535. This gives enough room to allocate the two option codes needed.

allocated to the IMS terminal. In this case, no interaction with DNS is needed, because the IMS terminal directly gets one or more IPv6 addresses.

These two ways are not mutually exclusive. It is possible, although not required, that an IMS terminal requests both the *SIP servers domain name list* and the *SIP servers IPv6 address list*. The DHCP server may be configured to answer both or just one of the options, but if the DHCP server answers with both lists the IMS terminal should give precedence to the *SIP servers domain name list*. Handling of all these conflicts is described in RFC 3263 [214].

Another way to provide the address of the P-CSCF relies in some means of configuration. This can be, for example, an SMS sent to the terminal for the purpose of configuration; or the Client Provisioning [170] or Device Management [174] specified by the Open Mobile Alliance (OMA).

Eventually, the IMS terminal discovers the IP address of its P-CSCF and can send SIP signaling to its allocated P-CSCF. The P-CSCF takes care of forwarding the SIP signaling to the next SIP hop. The P-CSCF allocated to the IMS terminal does not change until the next P-CSCF discovery procedure. This procedure typically takes place when the terminal is switched on or during severe error conditions. The important aspect to highlight is that the IMS terminal does not need to worry about possible changes of address of the P-CSCF, because its address is not variable.

5.5 IMS-level Registration

Once the IMS terminal has followed the procedures of getting access to an IP Connectivity Access Network, has acquired an IPv6 address or built an IPv6 address, and has discovered the IPv4 or IPv6 address of its P-CSCF, the IMS terminal can begin registration at the IMS level.

IMS-level registration is the procedure where the IMS user requests authorization to use the IMS services in the IMS network. The IMS network authenticates and authorizes the user to access the IMS network.

IMS-level registration is accomplished by a SIP REGISTER request. We explained in Section 4.1.4 that a SIP registration is the procedure whereby a user binds his *public URI* to a URI that contains the host name or IP address of the terminal where the user is logged in. Unlike regular SIP procedures, registration with the IMS is mandatory before the IMS terminal can establish a session.

The IMS registration procedure uses a SIP REGISTER request. However, this procedure is heavily overloaded in the IMS, for the sake of fulfilling the 3GPP requirement of a minimum number of round trips. The goal is achieved and the procedure completes after two round trips, as illustrated in Figure 5.4.[6]

5.5.1 IMS Registration with an ISIM

We explained in Section 3.6 that in order to authenticate users, when they access the IMS network, the IMS terminal needs to be equipped with a UICC. The UICC can include an ISIM application, a USIM application, or both. The parameters stored in both applications are completely different, since the ISIM is IMS specific and the USIM was already available for circuit-switched and packet-switched networks before the IMS was designed. Although

---

[6]Note that, for the sake of simplicity, Figure 5.4 does not show a Subscriber Location Function (SLF). An SLF is needed if there is more than one HSS in the home network of the subscriber.

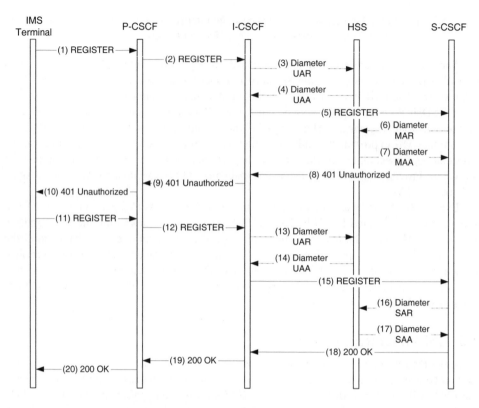

Figure 5.4: Registration at the IMS level

the registration procedure is quite similar, independently of the presence of an ISIM or USIM, there are certainly detailed differences. In this section we describe access to the IMS with an ISIM application in the UICC. Section 5.5.2 describes the registration procedures when the UICC only contains a USIM.

The IMS registration procedure satisfies the following requirements in two round trips:

(a) the user binds a Public User Identity to a contact address – this is the main purpose of a SIP REGISTER request;

(b) the home network authenticates the user;

(c) the user authenticates the home network;

(d) the home network authorizes the SIP registration and the usage of IMS resources;

(e) in case the P-CSCF is located in a visited network, the home network verifies that there is an existing roaming agreement between the home and the visited network and authorizes the usage of the P-CSCF;

(f) the home network informs the user about other possible identities that the home network operator has allocated exclusively to that user;

(g) the IMS terminal and the P-CSCF negotiates the security mechanism that will be in place for subsequent signaling;

(h) the P-CSCF and the IMS terminal establish a set of security associations that protect the integrity of SIP messages sent between the P-CSCF and the terminal;

(i) both the IMS terminal and the P-CSCF upload to each other the algorithms used for compression of SIP messages.

In this section we focus on a few aspects of the IMS registration procedure. Section 9.1.2 describes the security aspects that relate to registration.

Before creating the initial SIP REGISTER request, the IMS terminal retrieves from ISIM the Private User Identity, a Public User Identity, and the home network domain URI. Then, the IMS terminal creates a SIP REGISTER request and includes the following four parameters.

The registration URI: this is the SIP URI that identifies the home network domain used to address the SIP REGISTER request. This is a URI that typically points to the home network, but it could be any subdomain of the home network. The registration URI is included in the *Request-URI* of the REGISTER request.

The Public User Identity: this is a SIP URI that represents the user ID under registration. In SIP, it is known as the SIP *Address-of-Record* (i.e., the SIP URI that users print in their business cards). It is included in the To header field value of the REGISTER request.

The Private User Identity: this is an identity that is used for authentication purposes only, not for routing. It is equivalent to what in GSM is known as IMSI (International Mobile Subscriber Identity); it is never displayed to the user. It is included in the username parameter of the Authorization header field value included in the SIP REGISTER request.

The Contact address: this is a SIP URI that includes the IP address of the IMS terminal or the host name where the user is reachable. It is included in the SIP Contact header field value of the REGISTER request.

According to Figure 5.4 the IMS creates a SIP REGISTER request including the above mentioned information. Figure 5.5 shows an example of the REGISTER request that the IMS terminal sends to the P-CSCF. It must be noted that the P-CSCF may be either located in a visited or a home network. So, in general, the P-CSCF may not be located in the same network as the home network and needs to locate an entry point into the home network by executing the DNS procedures specified in RFC 3263 [214]. These procedures provide the P-CSCF with the SIP URI of an I-CSCF. That I-CSCF is located at the entrance to the home network. The P-CSCF inserts a P-Visited-Network-ID that contains an identifier of the network where the P-CSCF is located. The home network requires this header field to validate the existence of a roaming agreement between the home and visited networks. The P-CSCF also inserts a Path header field with its own SIP URI to request the home network to forward all SIP requests through this P-CSCF. Eventually, the P-CSCF forwards the SIP REGISTER request to an I-CSCF in the home network, (2) in Figure 5.4.

The I-CSCF does not keep a registration state, mainly because it is typically configured in DNS to be serving a load-balancing function. When a SIP proxy needs to contact a SIP proxy

```
REGISTER sip:home1.net SIP/2.0
Via: SIP/2.0/UDP [1080::8:800:200C:417A];comp=sigcomp;
      branch=z9hG4bK9h9ab
Max-Forwards: 70
P-Access-Network-Info: 3GPP-UTRAN-TDD;
                          utran-cell-id-3gpp=C359A3913B20E
From: <sip:alice@home1.net>;tag=s8732n
To: <sip:alice@home1.net>
Contact: <sip:[1080::8:800:200C:417A];comp=sigcomp>
          ;expires=600000
Call-ID: 23fi57lju
Authorization: Digest username="alice_private@home1.net",
                realm="home1.net", nonce="",
                uri="sip:home1.net", response=""
Security-Client: ipsec-3gpp; alg=hmac-sha-1-96;
                  spi-c=3929102; spi-s=0293020;
                  port-c:3333; port-s=5059
Require: sec-agree
Proxy-Require: sec-agree
Cseq: 1 REGISTER
Supported: path
Content-Length: 0
```

Figure 5.5: (1) REGISTER

located in another network, it gets a different IP address of an I-CSCF, because of the DNS load-balancing mechanisms. As a consequence, I-CSCFs do not keep any state associated to registration. In particular, I-CSCFs are not aware of whether an S-CSCF is allocated to the user and what the address of such an S-CSCF would be.

In order to carry out a first-step authorization and to discover whether there is an S-CSCF already allocated to the user, the I-CSCF sends a Diameter User-Authentication-Request (UAR) to the HSS, (3) in Figure 5.4. The I-CSCF transfers to the HSS the Public and Private User Identities and the visited network identifier, all of which are extracted from the SIP REGISTER request. The HSS authorizes the user to roam the visited network and validates that the Private User Identity is allocated to the Public User Identity under registration. The HSS answers with a Diameter User-Authentication-Answer (UAA) (3). The HSS also includes the SIP URI of a previously allocated S-CSCF in the Diameter UAA message, if there was an S-CSCF already allocated to the user. But if this was the first registration (e.g., after the user switched the IMS terminal on) there will most likely not be an S-CSCF allocated to the user. Instead, the HSS returns a set of S-CSCF capabilities that are the input for the I-CSCF when selecting the S-CSCF.

Let's assume for the time being that the user switches his IMS terminal on and the S-CSCF is unallocated as yet. The I-CSCF needs to perform an S-CSCF selection procedure based on the S-CSCF capabilities that the HSS returned in the Diameter UAA message. These capabilities are divided into mandatory capabilities, or capabilities that the chosen S-CSCF has to fulfill, and optional capabilities, or capabilities that the chosen S-CSCF may or may not fulfill. The standard does not indicate *what* these capabilities are and *how* they are specified. Instead, capabilities are represented by an integer and have semantics only in a particular home network. As an example, let's assume that operator A assigns the semantics of *capability 1* to the S-CSCF that provides detailed charging information, whereas *capability 2* indicates support in the S-CSCF for SIP calling preferences. Then, the operator can configure the user data in the HSS to indicate that for this particular subscriber it is mandatory that the S-CSCF supports capability 2 and, optionally, that it may support capability 1. Because the Diameter interface defined between the I-CSCF and the HSS is an intra-operator interface, mapping the capabilities to the semantics is a matter of operator configuration. In different networks, capabilities 1 and 2 will have different semantics.

S-CSCF selection is based on the capabilities received from the HSS in the Diameter UAA. The I-CSCF has a configurable table of S-CSCFs operating in the home network and the capabilities supported by each one. This allows the I-CSCF to choose an appropriate S-CSCF for this particular user. Then, the I-CSCF continues with the process by proxying the SIP REGISTER request to the chosen S-CSCF, (5) in Figure 5.4. An example of such a REGISTER request is shown in Figure 5.6. The S-CSCF receives the REGISTER request and authenticates the user. Initial registrations are always authenticated in the IMS. Other registrations may or may not be authenticated, depending on a number of security issues. Only REGISTER requests are authenticated in the IMS. Other SIP requests, such as INVITE, are never authenticated by the IMS.

```
REGISTER sip:home1.net SIP/2.0
Via: SIP/2.0/UDP icscf1.home1.net;branch=z9hG4bKea1dof,
      SIP/2.0/UDP pcscf1.visited1.net;branch=z9hG4bKoh2qrz,
      SIP/2.0/UDP [1080::8:800:200C:417A];comp=sigcomp;
                   branch=z9hG4bK9h9ab
Max-Forwards: 68
P-Access-Network-Info: 3GPP-UTRAN-TDD;
                        utran-cell-id-3gpp=C359A3913B20E
From: <sip:alice@home1.net>;tag=s8732n
To: <sip:alice@home1.net>
Contact: <sip:[1080::8:800:200C:417A];comp=sigcomp>
          ;expires=600000
Call-ID: 23fi571ju
Authorization: Digest username="alice_private@home1.net",
                realm="home1.net", nonce="",
                uri="sip:home1.net", response="",
                integrity-protected="no"
Require: path
Supported: path
Path: <sip:term@pcscf1.visited1.net;lr>
P-Visited-Network-ID: "Visited 1 Network"
P-Charging-Vector: icid-value="W34h6dlg"
Cseq: 1 REGISTER
Content-Length: 0
```

Figure 5.6: (5) REGISTER

The S-CSCF then contacts the HSS for a double purpose: on the one hand, the S-CSCF needs to download authentication data to perform authentication for this particular user. On the other hand, the S-CSCF needs to save the S-CSCF URI in the HSS, so that any further query to the HSS for the same user will return routing information pointing to this S-CSCF. For this purpose the S-CSCF creates a Diameter Multimedia-Auth-Request (MAR) message, (6) in Figure 5.4. The HSS stores the S-CSCF URI in the user data and answers in a Diameter Multimedia-Auth-Answer (MAA) message, (7) in Figure 5.4. In 3GPP IMS, users are authenticated by the S-CSCF with data provided by the HSS. These authentication data are known as *authentication vectors*. The HSS includes one or more authentication vectors in the Diameter MAA message, so that the S-CSCF can properly authenticate the user. Then, the S-CSCF creates a SIP 401 (Unauthorized) response, (8) in Figure 5.4. This response includes a challenge in the WWW-Authenticate header field that the IMS terminal should answer.

The SIP 401 (Unauthorized) response is forwarded, according to regular SIP procedures, via the I-CSCF and P-CSCF. An example of such a response is shown in Figure 5.7. When the IMS terminal receives the SIP 401 (Unauthorized) response, it realizes that there is a challenge included and produces an appropriate response to that challenge. The response to the challenge (sometimes known as credentials) is included in a new SIP REGISTER request, (11) in Figure 5.4. The actual contents of the credentials depend on the IMS network. If we are dealing with a 3GPP IMS terminal, then the terminal stores authentication information in a smart card (UICC, Universal Integrated Circuit Card). The IMS terminal extracts or derives the parameters stored in the smart card to build the credentials and does so transparently to the user. Chapter 9 is devoted to security in IMS networks.

```
SIP/2.0 401 Unauthorized
Via: SIP/2.0/UDP [1080::8:800:200C:417A];comp=sigcomp;
                 branch=z9hG4bK9h9ab
From: <sip:alice@home1.net>;tag=s8732n
To: <sip:alice@home1.net>;tag=409sp3
Call-ID: 23fi57lju
WWW-Authenticate: Digest realm="home1.net",
                  nonce="dcd98b7102dd2f0e8b11d0f600bfb0c093",
                  algorithm=AKAv1-MD5
Security-Server: ipsec-3gpp; q=0.1; alg=hmac-sha-1-96;
                 spi-c=909767; spi-s=421909;
                 port-c:4444; port-s=5058
Cseq: 1 REGISTER
Content-Length: 0
```

Figure 5.7: (10) 401 Unauthorized

In response, the IMS terminal sends a new SIP REGISTER request to the P-CSCF, (11) in Figure 5.4. An example of this new SIP REGISTER request is displayed in Figure 5.8.

The P-CSCF does the same operation as for the first REGISTER request; that is, it determines the entry point in the network stored in the *Request-URI* of the REGISTER request and finds an I-CSCF in the home network. It must be noted that this I-CSCF, due to DNS load-balancing mechanisms, may not be the same I-CSCF that the first REGISTER request, (2) in Figure 5.4, traversed.

The I-CSCF sends a new Diameter UAR message, (13) in Figure 5.4, for the same reasons as explained before. The difference in this situation is that the Diameter UAA message (14)

```
REGISTER sip:home1.net SIP/2.0
Via: SIP/2.0/UDP [1080::8:800:200C:417A]:5059;comp=sigcomp;
                branch=z9hG4bK9h9ab
Max-Forwards: 70
P-Access-Network-Info: 3GPP-UTRAN-TDD;
                        utran-cell-id-3gpp=C359A3913B20E
From: <sip:alice@home1.net>;tag=s8732n
To: <sip:alice@home1.net>
Contact: <sip:[1080::8:800:200C:417A]:5059;comp=sigcomp>
         ;expires=600000
Call-ID: 23fi571ju
Authorization: Digest username="alice_private@home1.net",
               realm="home1.net",
               nonce="dcd98b7102dd2f0e8b11d0f600bfb0c093",
               algorithm=AKAv1-MD5,
               uri="sip:home1.net",
               response="6629fae49393a05397450978507c4ef1"
Security-Verify: ipsec-3gpp; q=0.1; alg=hmac-sha-1-96;
                 spi-c=909767; spi-s=421909;
                 port-c:4444; port-s=5058
Require: sec-agree
Proxy-Require: sec-agree
Cseq: 2 REGISTER
Supported: path
Content-Length: 0
```

Figure 5.8: (11) REGISTER

includes routing information: the SIP URI of the S-CSCF allocated to the user. The HSS stored this URI when it received a Diameter MAR message (6). Therefore, no matter whether the I-CSCF is the same I-CSCF the first REGISTER request traversed or not, the second REGISTER request ends up in the same S-CSCF, the one that was allocated to the user at the time of the registration. The S-CSCF receives the REGISTER request that includes the user credentials (15). An example of this REGISTER request is displayed in Figure 5.9. The S-CSCF then validates these credentials against the authentication vectors provided by the HSS in a Diameter MAA message (7) in Figure 5.4. If authentication is successful, then the S-CSCF sends a Diameter SAR message to the HSS (16) for the purpose of informing the HSS that the user is now registered and to download the user profile (17). The user profile is an important piece of information that includes, among other things, the collection of all the Public User Identities allocated for authentication of the Private User Identity. It also indicates to the S-CSCF which of these Public User Identities are automatically registered in the S-CSCF in a *set of implicitly registered Public User Identities*. Additionally, the user profile also contains the *initial filter criteria*, which is the collection of triggers that determine when a SIP request is forwarded to the Application Server that will be providing the service.

At this stage the S-CSCF has stored the contact URI for this user, as it was present in the Contact header field of the SIP REGISTER request. It has also stored the list of URIs included in the Path header field. This list always includes the P-CSCF URI and

```
REGISTER sip:home1.net SIP/2.0
Via: SIP/2.0/UDP icscf1.home1.net;branch=z9hG4bKea1dof,
Via: SIP/2.0/UDP pcscf1.visited1.net;branch=z9hG4bKoh2qrz,
Via: SIP/2.0/UDP [1080::8:800:200C:417A]:5059;comp=sigcomp;
                 branch=z9hG4bK9h9ab
Max-Forwards: 68
P-Access-Network-Info: 3GPP-UTRAN-TDD;
                       utran-cell-id-3gpp=C359A3913B20E
From: <sip:alice@home1.net>;tag=s8732n
To: <sip:alice@home1.net>
Contact: <sip:[1080::8:800:200C:417A]:5059;comp=sigcomp>
         ;expires=600000
Call-ID: 23fi571ju
Authorization: Digest username="alice_private@home1.net",
               realm="home1.net",
               nonce="dcd98b7102dd2f0e8b11d0f600bfb0c093",
               algorithm=AKAv1-MD5,
               uri="sip:home1.net",
               response="6629fae49393a05397450978507c4ef1",
               integrity-protected="yes"
Require: path
Supported: path
Path: <sip:term@pcscf1.visited1.net;lr>
P-Visited-Network-ID: "Visited 1 Network"
P-Charging-Vector: icid-value="W34h6dlg"
Cseq: 2 REGISTER
Content-Length: 0
```

Figure 5.9: (15) REGISTER

may optionally include an I-CSCF URI. Later, the S-CSCF will route initial SIP requests addressed to the user via the list of URIs included in the `Path` header field and the contact address in this order.

Last, but not least, the S-CSCF sends a 200 (OK) response to the REGISTER request, to indicate the success of the REGISTER request, (18) in Figure 5.4. An example of this response is displayed in Figure 5.10. The 200 (OK) response includes a `P-Associated-URI` header field that contains the list of URIs allocated to the user (not to be confused with the list of implicitly registered URIs). It also contains a `Service-Route` header field that includes a list of SIP server URIs. Future SIP requests that the IMS terminal sends will be routed via these SIP servers, in addition to the outbound proxy (P-CSCF). In the IMS the `Service-Route` header field value always contains the address of the S-CSCF of the user and may also contain the address of an I-CSCF in the home network.

The 200 (OK) response traverses the same I-CSCF and P-CSCF that the REGISTER request traversed. Eventually, the IMS terminal gets the 200 (OK) response, (20) in Figure 5.4. At this stage the registration procedure is complete. The IMS terminal is registered with the IMS for the duration of time indicated in the `expires` parameter of the `Contact` header field.

```
SIP/2.0 200 OK
Via: SIP/2.0/UDP [1080::8:800:200C:417A]:5059;comp=sigcomp;
                  branch=z9hG4bK9h9ab
Path: <sip:term@pcscf1.visited1.net;lr>
Service-Route: <sip:orig@scscf1.home1.net;lr>
From: <sip:alice@home1.net>;tag=s8732n
To: <sip:alice@home1.net>;tag=409sp3
Call-ID: 23fi571ju
Contact: <sip:[1080::8:800:200C:417A]:5059;comp=sigcomp>
          ;expires=600000
Cseq: 2 REGISTER
Date: Wed, 21 January 2004 18:19:20 GMT
P-Associated-URI: <sip:alice-family@home1.net>,
                  <sip:alice-business@home1.net>,
                  <sip:+1-212-555-1234@home1.net;user=phone>
Content-Length: 0
```

Figure 5.10: (20) 200 OK

As the reader may have noticed, the registration process with the IMS is based on SIP REGISTER requests that are authenticated. The procedure is overloaded with some extra functionality. For instance, the SIP Path header field extension (specified in RFC 3327 [235]) informs the S-CSCF of a P-CSCF (and perhaps an I-CSCF) via which SIP requests addressed to the IMS terminal should be routed. In the opposite direction the SIP Service-Route header field extension (specified in RFC 3608 [236]) provides the IMS terminal with a sequence of proxies via which future SIP requests have to be routed (e.g., S-CSCF), in addition to the outbound SIP server (P-CSCF). We also indicated the existence of a SIP P-Associated-URI header field (specified in RFC 3455 [109]). The S-CSCF populates the P-Associated-URI header field with a list of additional URIs that the home network operator has reserved or allocated to the user.

We want to stress that the IMS terminal is able to register a Public User Identity at any time. For instance, when the IMS application in the terminal is switched on, the IMS terminal may be configured to register one Public User Identity. Other Public User Identities may be registered later, at any time, upon explicit indication of the user. For example, this allows us to register independently a personal Public User Identity from a business Public User Identity.

5.5.2   IMS Registration with a USIM

In case the IMS terminal is equipped with a UICC that does not contain an ISIM application, perhaps because the card was acquired before the IMS service came into operation, the user can still register with the IMS network, but there are a few problems.

The first problem is that the IMS terminal is unable to extract or derive the Private User Identity, a Public User Identity and the home network domain URI where to address the SIP REGISTER request, because all of these parameters are stored in the ISIM, not the USIM. However, the IMS terminal can access a USIM. Of special interest in the USIM is the IMSI, which is a collection of a maximum of 15 decimal digits that globally represent the identity of a mobile subscriber, including their country and mobile network operator. The home operator allocates an IMSI to each subscriber.

Figure 5.11 depicts the structure of an IMSI. Beginning from the left side the first three digits form the *Mobile Country Code (MCC)*. The MCC represents the country of operation of the home network. The MCC is followed by two or three digits that constitute the *Mobile Network Code (MNC)*. The MNC represents the home operator within the country of the MCC. The remaining digits constitute the *Mobile Subscriber Identification Number (MSIN)*.

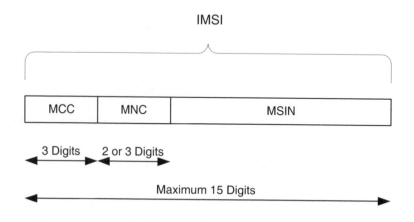

Figure 5.11: Structure of the IMSI

The IMSI is never used for routing calls in cellular networks, it is just used for identification of subscribers and their data stored in the network, authentication, and authorization purposes.

When an IMS terminal is loaded with a UICC that does not contain an ISIM, the terminal extracts the IMSI from the USIM in order to build a *temporary Private User Identity*, a *temporary Public User Identity*, and a *home network domain URI* that allows it to build a SIP REGISTER request and route it to the home network. These three parameters are only used during registration, re-registration, and deregistration procedures. When the user is eventually registered the S-CSCF sends a collection of the regular Public User Identities allocated to the user. The IMS terminal only uses these Public User Identities for any SIP traffic other than REGISTER requests. Consequently, the temporary identities are never known or used outside the home network (e.g., in a session setup).

5.5.2.1   Temporary Private User Identity

The temporary Private User Identity is derived from the IMSI. A regular Private User Identity has the format `username@realm`. A temporary Private User Identity has the same format. On building a temporary Private User Identity the IMS terminal inserts the complete IMSI as the `username` of the Private User Identity. The `realm` gets split into subrealms separated by dots (like a DNS subdomain), where the first subrealm is the MNC of the IMSI, the next subrealm is the MCC of the IMSI and the rest is the fixed string `.imsi.3gppnetwork.org`. As an example, let's imagine an IMSI: 2483235551234, where the MCC is 248, the MNC is 323, and the MSIN is 5551234. According to the explanation above the temporary Private User Identity is `2483235551234@323.248.imsi.3gppnetwork.org`.

5.5.2.2 Temporary Public User Identity

An IMS terminal without an ISIM also needs to build a temporary Public User Identity to be registered. A regular Public User Identity during registration is a SIP URI that takes the format `sip:user@domain`. It is very simple to build a temporary Public User Identity, because it takes the same format as the temporary Private User Identity, but now, it is prepended by the string: "sip:" since the identity is a SIP URI. So, if we take as an example the same IMSI that we chose to illustrate a temporary Private User Identity, the corresponding temporary Public User Identity is `sip:2483235551234@323.248.imsi.3gppnetwork.org`.

5.5.2.3 Home Network Domain URI

When an IMS terminal equipped only with a USIM needs to build a home network domain URI for inclusion in the *Request-URI* of the SIP REGISTER request, the terminal just removes the "user" part of the temporary Public User Identity. According to the example that we have been following the home network domain URI is set to:

`sip:323.248.imsi.3gppnetwork.org`

5.5.2.4 Registration Flow

Registration flow is the same no matter whether the IMS terminal is provided with an ISIM or a USIM application in the UICC. Figure 5.4 was discussed when we described registration with an ISIM and is still valid with a USIM. However, the contents of the messages change since the messages convey temporary Private and Public User Identities.

Figure 5.12 shows an example of a REGISTER request when the IMS terminal is equipped with a USIM in the UICC. The *Request-URI* and the `uri` parameter of the `Authorization` header are set to the home network domain derived from the IMSI. The `From` and `To` headers are set to the temporary Public User Identity derived from the IMSI. The `username` parameter in the `Authorization` header is set to the temporary Private User Identity also derived from the IMSI.

When the P-CSCF receives the REGISTER request (1), it performs its regular procedures to discover an I-CSCF in the home network. In this case, since the *Request-URI* of the REGISTER request is set to `323.248.imsi.3gppnetwork.org`, the P-CSCF uses this domain as the input to the DNS procedures (which are specified in RFC 3263 [214]). This requires that home network operators have configured appropriately not only the DNS entries corresponding to their own domain name but also the DNS entries corresponding to a DNS domain that follows the `imsi.3gppnetwork.org` pattern.

The I-CSCF queries the HSS with a Diameter UAR message (3) which contains the temporary Private and Public User Identities extracted from the REGISTER request. Eventually, the I-CSCF forwards the REGISTER request (5) to the S-CSCF. The S-CSCF downloads the authentication vectors from the HSS with a Diameter MAA message (7). These vectors are synchronized with the USIM in the UICC, since there is no ISIM present. The S-CSCF challenges the IMS terminal in a 401 (Unauthorized) response (8), which is forwarded to the IMS terminal via the I-CSCF (9) and the P-CSCF (10). Figure 5.13 shows an example of the contents of the 401 (Unauthorized) response (10).

The IMS terminal computes the challenge within the USIM, builds a new SIP REGISTER request (11) that contains an answer to the challenge, and sends it again to the P-CSCF, as shown in Figure 5.14. The message is forwarded to the S-CSCF.

```
REGISTER sip:323.248.imsi.3gppnetwork.org SIP/2.0
Via: SIP/2.0/UDP [1080::8:800:200C:417A];comp=sigcomp;
     branch=z9hG4bK9h9ab
Max-Forwards: 70
P-Access-Network-Info: 3GPP-UTRAN-TDD;
                       utran-cell-id-3gpp=C359A3913B20E
From: <sip:2483235551234@323.248.imsi.3gppnetwork.org>;tag=4fa3
To: <sip:2483235551234@323.248.imsi.3gppnetwork.org>
Contact: <sip:[1080::8:800:200C:417A];comp=sigcomp>
          ;expires=600000
Call-ID: 23fi571ju
Authorization: Digest
        username="2483235551234@323.248.imsi.3gppnetwork.org",
        realm="323.248.imsi.3gppnetwork.org", nonce="",
        uri="sip:323.248.imsi.3gppnetwork.org", response=""
Security-Client: ipsec-3gpp; alg=hmac-sha-1-96;
                 spi-c=3929102; spi-s=0293020;
                 port-c:3333; port-s=5059
Require: sec-agree
Proxy-Require: sec-agree
Cseq: 1 REGISTER
Supported: path
Content-Length: 0
```

Figure 5.12: (1) REGISTER

```
SIP/2.0 401 Unauthorized
Via: SIP/2.0/UDP [1080::8:800:200C:417A];comp=sigcomp;
                branch=z9hG4bK9h9ab
From: <sip:2483235551234@323.248.imsi.3gppnetwork.org>;tag=4fa3
To: <sip:2483235551234@323.248.imsi.3gppnetwork.org>;tag=409sp3
Call-ID: 23fi571ju
WWW-Authenticate: Digest realm="323.248.imsi.3gppnetwork.org",
                  nonce="dcd98b7102dd2f0e8b11d0f600bfb0c093",
                  algorithm=AKAv1-MD5
Security-Server: ipsec-3gpp; q=0.1; alg=hmac-sha-1-96;
                 spi-c=909767; spi-s=421909;
                 port-c:4444; port-s=5058
Cseq: 1 REGISTER
Content-Length: 0
```

Figure 5.13: (10) 401 Unauthorized

```
REGISTER sip:323.248.imsi.3gppnetwork.org SIP/2.0
Via: SIP/2.0/UDP [1080::8:800:200C:417A]:5059;comp=sigcomp;
                    branch=z9hG4bK9h9ab
Max-Forwards: 70
P-Access-Network-Info: 3GPP-UTRAN-TDD;
                          utran-cell-id-3gpp=C359A3913B20E
From: <sip:2483235551234@323.248.imsi.3gppnetwork.org>;tag=4fa3
To: <sip:2483235551234@323.248.imsi.3gppnetwork.org>
Contact: <sip:[1080::8:800:200C:417A]:5059;comp=sigcomp>
          ;expires=600000
Call-ID: 23fi57lju
Authorization: Digest
        username="2483235551234@323.248.imsi.3gppnetwork.org",
        realm="323.248.imsi.3gppnetwork.org",
        nonce="dcd98b7102dd2f0e8b11d0f600bfb0c093",
        algorithm=AKAv1-MD5,
        uri="sip:323.248.imsi.3gppnetwork.org",
        response="6629fae49393a05397450978507c4ef1"
Security-Verify: ipsec-3gpp; q=0.1; alg=hmac-sha-1-96;
                  spi-c=909767; spi-s=421909;
                  port-c:4444; port-s=5058
Require: sec-agree
Proxy-Require: sec-agree
Cseq: 2 REGISTER
Supported: path
Content-Length: 0
```

Figure 5.14: (11) REGISTER

When the S-CSCF gets the REGISTER request (15) it verifies the response and providing that it was correct it contacts the HSS to download the user profile. Part of the user profile contains the list of Public User Identities that are equivalent, or alias to it, and some of them are implicitly registered when the user registers any of them. The HSS also sends an indication of whether the temporary Public User Identity derived from the IMSI is barred, so that the user cannot originate or receive any SIP traffic with that temporary Public User Identity other than for registrations.

The S-CSCF inserts all the non-barred Public User Identities that are associated with this registered Public User Identity in the P-Associated-URI header field value. The header field does not indicate which of those identities have been automatically registered, it only indicates the Public User Identities the home network has reserved for this user. The S-CSCF sends the 200 (OK) response, (18) in Figure 5.4, via the I-CSCF, P-CSCF to the IMS terminal (20). Figure 5.15 shows an example of the 200 (OK) response, (20) in Figure 5.4, that the IMS terminal receives. In this example, the temporary Public User Identity is barred, since it is not included in the P-Associated-URI header field.

Once the registration process is complete the IMS terminal has got a set of regular Public User Identities that can be used for subscriptions, establishing sessions, etc. The temporary identities are used only for registration purposes.

```
SIP/2.0 200 OK
Via: SIP/2.0/UDP [1080::8:800:200C:417A]:5059;comp=sigcomp;
                branch=z9hG4bK9h9ab
Path: <sip:term@pcscf1.visited1.net;lr>
Service-Route: <sip:orig@scscf1.home1.net;lr>
From: <sip:2483235551234@323.248.imsi.3gppnetwork.org>;tag=4fa3
To: <sip:2483235551234@323.248.imsi.3gppnetwork.org>;tag=409sp3
Call-ID: 23fi57lju
Contact: <sip:[1080::8:800:200C:417A]:5059;comp=sigcomp>
         ;expires=600000
Cseq: 2 REGISTER
Date: Wed, 21 January 2004 18:19:20 GMT
P-Associated-URI: <sip:alice@home1.net>,
                  <sip:alice-family@home1.net>,
                  <sip:alice-business@home1.net>,
                  <sip:+1-212-555-1234@home1.net;user=phone>
Content-Length: 0
```

Figure 5.15: (20) 200 OK

5.6   Subscription to the `reg` Event State

Let's consider for a second the operation of SIP in a general Internet context rather than in the IMS. In SIP a User Agent can register with a registrar. The registrar accepts the registration and creates a registration state. When registering the SIP the User Agent publishes its contact address (i.e., the location where the user is available). The registrar stores this information for a determined length of time, according to the expiration timer of the SIP REGISTER request. Now, let's imagine for a second that the registrar gracefully shuts down, or simply that the operator of the registrar wants, for whatever reason, to clear the registration stage of the SIP User Agent in the registrar. Would the SIP User Agent be informed about this hypothetical deregistration? The answer is no, because basic SIP functionality does not offer a mechanism for the UA to be informed that a deregistration has occurred.

In the IMS there is a clear requirement, perhaps inherited from the GSM age, for the user to be informed whether he is reachable or not (i.e., whether the terminal is under radio coverage and whether the user is registered with the network or not). Most existing GSM phones provide information to the user on whether the phone is under radio coverage or not and whether the terminal is registered to the network. The core SIP specified in RFC 3261 [215] does not offer a solution to this requirement. The solution that the IETF worked out and the IMS adopted was to create a registration package for the SIP event framework. The solution (specified in RFC 3680 [201]), allows an IMS terminal to subscribe to its registration-state information stored in the S-CSCF. When the IMS terminal has completed its registration, it sends a SUBSCRIBE request for the `reg` event. This request is addressed to the same Public User Identity that the SIP User Agent just registered. The S-CSCF receives the request and installs that subscription (i.e., the S-CSCF takes the role of a notifier, according to RFC 3265 [198]). According to the same RFC, the S-CSCF sends a NOTIFY request to the user (e.g., the IMS terminal). This request includes an XML document in its body (specified in RFC 3680 [201]) that contains a list of all the Public User Identities allocated to the user, along with the user's registration state. The IMS terminal now knows whether the user is registered or not, and with which Public User Identities the user is registered (remember that the user may be registered with the IMS network with more than a single Public User Identity). The IMS terminal may decide to display a list of all the Public User Identities that the operator has allocated to the user and it may display icons showing whether that Public User Identity is registered with the network or not.

In case the S-CSCF has to shut down or if the operator needs to manually deregister a Public User Identity, the registration information will be changed. This will provoke the S-CSCF into informing each subscriber of the reg event of that user.

In addition to the IMS terminal subscribing to its own registration state, the P-CSCF also subscribes to the registration state of the user. This registration allows the P-CSCF to be informed in real time about which Public User Identities are registered. In case there is administrative action taking place on one or all of them the P-CSCF can delete any state it may have.[7]

Other entities that may require a subscription to the reg state of a user are Application Servers. However, this subscription is not compulsory, as it depends on the type of service offered to the user. For instance, an Application Server offering a welcome message when a user switches on their IMS terminal may subscribe to the reg state of the user, so that when the user connects to the IMS network the Application Server is informed that the user has now switched their IMS terminal on. In response, the Application Server sends an instant message to the user, giving a welcome message or any other information of interest to the user.

Figure 5.16 shows the complete registration flow.[8] This figure expands the contents of Figure 5.4 by including additionally the subscriptions to the reg event of both the P-CSCF and the IMS terminal. Messages 1–20 in Figure 5.16 have already been described previously when dealing with Figure 5.4.

Figure 5.17 shows the contents of the SUBSCRIBE request that the IMS terminal sends to the user's registration stage. You may notice that the *Request-URI* contains the Public User Identity that was just registered in the previous REGISTER requests. The IMS terminal sends this SUBSCRIBE request to its P-CSCF. The P-CSCF honors the Route header field and proxies the request to the S-CSCF, (26) in Figure 5.16. The S-CSCF acts as a SIP notifier and, as it is also a registrar for the Public User Identity of interest, accepts the subscription by sending a 200 (OK) response (27). The P-CSCF forwards the response (28) to the IMS terminal. Additionally, the S-CSCF sends a NOTIFY request (29) that contains a well-formed and valid XML document that contains registration information. The P-CSCF forwards the NOTIFY request to the IMS terminal. Figure 5.18 shows an example of this NOTIFY request, as received at the IMS terminal. If we focus on the XML registration information document we can observe that there is a set of Public User Identities allocated to this user. In this case, these Public User Identities are sip:alice@home1.net, sip:alice-family@home1.net and tel:+1-212-555-1234. They are indicated in the aor attribute of the registration element. The S-CSCF also indicates in the state attribute that all of these Public User Identities are active. The contact address of each Public User Identity is also supplied. In this example, each contact element contains a uri element that includes the address of the different identities that point to the same SIP URI. This URI contains the IP address allocated to the IMS terminal. An important piece of information is included in the event attribute

---

[7]Although the P-CSCF does not keep a registration state that relates to SIP registrations, it keeps states that relate to the compression of signaling (SigComp) and the establishment of security associations.

[8]Note that the SLF, which is required if there is more than one HSS in the home network, is not shown.

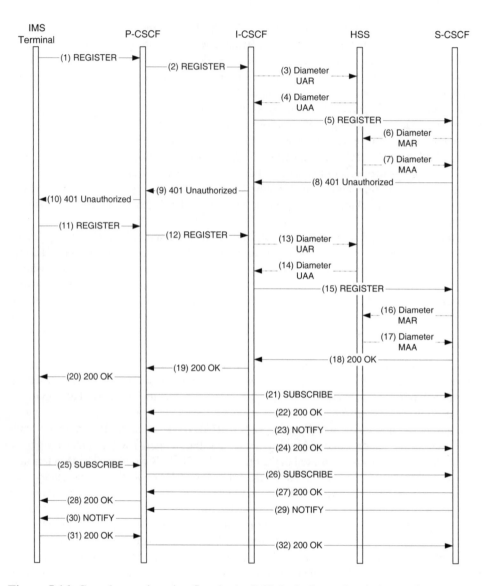

Figure 5.16: Complete registration flow in the IMS, including subscription to the reg event state

of the contact address. The event attribute indicates which event triggered the change in state. In this example the S-CSCF indicates that a SIP registration made the first Public User Identity change to active, because the event attribute of the Public User Identity of sip:alice@home1.net was set to registered. However, the event attribute of the other two Public User Identities was set to created, indicating that these two Public User Identities were created administratively. These two administratively created Public User Identities form what is know as the *set of implicitly created Public User Identities*. This is an interesting feature of IMS that allows a user to configure his registration so that whenever one Public

```
SUBSCRIBE sip:alice@home1.net SIP/2.0
Via: SIP/2.0/UDP [1080::8:800:200C:417A]:5059;comp=sigcomp;
                 branch=z9hG4bK9h9ab
Max-Forwards: 70
Route: <sip:pcscf1.visited1.net:5058;lr;comp=sigcomp>,
       <sip:orig@scscf1.home1.net;lr>
P-Preferred-Identity: "Alice Bell" <sip:alice@home1.net>
Privacy: none
P-Access-Network-Info: 3GPP-UTRAN-TDD;
                       utran-cell-id-3gpp=C359A3913B20E
From: <sip:alice@home1.net>;tag=d9211
To: <sip:alice@home1.net>
Call-ID: b89rjhnedlrfjflslj40a222
Require: sec-agree
Proxy-Require: sec-agree
Cseq: 61 SUBSCRIBE
Event: reg
Expires: 600000
Accept: application/reginfo+xml
Security-Verify: ipsec-3gpp; q=0.1; alg=hmac-sha-1-96;
                 spi-c=98765432; spi-s=909767;
                 port-c=5057; port-s=5058
Contact: <sip:[1080::8:800:200C:417A]:5059;comp=sigcomp>
Content-Length: 0
```

Figure 5.17: (25) SUBSCRIBE

User Identity is registered other Public User Identities are automatically registered. The same applies for deregistration: if a Public User Identity is deregistered the set of implicitly registered Public User Identities are automatically deregistered. This feature can be used to speed up the registration process, as now this process is independent of the number of identities allocated to the user.

5.7 Basic Session Setup

This section explores the process of establishing a basic session in the IMS. For the sake of clarity we will assume that an IMS terminal is establishing a session toward another IMS terminal. As both terminals are IMS terminals they will support the same sort of capabilities.

For the sake of simplicity, we assume that neither the calling party nor the called party have any services associated with the session. We describe in Section 5.8.3 how an Application Server is involved and how Application Servers can provide valuable services to the user.

Figures 5.19 and 5.20 contain a description flow chart of the signaling involved in a basic session setup. At first glance the reader may think that there are a lot of SIP messages flowing around. Additionally, there are many nodes involved in setting up the session. The evaluation is probably right, SIP is a rich protocol in its expression at the cost of a high number of messages.

```
NOTIFY sip:[1080::8:800:200C:417A]:5059;comp=sigcomp SIP/2.0
Via: SIP/2.0/UDP pcscf1.visited1.net:5058;comp=sigcomp
                ;branch=z9hG4bKoh2qrz
Via: SIP/2.0/UDP scscf1.home1.net;branch=z9hG4bKs1pp0
Max-Forwards: 69
Route: <sip:pcscf1.home1.net;lr>
From: <sip:alice@home1.net>;tag=d9211
To: <sip:alice@home1.net>;tag=151170
Call-ID: b89rjhnedlrfjflslj40a222
Cseq: 42 NOTIFY
Subscription-State: active;expires=600000
Event: reg
Content-Type: application/reginfo+xml
Contact: <sip:scscf1.home1.net>
Content-Length: 873

<?xml version="1.0"?>
<reginfo xmlns="urn:ietf:params:xml:ns:reginfo"
                version="1" state="full">
    <registration aor="sip:alice@home1.net"
                id="11a" state="active">
        <contact id="542" state="active" event="registered"
                duration-registered="0">
            <uri>sip:[1080::8:800:200C:417A]</uri>
        </contact>
    </registration>
    <registration aor="sip:alice-family@home1.net"
                id="11b" state="active">
        <contact id="543" state="active" event="created"
                duration-registered="0">
            <uri>sip:[1080::8:800:200C:417A]</uri>
        </contact>
    </registration>
    <registration aor="tel:+1-212-555-1234"
                id="11c" state="active">
        <contact id="544" state="active" event="created"
                duration-registered="0">
            <uri>sip:[1080::8:800:200C:417A]</uri>
        </contact>
    </registration>
</reginfo>
```

Figure 5.18: (30) NOTIFY

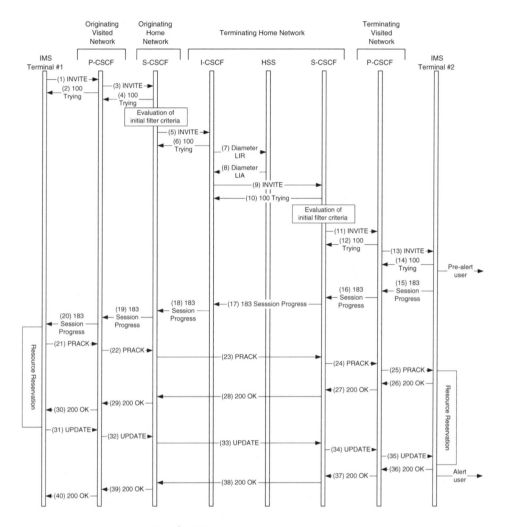

Figure 5.19: Basic session setup, part 1

But let's focus on Figures 5.19 and 5.20. We are assuming that both users are roaming to a network outside their respective home networks, such as when both users are outside their respective countries. This leads to having two different visited networks in the figures. We also assume that each of the users has a different business relationship with his respective operator; therefore, there are two different home networks in the figures. Additionally, we assume that the P-CSCF is located in a visited network. When we consider all the roaming/non-roaming scenarios, different home networks, etc., this is the most complete and complicated case. Once the reader is familiar with this scenario, variations of it are just subsets or simplifications of it.

For the sake of clarity we refer to the *originating P-CSCF* and *originating S-CSCF* as the P-CSCF and S-CSCF that are serving the caller. Similarly, we refer to the *terminating P-CSCF* and *terminating S-CSCF* as the P-CSCF and S-CSCF that are serving the callee.

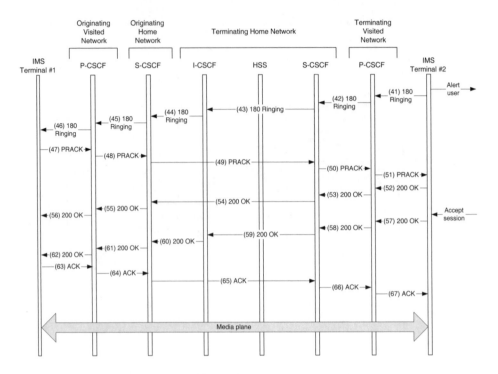

Figure 5.20: Basic session setup, part 2

The first observation the reader might have noticed is the complete separation of the signaling and media planes. Signaling (i.e., SIP), traverses a set of CSCFs, whereas media is sent end to end (i.e., from an IMS terminal to another IMS terminal) only traversing IP routers and, if applicable, GGSNs.

Another observation is that all SIP signaling traverses both the originating P-CSCF and the originating S-CSCF in all circumstances. This is a significant difference with respect to other cellular networks where, if the user is roaming, signaling does not traverse the home network. The P-CSCF must be present in all the signaling exchanged with the terminal because it compresses/decompresses SIP in the interface toward the terminal. The S-CSCF is traversed in all requests to allow the triggering of services that the user may have requested. The S-CSCF plays an important role in service provision by involving one or more Application Servers that implement the service logic. Since the S-CSCF is always located in the home network, services are always available to the user regardless of whether the user is roaming or not. We describe the triggering of services in Section 5.8.3. Note, however, that traversing the CSCFs only affects the signaling plane. It does not affect, in principle, the media plane, which is transmitted end-to-end.

If we continue with the examination of the figures, we discover that the flow follows the model described in "Integration of Resource Management and SIP" (RFC 3312 [67]), sometimes known as *preconditions*. We already explored the model in Section 4.14.

While the use of preconditions was mandatory in 3GPP Release 5, Release 6 allows session establishment with no preconditions when the remote terminal does not support them or when a particular service does not require them. Push-to-talk over Cellular (PoC), which is described in Chapter 20, is an example of a service that does not use preconditions.

For regular sessions between two IMS terminals the requirement to use the preconditions call flow model comes from the fact that in cellular networks radio resources for the media plane are not always available.[9] If the preconditions extension was not used, when the called party accepts the call, the radio bearers at both the calling and called party would not be ready. This means that the terminals would not be able to transmit and receive media-plane traffic (e.g., Real-Time Transport Protocol). As a consequence, the first words the callee says may be lost.

If we take a look at Figure 5.19 we can see a Diameter interaction between the I-CSCF and the HSS in the terminating network. This is required because the I-CSCF needs to discover the address of the S-CSCF serving the destination user. However, it must be noted that there is no interaction between the HSS and any other node in the originating network. In other words, the flow is asymmetric, considering the originating and terminating sides of it. The advantage is that the HSS is relieved of involvement in extra messages that are not required in the originating call leg.[10]

By examining Figures 5.19 and 5.20 we realize that all SIP signaling traverses the originating and terminating P-CSCF and S-CSCF nodes (four nodes in total). This requires that each of these nodes inserts a `Record-Route` header field that contains the SIP URI of the node. This guarantees that future signaling (e.g., a PRACK or BYE request) will visit that node. However, the I-CSCF in the terminating network plays a different role and may or may not insert a `Record-Route` header field pointing to itself. This means that the INVITE request and all its responses (both provisional and the final one) will visit the I-CSCF, but other requests such as PRACK, ACK, or BYE, will not visit the I-CSCF.

5.7.1 The IMS Terminal Sends an INVITE Request

Figure 5.21 shows an example of the INVITE request (1) that the IMS terminal sends to the network. We will devote a few pages to this INVITE request and analyze it in depth.

The *Request-URI* contains the Public User Identity of the intended destination. In this case the *Request-URI* points to Bob's identity that belongs to an operator known as home2.net.

Figure 5.22 reproduces the contents of the `Via` header field displayed in Figure 5.21. The `Via` header field contains the IP address and port number where the IMS terminal is supposed to receive the responses to the INVITE request. The port number is of importance because it is the port number bound to the security association that the P-CSCF has established toward the IMS terminal (we describe security associations in detail in Section 9.1.4). The P-CSCF will forward responses to the INVITE request to the IP address and port number declared in the `Via` header field. If the IMS terminal fails to include a port number bound to the security association at the P-CSCF the IPsec layer at the P-CSCF will discard the packet. In addition, the header also indicates the willingness of IMS terminals to receive compressed signaling using Signaling Compression (SigComp, RFC 3320 [191]). The contents of the `Contact` header field are quite similar, as they indicate the SIP URI where the IMS terminal is willing to receive subsequent requests belonging the same SIP dialog as the INVITE request.

---

[9]It must be noted that radio resources for the signaling plane are always allocated, so that sessions and other types of SIP signaling can be addressed to the IMS terminal. However, the situation is not the same for the radio resources that the media plane will use.

[10]Note that the HSS in typical implementations is a high-capacity node, serving hundreds of thousands of users. Saving a couple of messages on each session attempt is a significant saving in capacity at the HSS.

```
INVITE sip:bob@home2.net SIP/2.0
Via: SIP/2.0/UDP [1080::8:800:200C:417A]:5059;
                comp=sigcomp;branch=z9hG4bK9h9ab
Max-Forwards: 70
Route: <sip:pcscf1.visited1.net:5058;lr;comp=sigcomp>,
       <sip:orig@scscf1.home1.net;lr>
P-Preferred-Identity: "Alice Smith" <sip:alice@home1.net>
Privacy: none
P-Access-Network-Info: 3GPP-UTRAN-TDD;
                       utran-cell-id-3gpp=C359A3913B20E
From: <sip:alice@home1.net>;tag=ty20s
To: <sip:bob@home2.net>
Call-ID: 3s09cs03
Cseq: 127 INVITE
Require: precondition, sec-agree
Proxy-Require: sec-agree
Supported: 100rel
Security-Verify: ipsec-3gpp; q=0.1;
                 alg=hmac-sha-1-96;
                 spi-c=98765432; spi-s=909767;
                 port-c=5057; port-s=5058
Contact: <sip:[1080::8:800:200C:417A]:5059;comp=sigcomp>
Allow: INVITE, ACK, CANCEL, BYE, PRACK, UPDATE, REFER, MESSAGE
Content-Type: application/sdp
Content-Length: 569

v=0
o=- 1073055600 1073055600 IN IP6 1080::8:800:200C:417A
s=-
c=IN IP6 1080::8:800:200C:417A
t=0 0
m=video 8382 RTP/AVP 98 99
b=AS:75
a=curr:qos local none
a=curr:qos remote none
a=des:qos mandatory local sendrecv
a=des:qos none remote sendrecv
a=rtpmap:98 H263
a=fmtp:98 profile-level-id=0
a=rtpmap:99 MP4V-ES
m=audio 8283 RTP/AVP 97 96
b=AS:25.4
a=curr:qos local none
a=curr:qos remote none
a=des:qos mandatory local sendrecv
a=des:qos none remote sendrecv
a=rtpmap:97 AMR
a=fmtp:97 mode-set=0,2,5,7; maxframes=2
a=rtpmap:96 telephone-event
```

Figure 5.21: (1) INVITE

The Via header field also indicates the transport protocol that is used to transport SIP messages to the next hop. SIP supports any transport protocol (e.g. UDP, TCP, or SCTP). So, every node is free to choose the more appropriate transport protocol according to the guidelines given in RFC 3261 [215].

```
Via: SIP/2.0/UDP [1080::8:800:200C:417A]:5059;
                comp=sigcomp;branch=z9hG4bK9h9ab
```

Figure 5.22: The Via header field

The reader can observe that there is a preloaded Route header field (Figure 5.23). The value of the Route header field points to the P-CSCF in the visited network and the S-CSCF in the home network. Typically, in SIP the Route header fields are learnt during the initial INVITE transaction. In normal circumstances the SIP UA obtains the route out of the Record-Route header field value, and the set of proxies to traverse will be operational for subsequent transactions that belong to the dialog created with the INVITE request.

```
Route: <sip:pcscf1.visited1.net:5058;lr;comp=sigcomp>,
       <sip:orig@scscf1.home1.net;lr>
```

Figure 5.23: The Route header field

In the IMS, the first messages need to traverse a set of SIP proxies, which serve the IMS terminal. The SIP UA needs to create a list of proxies to be traversed, and place this list into the Route header field. This set of proxies is created from the concatenation of the outbound SIP proxy, learnt during the P-CSCF discovery procedure (see Section 5.4), and the value of the Service-Route header field received in the 200 (OK) response for a REGISTER request. We already saw in Figure 5.10 what this 200 (OK) response looks like. In our example the Service-Route header field contained a single node to visit, the S-CSCF in the home network. So, when the IMS terminal creates a request other than a subsequent request within an existing dialog, it inserts the preloaded Route header field pointing to the mentioned nodes.

If we continue with the examination of the relevant header fields in Figure 5.21, we can observe the presence of a P-Preferred-Identity header field with the value set to Alice Smith and a SIP URI. This header field is a SIP extension (which is specified in the Private "Extensions to the Session Initiation Protocol (SIP) for Asserted Identity within Trusted Networks", RFC 3325 [147]). The reason for this header field is that a user may have several Public User Identities. When the user initiates a session, they need to indicate which one of their identities should be used for this session. The user is identified with this identity in the network charging records and, if applicable, at the callee. In addition, the identity is also used to trigger services; so, different identities may trigger different services. The user chooses a preferred identity and inserts the value in the P-Preferred-Identity header field of the INVITE request. When the P-CSCF receives the INVITE request, it verifies that the INVITE request was received within an already established security association with the terminal that registered the identity of the P-Preferred-Identity. The P-CSCF also verifies the possible identities that the user can use in requests received through that security association. Providing that all the verifications are correct, the P-CSCF, when forwarding the INVITE request toward the home network, replaces the P-Preferred-Identity with a P-Asserted-Identity header field that contains the same value. If the P-CSCF considers that the preferred identity is not allocated to the user or not registered to the IMS network, the P-CSCF removes the `P-Preferred-Identity` header field and inserts a `P-Asserted-Identity` that contains a value the P-CSCF considers to be one of the valid identities of the user. Additionally, if the user didn't indicate a `P-Preferred-Identity` header field, then the P-CSCF chooses a default identity and inserts it in a `P-Asserted-identity` header field. In any case the SIP request that the P-CSCF forwards always includes a `P-Asserted-Identity` header field that includes an authenticated Public User Identity value.

The next header field that we see in Figure 5.21 is the `P-Access-Network-Info` header field. This header field is an extension created by the IETF as a consequence of 3GPP requirements (this header field is documented in the *P-Header extensions to SIP for 3GPP*, RFC 3455 [109]). Figure 5.24 shows an example of its utilization. The header field provides two types of access information.

(a) The type of layer 2/3 technology used by the IMS terminal. In our example, the IMS terminal indicates that is using the UTRAN (UMTS Terrestrial Radio Access Network). This information may be valuable to Application Servers, as they may use it to customize the service depending on the characteristics of the access network. The available bandwidth to the terminal is also determined by the type of access network. For instance, a wireless LAN access typically provides higher rates than the GERAN (GSM/Edge Radio Access Network).

(b) The identity of the radio cell where the IMS terminal is connected. This parameter is optional, as it may or may not be available in the SIP application layer. But, when the SIP application layer can read the cell identity from the lower radio layers the IMS terminal will insert it. The base station transceiver broadcasts its cell identity over the radio channels, and the radio layer at the terminal stores the last received cell identity. The IMS application reads that information and sticks it in the `P-Access-Network-Info` header field. On the other hand, the cell ID implicitly contains some rough location information that may be used to provide a personalized service to the user (e.g., an announcement giving a list of the closest Italian restaurants in the same location area).

```
P-Access-Network-Info: 3GPP-UTRAN-TDD;
                       utran-cell-id-3gpp=C359A3913B20E
```

Figure 5.24: The `P-Access-Network-Info` header field

Due to the sensitive private information contained in the `P-Access-Network-Info` header field the header field is transmitted from the IMS terminal until the last hop in the home network, but it is never transferred to the callee's home network. This scheme allows Application Servers in the home network to receive the header field when they process a SIP message. Referring to Figure 5.19, the INVITE request in steps (1)–(4) contains the header field, but the INVITE request in step (5) does not contain the header field, as this INVITE request is destined for the callee's home network.

The `Privacy` header field in Figure 5.21 shows the willingness of the user to reveal certain privacy information to nontrusted parties (such as the called party). In this example the user does not have any requirement regarding the privacy of the session. In other cases the user may have indicated that they do not want to reveal the contents of some header fields to nontrusted parties. One of these header fields is the `P-Asserted-Identity`, which as we previously described, is inserted not by the IMS terminal but by the P-CSCF.

The `From`, `To`, and `Call-ID` header fields are set according to the SIP procedures described in the SIP core specification, RFC 3261 [215]. It must be noted that these header fields transport end-to-end information and that network nodes do not assert, inspect, or take any action on the value of these header fields. It is especially important to highlight that the `From` header field is not policed, so the IMS terminal can insert any value there, including a SIP URI that does not belong to the user. For the purpose of identifying the user the `P-Asserted-Identity` header field is used instead.

The `Require`, `Proxy-Require`, and `Supported` header fields are shown in Figure 5.25. These header fields include those values that declare the mandatory or optional requirements of certain capabilities or SIP extensions. The mandatory capabilities are required in the `Require` and `Proxy-Require` header fields, whereas the optional capabilities are declared in the `Supported` header field. In the example in Figure 5.25 the mandatory capabilities that have to be supported are the `precondition` extension and the SIP security agreement `sec-agree`, whereas the optional requirement is the reliability of provisional responses `100rel`. We describe the SIP security agreement in Chapter 9.

```
Require: precondition, sec-agree
Proxy-Require: sec-agree
Supported: 100rel
```

Figure 5.25: The `Require`, `Proxy-Require` and `Supported` header fields

The `precondition` value in the `Require` header field is an indication to the remote terminal that it has to use the SIP preconditions extension. All IMS terminals implement this extension, which determines the call flow. In particular, because the preconditions extension is used the call flow contains 183 (Session Progress) responses, PRACK, and UPDATE requests. We described the generalities of the preconditions extension in Section 4.14.

In the IMS we are interested in the quality of service aspects of this extension. The problem in cellular networks is that resources are not preallocated before its usage is needed. As such, the calling party cannot freely request the network to establish a session comprising high-bandwidth audio and video if, at the end of the SIP/SDP negotiation, the remote party just supports a low-bandwidth codec or, even worse, declines the session establishment. The preconditions extension allows us to have a first exchange of SDP in an INVITE request and a 183 (Session Progress) response. After that, both parties are aware of the capabilities and willingness of the remote party to establish a particular set of media streams with a particular set of codecs. This is the moment when the IMS terminal can request a guaranteed quality of service from the network to establish the appropriate media streams.

The preconditions extension defines two models of operation with respect to quality of service. In the IMS the so-called local segmentation model for quality of service reservation is used. Each of the terminals is responsible for maintaining the appropriate resource reservation in its respective local segment. In the case of a GPRS access network, the local segment is defined between the IMS terminal and the GGSN. A particular quality of service is requested using the PDP Context Activation procedure.

The `Supported` header field indicates all those SIP extensions that the IMS terminal supports and are appropriate for the invited party to use in the response. In the example we saw that the IMS terminal declares its support for the reliable provisional responses extension 100rel. All IMS terminals support this extension which, in any case, is already required by the preconditions extension.

The Contact header field in Figure 5.26 indicates the SIP URI where the IMS terminal wants to receive further SIP requests pertaining to the same dialog. Typically, the IMS terminal is not aware of its own host name (unless it does a reverse DNS query to find its own host name), and it is really not required. So, this URI typically contains an IP address rather than a host name. The port number that follows the IP address refers to the port number bound to the security association that the P-CSCF has established toward the IMS terminal. If the IMS terminal fails to indicate a port number that falls into the security association that the P-CSCF has established toward the IMS terminal, the IPsec layer in the P-CSCF discards packets and the IMS terminal does not receive the desired SIP signaling. Additionally, the URI declares the willingness to provide signaling compression, because of the inclusion of the comp=sigcomp parameter.

```
Contact: <sip:[1080::8:800:200C:417A]:5059;comp=sigcomp>
```

Figure 5.26: The Contact header field

The next header field the IMS terminal includes in the INVITE request in Figure 5.21 is the Allow header field, reproduced in Figure 5.27. The Allow header field is optional in INVITE requests, although highly recommended, as it gives a hint to the peer terminal about the SIP methods the IMS terminal supports. In our example in Figure 5.27 the IMS terminal declared its support for the core SIP methods (e.g., INVITE, ACK, CANCEL, BYE), the methods supported by the preconditions extension (RFC 3312 [67]) (e.g., PRACK, UPDATE), the REFER method (RFC 3515 [229]), and the MESSAGE method (RFC 3428 [76]). When the session is established, the peer terminal grants support for REFER or MESSAGE methods that, otherwise, could or could not be supported, as they are extensions to the SIP core protocol.

```
Allow: INVITE, ACK, CANCEL, BYE, PRACK, UPDATE, REFER, MESSAGE
```

Figure 5.27: The Allow header field

Content-Type and Content-Length are the last two header fields in Figure 5.21. The values of these header fields depend on the type of body included in the INVITE request. SDP (Session Description Protocol, whose core protocol is specified in RFC 2327 [115][11]) is the only protocol used in the IMS to describe sessions. Furthermore, although it is possible that for INVITE requests not to include any body at all, IMS-compliant terminals, when creating INVITE requests, always insert an SDP offer describing the session they are trying to establish.

Figure 5.28 reproduces the Content-Type and Content-Length header fields. These headers indicate that the accompanying body is an SDP body of a certain length. If we take a look at the SDP body, we can see a number of lines that indicate the type of session that the IMS terminal wants to create. In particular, the c= line indicates the IP address where the terminal wants to receive media streams. In this example the IMS terminal wants to establish

---

[11]There is work in progress in the IETF to revise SDP, RFC 2327, into a new document that will contain some clarifications about the protocol, support for IPv6, etc. The document will be allocated a new RFC number.

two media streams, indicated by the presence of two m= lines: one is an audio stream, the other is a video stream. It is worth noting that IMS terminals always include a b= line per audio or video media stream, indicating the bandwidth requirements for that particular media stream. The b= line is otherwise optional in SDP, but IMS terminals include it to give a hint to the network nodes that need to control resources (such as bandwidth allocation).

```
Content-Type: application/sdp
Content-Length: 569

v=0
o=- 1073055600 1073055600 IN IP6 1080::8:800:200C:417A
s=-
c=IN IP6 1080::8:800:200C:417A
t=0 0
m=video 8382 RTP/AVP 98 99
b=AS:75
a=curr:qos local none
a=curr:qos remote none
a=des:qos mandatory local sendrecv
a=des:qos none remote sendrecv
a=rtpmap:98 H263
a=fmtp:98 profile-level-id=0
a=rtpmap:99 MP4V-ES
m=audio 8283 RTP/AVP 97 96
b=AS:25.4
a=curr:qos local none
a=curr:qos remote none
a=des:qos mandatory local sendrecv
a=des:qos none remote sendrecv
a=rtpmap:97 AMR
a=fmtp:97 mode-set=0,2,5,7; maxframes=2
a=rtpmap:96 telephone-event
```

Figure 5.28: The Content-Type and Content-Length header fields and the SDP body

The m= lines in SDP include the port number where the IMS terminal wants to receive the media stream and a list of codecs that the terminal is willing to support for this media stream and for this session. The first m= line in Figure 5.28 indicates the desire to establish a video stream and includes the supported and desired video codecs for this session. In this example the IMS terminal indicates support for the H.263 [137] and MPEG-4 Visual [129] codecs. The H.263 [137] baseline is the mandatory codec for 3GPP IMS terminals. 3GPP also recommends IMS terminals to implement H.263 version 2 Interactive and Streaming Wireless Profile (Profile 3) Level 10 [142] and MPEG-4 Visual Simple Profile at Level 0 [129]. Other codecs may be supported at the discretion of the IMS terminal manufacturer. 3GPP2 supports a different set of codecs that include EVRC (Enhanced Variable Rate Codec) and SMV (Selectable Mode Vocoder).

The second m= line in Figure 5.28 indicates the wish to establish an audio stream and includes the supported and desired audio codecs for this session. In this case the IMS terminal has indicated the support for the AMR narrowband speech codec (specified in 3GPP TS 26.071 [5]) and the telephone-event RTP payload format (used to transmit key presses during a session), as specified in RFC 2833 [227]. Both the AMR and the telephone-event RTP payload format are mandatory audio codecs for a 3GPP IMS terminal. If the IMS terminal supports wideband speech codecs, the mandatory codec to implement is the AMR wideband codec, as specified in 3GPP TS 26.171 [38], working at a 16 kHz sampling frequency.

We also observe that the SDP in Figure 5.28 indicates the use of the precondition extension, observable through the presence of a few `a=curr:qos` and `a=des:qos` lines directly below each `m=` line. These attributes indicate the current and desired local quality of service bearers for the respective media streams. Typically the IMS terminal has not established a radio bearer to send and receive the media streams when it sends the INVITE request; therefore, the quality of service is set to `none`.

This concludes the examination of the INVITE request that the IMS terminals sends out. But what happens to that request? How does it reach the peer terminal? What other transformations occur en route to its final destination? We analyze them in the next section.

5.7.2   The Originating P-CSCF Processes the INVITE Request

So far we have analyzed all the SIP header fields and SDP that the IMS terminal populates. When this INVITE is ready the terminal sends it to the P-CSCF discovered during the P-CSCF discovery procedures. The P-CSCF receives the INVITE request through an established security association (we describe security associations in Section 9.1.4). Assuming that the security functions are correctly performed the P-CSCF takes a few actions that we describe in the following paragraphs.

First, the P-CSCF verifies that the IMS terminal is acting correctly, according to IMS routing requirements. For instance, the P-CSCF verifies that the `Route` header is correctly populated; in other words, it contains the contents of the `Service-Route` header field that the IMS terminal received in a 200 (OK) response to a REGISTER request (during the last registration procedure). This guarantees that the SIP signaling will traverse the S-CSCF in the home network, as requested during registration in the mentioned `Service-Route` header. If the P-CSCF considers that the `Route` header field does not include the value of the `Service-Route` header field received during the registration process, the P-CSCF either replaces the `Route` header according to what the P-CSCF considers to be correct, or the P-CSCF answers the INVITE request with a 400 (Bad Request) response and does not forward the INVITE request.

The P-CSCF inspects the SDP offer. The P-CSCF is configured with the local policy of the network where the P-CSCF is operating. Such a policy may indicate that some media parameters are not allowed in the network. For instance, the policy may indicate that G.711 codecs are not allowed. This is likely to be prohibited, since G.711 requires a bandwidth of 64 kb/s, whereas AMR requires only a maximum of 14 kb/s. Basically, if an IMS terminal uses G.711 as an audio codec, then it is using a radio spectrum that could potentially be used by four users. As radio is a scarce resource and since AMR is mandated in all the 3GPP IMS implementations, it sounds natural that operators will prevent IMS terminal misuse of such expensive resources. The local policy is dependent on each operator's needs, network topology, charging models, roaming agreements, etc.

If the P-CSCF, when policing the SDP, discovers a media parameter that does not fit in the current local policy, the P-CSCF generates a 488 (Not Acceptable Here) response containing an SDP body and does not forward the INVITE request. The SDP body included in the 488 response indicates the media types, codecs, and other SDP parameters which are allowed according to local policy.

Then, the P-CSCF looks to see whether a `P-Preferred-Identity` header is included in the INVITE request. If it is, as in our example in Figure 5.21, then the P-CSCF verifies whether the value in that header corresponds to one of the implicitly or explicitly registered Public User Identities. The P-CSCF is aware of the security association related to this request, so only that IMS terminal, and not an impersonating agent, is at the other side of the security association sending the INVITE request. The P-CSCF learnt, during the registration procedure (see Section 5.6 for a complete description of the subscription to the `reg` state), all the Public User Identities registered to the user. The P-CSCF deletes the `P-Preferred-Identity` header in the INVITE request and inserts a `P-Asserted-Identity` header following the procedures specified in RFC 3325 [147]. The value of the `P-Asserted-Identity` header field is set to a valid SIP registered Public User Identity of the user; for example, the same value of the `P-Preferred-Identity` header field or any other registered Public User Identity of that user, if the one included in the `P-Preferred-Identity` is not registered or the `P-Preferred-Identity` is not present in the request. It must be noted that, in asserting the identity of the user, the P-CSCF never takes into account the `From` header field. The `From` header field is considered an end-to-end header.

Once done the P-CSCF removes and modifies the headers that relate to the security agreement. These procedures are described in Section 9.1.4. The P-CSCF also inserts charging headers. This procedure is described in detail in Chapter 7.

The P-CSCF always records the route, because all the SIP requests and responses have to traverse the P-CSCF, so that the P-CSCF can apply compression, security, and policing, as mentioned previously. When a SIP proxy wants to remain in the path for a subsequent exchange of requests pertaining to the same SIP dialog, it inserts a `Record-Route` header field with is own SIP URI as a value. And this is exactly what the P-CSCF does: it inserts a `Record-Route` header field that contains the P-CSCF's SIP URI. The reader must note that this `Record-Route` header does not contain the `comp=sigcomp` parameter or the ports derived from the security associations. This is because the INVITE request is transmitted toward the core network, where there is no need for SIP compression or IPsec connections.

In addition to recording the route the P-CSCF modifies the `Via`, `Route`, and `Max-Forwards` header field values as per regular SIP procedures.

Figure 5.29 shows the INVITE request that the P-CSCF forwards to the S-CSCF. This INVITE corresponds to step (3) in Figure 5.19. The P-CSCF forwards the request to the next hop included in the `Route` header. In this example that node is an S-CSCF, but it could have been an I-CSCF in the home network, if that was decided by the home network operator.

5.7.3 *The Originating S-CSCF Processes the INVITE Request*

The S-CSCF allocated to the caller receives the INVITE request and examines the `P-Asserted-Identity` header to identify the user who originated the INVITE request. The S-CSCF downloaded the user profile at registration. Among other information, the user profile contains the so-called filter criteria. The filter criteria contain the collection of triggers that determine whether a request has to traverse one or more Application Servers that provide services to the user. It must be noted that the S-CSCF does not execute services, but triggers services to be executed by Application Servers.

```
INVITE sip:bob@home2.net SIP/2.0
Via: SIP/2.0/UDP pcscf1.visited1.net;branch=z9hG4bKoh2qrz
Via: SIP/2.0/UDP [1080::8:800:200C:417A]:5059;
               comp=sigcomp;branch=z9hG4bK9h9ab
Max-Forwards: 69
Route: <sip:orig@scscf1.home1.net;lr>
Record-Route: <sip:pcscf1.visited1.net;lr>
P-Asserted-Identity: "Alice Smith" <sip:alice@home1.net>
Privacy: none
P-Access-Network-Info: 3GPP-UTRAN-TDD; utran-cell-id-3gpp=C359A3913B20E
P-Charging-Vector: icid-value="W34h6dlg"
From: <sip:alice@home1.net>;tag=ty20s
To: <sip:bob@home2.net>
Call-ID: 3s09cs03
Cseq: 127 INVITE
Require: precondition
Supported: 100rel
Contact: <sip:[1080::8:800:200C:417A]:5059;comp=sigcomp>
Allow: INVITE, ACK, CANCEL, BYE, PRACK, UPDATE, REFER, MESSAGE
Content-Type: application/sdp
Content-Length: 569

v=0
o=- 1073055600 1073055600 IN IP6 1080::8:800:200C:417A
s=-
c=IN IP6 1080::8:800:200C:417A
t=0 0
m=video 8382 RTP/AVP 98 99
b=AS:75
a=curr:qos local none
a=curr:qos remote none
a=des:qos mandatory local sendrecv
a=des:qos none remote sendrecv
a=rtpmap:98 H263
a=fmtp:98 profile-level-id=0
a=rtpmap:99 MP4V-ES
m=audio 8283 RTP/AVP 97 96
b=AS:25.4
a=curr:qos local none
a=curr:qos remote none
a=des:qos mandatory local sendrecv
a=des:qos none remote sendrecv
a=rtpmap:97 AMR
a=fmtp:97 mode-set=0,2,5,7; maxframes=2
a=rtpmap:96 telephone-event
```

Figure 5.29: (3) INVITE

The S-CSCF always evaluates the initial filter criteria in initial SIP requests.[12] For instance, the S-CSCF evaluates the filter criteria when it receives an initial INVITE or SUBSCRIBE request, because INVITE and SUBSCRIBE requests create a dialog. The S-CSCF also evaluates the filter criteria when it receives OPTIONS or MESSAGE requests outside any other dialog (they can be sent within or outside an existing dialog). However, the S-CSCF does not evaluate the filter criteria when it receives a PRACK or UPDATE request, as this is a subsequent request sent within a SIP dialog.

Having said that, we will not stop now to find out how filter criteria work. We assume for the time being that our user, named `sip:user1_public1@home1.net`, does not have any services associated with the origination of a SIP session. We will revisit this assumption in Section 5.8.4 when we describe how filter criteria work and how Application Servers are involved in the signaling path.

The S-CSCF, like the P-CSCF, polices SDP. Like the P-CSCF, the S-CSCF polices those SDP media parameters that are not set according to local policy. Unlike the P-CSCF the S-CSCF uses the HSS user-related information (the user profile). The S-CSCF also policies the SDP for user-related information. This allows operators, for instance, to offer cheap subscriptions that do not allow the use of certain media streams (e.g., video) or certain premium codecs. In case the request does not fit with the policy the S-CSCF may not process the INVITE request and answer it with a 488 (Not Acceptable Here) response indicating the media types, codecs, and other SDP parameters which are allowed according to the policy.

The S-CSCF of the originating user is the first node that tries to route the SIP request based on the destination (the callee) in the *Request-URI*. The originating P-CSCF (and I-CSCF if it was traversed) were never interested in the destination of the session setup nor did they look at the *Request-URI* field of the request. However, the S-CSCF in the originating home network is the first node that takes a look at the destination (*Request-URI*). In doing so the S-CSCF may find two different types of contents for the *Request-URI*: a SIP URI or a TEL URI.

If the S-CSCF finds a SIP URI in the *Request-URI*, then regular SIP routing procedures apply. Basically, given a domain name the S-CSCF has to find a SIP server in that domain name. The procedures are normatively described in RFC 3263 [214]. In our example the *Request-URI* is set to `sip:bob@home2.net` and the S-CSCF has to find a SIP server (typically an I-CSCF) in the network named *home2.net*. The S-CSCF extracts the domain name *home2.net* and does a number of DNS queries, as shown in Figure 5.30. The first DNS query (1) tries to find the transport protocols supported by SIP services in *home2.net*. The result (2) reveals that *home2.net* is implementing SIP services on both TCP and UDP. Then the S-CSCF, as it is interested in UDP services, queries the DNS requesting information about SIP services over UDP, hence the *_sip._udp.home2.net* query (3). The result (4) returns the host names and port numbers of two different SIP servers, *icscf1* and *icscf2*. Having determined the supported transport protocol host name and port number, the S-CSCF queries the DNS to find the IP address of that SIP proxy (5). The result is returned in step (6). At this stage the S-CSCF has learnt all the information required to build the INVITE request and to forward it to the appropriate I-CSCF at the terminating home network.

Some may think that so many queries to the DNS will have a severe impact on the session setup time. DNS offers a way around this problem. On one side, DNS typically tries to guess future queries based on the responses offered by DNS. If the information is available

---

[12]In this context we refer to *initial requests* as those SIP requests that either create a dialog or are stand-alone requests (i.e., those that are not subsequent requests).

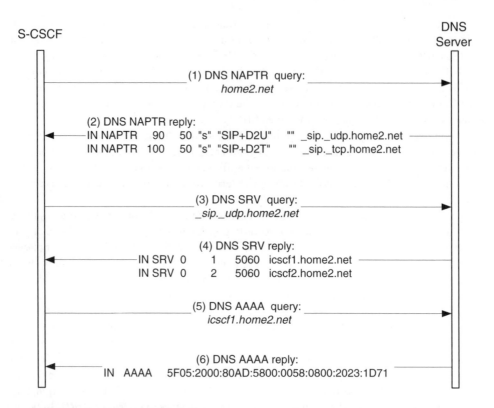

Figure 5.30: DNS procedures to locate a SIP server in the terminating home network

DNS returns not only the required information, but also the information that is likely to be requested at a later point. This avoids subsequent queries to the DNS. On the other hand, the responses offered by DNS are cached for a certain length of time. This means that, if within a certain configurable period of time, the S-CSCF has to query the DNS with the same input query (e.g., the same domain name), then all the data are locally cached and no interaction with DNS is required.

We indicated that the *Request-URI* of the INVITE request may contain either a SIP URI or a TEL URI. In the case where the *Request-URI* contains a TEL URI (specified in RFC 3966 [220]), there are two possibilities. The first possibility is that the telephone number contained in the TEL URI "belongs" to an IMS user, therefore, there is most likely a SIP URI allocated to that user that maps the TEL URI to a SIP URI. The other possibility is that the telephone number does not belong to an IMS user. This could be the case for a telephone number that belongs to a PSTN user or other circuit-switched network (e.g., GSM). The S-CSCF, being a SIP proxy server, implements routing based on SIP, and not on telephone numbers. So, the S-CSCF first tries to map the TEL URI into a SIP URI, typically by using another DNS service: the ENUM service (specified in RFC 2916 [100]).

DNS ENUM specifies a mechanism that consists of reversing the order of telephone number digits, adding a dot in between two digits, adding the domain name *.e164.arpa*, and performing a DNS NAPTR query with the resulting string. For instance, if the TEL URI is tel:+1-212-555-1234, the DNS query for the NAPTR record is sent to

`4.3.2.1.5.5.5.2.1.2.1.e164.arpa`. In the case where the telephone number is mapped to a SIP URI, DNS answers typically with a list of URIs including, but not restricted to, the SIP URI. The S-CSCF uses any of the SIP URIs as an entry to the routing process.

However, if the telephone number is not associated with a SIP URI, there is no mapping to DNS, so DNS returns a negative response, indicating that no record was available for that telephone number. This is an indication that the user is neither an IMS user nor a user defined at any other SIP domain. This may be the case when the user resides in the PSTN or a traditional circuit-switched network. As a result the S-CSCF is unable to forward the SIP request to its destination. Instead, the S-CSCF requires the services of a BGCF (Breakout Gateway Control Function). The BGCF is specialized in routing SIP requests based on telephone numbers, typically addressed to a circuit-switched network. We describe the interactions with circuit-switched networks in more detail in Section 5.9.1.

Once the S-CSCF has taken a determination of how to route the INVITE request, it adds a new value to the existing `P-Asserted-Identity` header field. We saw how the P-CSCF inserts a `P-Asserted-Identity` header field that contains, as a value, one of the multiple SIP URIs that the user registered. The S-CSCF receives this value and identifies the user. Before forwarding the INVITE request the S-CSCF adds a new value with a TEL URI associated with the user. The assumption here is that every user has a TEL URI associated to a SIP URI. The SIP URI enables SIP communications, whereas the TEL URI allows voice calls originated on or destined to a circuit-switched network, such as the PSTN. If the S-CSCF does not insert the TEL URI in the `P-Asserted-Identity` header field and the session terminates in the PSTN (perhaps due to a forwarding), then the PSTN network is unable to identify who is calling. Figure 5.31 shows the contents of the `P-Asserted-Identity` when the S-CSCF forwards the INVITE request to the final destination.

```
P-Asserted-Identity: "Alice Smith" <sip:alice@home1.net>,
                     <tel:+1-212-555-1234>
```

Figure 5.31: The `P-Asserted-Identity` header field

The S-CSCF also takes action on the `P-Access-Network-Info` header field. If the SIP request is forwarded outside the home domain, then the S-CSCF removes the header, since it may contain sensitive private information (e.g., the cell ID). If the request is forwarded to an AS that is part of the home network, then the S-CSCF keeps the `P-Access-Network-Info` header field in the request, so that the AS can read the contents.

In 3GPP Release 5 the S-CSCF always remains in the path for subsequent SIP signaling sent within the SIP dialog created by the INVITE request. 3GPP Release 6 adds further flexibility by allowing an S-CSCF not to remain in the path for subsequent signaling. This can be used, for example, if the SIP request is SUBSCRIBE request forwarded or terminated by an AS that is keeping charging records and perhaps other data. If the SIP request is an INVITE request, the S-CSCF typically remains in the path for subsequent requests and adds its own SIP URI to the `Record-Route` header field value.

5.7.4 The Terminating I-CSCF Processes the INVITE Request

Let's return to Figure 5.19 and follow up the INVITE request. The S-CSCF has found, through DNS procedures, a SIP server in the destination home network. This SIP server in the IMS is an I-CSCF. So the I-CSCF in the destination home network receives the INVITE request, (5) in Figure 5.19. The I-CSCF is not interested in the identity of the caller, identified in the P-Asserted-Identity header field. The I-CSCF is, however, interested in the callee, identified in the *Request-URI* of the INVITE request. The I-CSCF has to forward the SIP request to the S-CSCF allocated to the callee. For the sake of simplicity, let's assume that the callee is registered to the network and, as such, an S-CSCF is allocated to them. The I-CSCF is not aware of the address of the S-CSCF allocated to the callee, so the I-CSCF has to discover it. The S-CSCF saved its own address in the HSS during the registration procedure. Therefore, the I-CSCF queries the HSS with a Diameter Location-Information-Request (LIR) message (7).

The I-CSCF includes the address of the callee, copied from the *Request-URI* of the SIP request, in a Public-Identity Attribute Value Pair (AVP) of the Diameter LIR request. In the INVITE request example we have been following, the *Request-URI* is set to sip:bob@home2.net, and so is the Public-Identity AVP of the Diameter LIR request. The HSS receives the Diameter LIR request, inspects the Public-Identity AVP, searches for the data associated with that user, gets the stored S-CSCF address, and inserts it into a Server-Name AVP in a Diameter Location-Information-Answer (LIA) message, which is step (8) in Figure 5.19.

Once the I-CSCF receives the Diameter LIA message, it knows where to route the INVITE request (5), whose routing was temporarily suspended. The I-CSCF at this stage does not modify or add any SIP header field, other than the SIP routing header fields (e.g., Route, Max-Forwards, Via, etc.). So, the I-CSCF forwards the INVITE request to the S-CSCF that is assigned to the callee (9) in Figure 5.19.

The I-CSCF may or may not remain in the path for subsequent SIP signaling sent within the SIP dialog created by the INVITE request. It depends on the network topology and the configuration. For instance, the security policy in the network may dictate that SIP signaling has to traverse an I-CSCF, perhaps because the operator does not want to expose the S-CSCF to SIP signaling generated outside the home network. In such a case the I-CSCF adds its own SIP URI to the Record-Route header field value. Otherwise, the I-CSCF does not add itself to the Record-Route header field value, and once the INVITE transaction is complete the I-CSCF will not receive further SIP requests and responses belonging to the dialog created by the INVITE request.

5.7.5   The Terminating S-CSCF Processes the INVITE Request

The S-CSCF in the terminating network that takes care of the callee receives the INVITE request, (9) in Figure 5.19. The S-CSCF first identifies the callee by the *Request-URI* in the INVITE request. Then, the S-CSCF evaluates the initial filter criteria of the called user. This evaluation is the same as that undertaken by the S-CSCF in the originating side with respect to the calling user, with the only exception that in this case the terminating S-CSCF is looking for services that apply to sessions established toward the user, rather than sessions originated by the user. For the sake of simplicity, we assume at this stage that the callee does not have any service that requires the involvement of an Application Server.

In 3GPP Release 5 the terminating S-CSCF always remains in the path for subsequent SIP signaling sent within the SIP dialog created by the INVITE request. 3GPP Release 6 adds the possibility of the S-CSCF not remaining in the path. The procedure is most likely applicable to non-INVITE requests that are terminated in an Application Server. This may be the case with the presence service. However, for INVITE requests the terminating S-CSCF typically remains in the path; therefore, it adds its own SIP URI to the `Record-Route` header field value.

Then, the S-CSCF continues with the processing of the INVITE request. The goal is to forward the INVITE request to the callee's IMS terminal, by means of traversing a set of proxies determined during the registration process of the callee. This set of proxies will always include a P-CSCF and may include one or more I-CSCFs. In the example in Figure 5.19 we assume that SIP has to traverse only a P-CSCF.

The S-CSCF, therefore, creates a new *Request-URI* with the contents of the `Contact` header field value registered by the callee during registration. The S-CSCF also sets the value of the `Route` header field to that of the `Path` header field learnt during the same registration process. In our case, as the SIP requests have to traverse the P-CSCF that is assigned to the user, the `Route` header contains the P-CSCF URI learnt during user registration. Figure 5.32 shows the details of the `Route` header field.

`Route: <sip:pcscf2.visited2.net;lr>`

Figure 5.32: The `Route` header field

3GPP had a requirement for SIP that led to an extension in the form of the `P-Called-Party-ID` header field. Let's take a look at the problem description with the help of Figure 5.33. We saw in Section 5.5 that the IMS terminal can register a new Public User Identity at any time. Steps (1)–(4) in Figure 5.33 describe an IMS user registering two different Public User Identities, identified by the `To` header field value of the SIP REGISTER request. In step (1), Bob is registering the Public User Identity `sip:bob@home2.net`, whereas in step (3) Bob is registering the identity `sip:bob-business@home2.net`. In both REGISTER requests the `Contact` header field contains the same value: the SIP URI of the IMS terminal.

We have seen in previous paragraphs that the S-CSCF on receiving an INVITE request, such as step (5) of Figure 5.33, replaces the contents of the original *Request-URI* header field with more accurate information extracted from the `Contact` header during the registration of the user. This procedure in SIP parlance is known as *retarget*. So, the S-CSCF retargets the *Request-URI* in the INVITE request. When the INVITE request is delivered to the IMS terminal, as in step (6) of Figure 5.33, Bob does not know to which of the several Public User Identities this INVITE request corresponds. Was this INVITE sent toward Bob's personal or business identity? Perhaps the IMS terminal wants to apply a different alerting tone, depending on the callee's address, but this is not possible because the callee's address was lost during the retarget process.

The solution led to the creation of the `P-Called-Party-ID` header field in SIP, which is sketched in Figure 5.34. The `P-Called-Party-ID` header field is documented in 3GPP extensions to SIP (RFC 3455 [109]). The mechanism is quite simple: when the S-CSCF retargets, it also inserts a `P-Called-Party-ID` header field, whose value is set to the original *Request-URI*. Then, the S-CSCF forwards the INVITE request, (6) in Figure 5.34, including the `P-Called-Party-ID` header field. When the IMS terminal receives the INVITE request, it inspects the contents of the `P-Called-Party-ID` header field in order to determine to which of the various identities the INVITE request is destined. Then, the IMS terminal may apply any distinguished alerting tone or similar service.

Returning to our example in Figure 5.19 the S-CSCF eventually forwards the INVITE request to the P-CSCF, (11) in Figure 5.19.

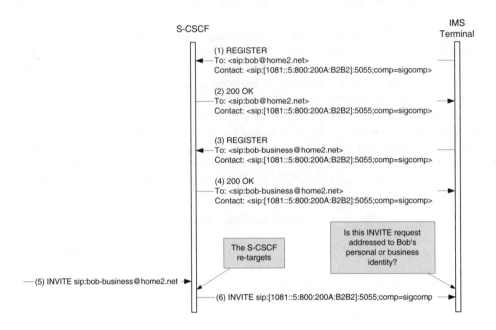

Figure 5.33: The S-CSCF re-targets

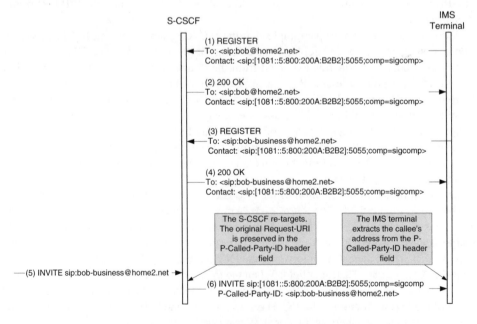

Figure 5.34: Usage of the P-Called-Party-ID header field

5.7.6   The Terminating P-CSCF Processes the INVITE Request

The P-CSCF receives the INVITE request, (11) in Figure 5.19 and does not need to take any routing decision, because the *Request-URI* already contains a SIP URI that includes the IP address (or host name) of the IMS terminal. The P-CSCF needs to identify the Public User Identity of the callee in order to find the proper security association established with the IMS terminal of the user. The P-CSCF extracts this Public User Identity from the P-Called-Party-ID header of the SIP INVITE request.

The P-CSCF always remains in the path for subsequent SIP signaling sent within the SIP dialog created by the INVITE request. Therefore, the P-CSCF adds its own SIP URI to the Record-Route header field value.

The terminating P-CSCF inspects the contents of the Privacy header field and, if needed, acts to remove the P-Asserted-Identity header field. For instance, if the Privacy header field is set to a value of id, then the P-CSCF removes the P-Asserted-Identity header field from the INVITE request that it sends to the callee's IMS terminal. This gives the originating user some degree of privacy, as the terminating user is unable to determine who the caller is. In the example we have been following, the caller set the Privacy header field to none, indicating that there is no requirement to keep any privacy; therefore, the P-CSCF keeps the P-Asserted-Identity header field in the INVITE request.

The P-CSCF inserts a P-Media-Authorization header field in the INVITE request. This mechanism is described in detail in Section 11.2.2.

The P-CSCF needs to carry out a number of functions that are related to charging, other security aspects, control of the edge router (e.g., GGSN), compression of SIP signaling, etc. But there are no other SIP-specific functions other than acting as a SIP proxy. We describe all of these functions in their respective sections.

Figure 5.35 shows an example of the INVITE request that the P-CSCF sends to the callee.

5.7.7   The Callee's Terminal Processes the INVITE Request

The INVITE request in Figure 5.35 is received at the callee's IMS terminal. The INVITE carries an SDP offer generated at the caller's terminal. The SDP offer indicates the IP address and port numbers where the caller wants to receive media streams, the desired and supported codecs for each of the media streams, etc.

The INVITE contains a request in the Require header field to follow the *precondition* call flow model. This model requires that the callee responds with a 183 (Session Progress) response that contains an SDP answer. The SDP answer contains the media streams and codecs that the callee is able to accept for this session. We expect IMS terminals to be configured to answer this INVITE request automatically without requiring any interaction with the user. For instance, users may have configured their IMS terminals to automatically accept sessions that require support for the AMR audio codec and the H.263 video codec. However, this is terminal implementation-dependent, and different terminal manufacturers may include different features. In any case, if automatic configuration is not available, the terminal has to alert the user to take a decision about how to answer the SDP with the preferred streams and codecs.

The callee's IMS terminal could start its resource reservation process at this stage, because it knows all the parameters needed (all included in the SDP) to start resource reservation. However, if the terminal started resource reservation at this stage, it would need to reserve resources for the higher demanding codec, when several codecs are possible within

```
INVITE sip:[1081::5:800:200A:B2B2]:5055;comp=sigcomp SIP/2.0
Via: SIP/2.0/UDP pcscf2.visited2.net:5056;comp=sigcomp;
               branch=z9hG4bK2a2qr
Via: SIP/2.0/UDP scscf2.home2.net;branch=z9hG4bKvp2yml
Via: SIP/2.0/UDP icscf2.home2.net;branch=z9hG4bKra1ar
Via: SIP/2.0/UDP scscf1.home1.net;branch=z9hG4bKs1pp0
Via: SIP/2.0/UDP pcscf1.visited1.net;branch=z9hG4bKoh2qrz
Via: SIP/2.0/UDP [1080::8:800:200C:417A]:5059;comp=sigcomp;
               branch=z9hG4bK9h9ab
Max-Forwards: 65
Record-Route: <sip:pcscf2.visited2.net:5056;lr;comp=sigcomp>
Record-Route: <sip:scscf2.home2.net;lr>
Record-Route: <sip:scscf1.home1.net;lr>
Record-Route: <sip:pcscf1.visited1.net;lr>
P-Asserted-Identity: "Alice Smith" <sip:alice@home1.net>,
                     <tel:+1-212-555-1234>
Privacy: none
P-Media-Authorization: 0020000100100101706466312e686f6d65312e6
e6574000c02013942563330373200
From: <sip:alice@home1.net>;tag=ty20s
To: <sip:bob@home2.net>
Call-ID: 3s09cs03
Cseq: 127 INVITE
Require: precondition
Supported: 100rel
Contact: <sip:[1080::8:800:200C:417A]:5059;comp=sigcomp>
Allow: INVITE, ACK, CANCEL, BYE, PRACK, UPDATE, REFER, MESSAGE
Content-Type: application/sdp
Content-Length: 569

v=0
o=- 1073055600 1073055600 IN IP6 1080::8:800:200C:417A
s=-
c=IN IP6 1080::8:800:200C:417A
t=0 0
m=video 8382 RTP/AVP 98 99
b=AS:75
a=curr:qos local none
a=curr:qos remote none
a=des:qos mandatory local sendrecv
a=des:qos none remote sendrecv
a=rtpmap:98 H263
a=fmtp:98 profile-level-id=0
a=rtpmap:99 MP4V-ES
m=audio 8283 RTP/AVP 97 96
b=AS:25.4
a=curr:qos local none
a=curr:qos remote none
a=des:qos mandatory local sendrecv
a=des:qos none remote sendrecv
a=rtpmap:97 AMR
a=fmtp:97 mode-set=0,2,5,7; maxframes=2
a=rtpmap:96 telephone-event
```

Figure 5.35: (13) INVITE

a particular media stream. For instance, the SDP in the INVITE request declares its support for two video codecs: H.263 and MPEG-4. It is not worth reserving resources for a codec that requires a higher bandwidth, when that codec may not be used at all during the session. If the session is charged depending on the reserved resources, then it is easier to justify the short period of time that the terminal needs to wait for a potential new SDP offer. If the SDP offer contains one codec per media stream, then resource reservation can start straight away.

The IMS terminal inspects the `P-Asserted-Identity` header field, if present, and extracts the identity of the caller. The P-CSCF may remove this header field if the privacy requirement indicates so. The IMS terminal also determines, by inspecting the value of the `P-Called-Party-ID` header field, which of the several identities of its user the INVITE request is addressed to. The combination of caller and callee identities may be used at this or a later stage to play a personalized ringing tone, display a picture of the calling party, etc.

Figure 5.36 shows an example of a 183 (Session Progress) provisional response that the callee's IMS terminal sends back to the originating side. The IMS terminal inserts a `P-Access-Network-Info` header field in a similar way to that the originating terminal used to insert the same header field in the INVITE request.

The terminal also inserts a `Require` header field with the value `100rel`. This is required because the *precondition* extension requires that provisional responses be transmitted reliably. If a reliable provisional responses extension is not used and if UDP is used as a transport protocol in between any two hops in the path, then there is no guarantee that the response ever arrives at the remote party. It can be lost, and the UAS that issued the provisional response will never notice it. To solve the problem the "Reliability of Provisional Responses in SIP" (specified in RFC 3262 [213]) guarantees that the reception of a provisional response is acknowledged by a PRACK request. If the callee does not receive a PRACK request to confirm the reception of the provisional response within a determined time, then it will retransmit the provisional response.

The *precondition* extension requires that SDP be included in a provisional response and that support for the reliable provisional responses mechanism is given to guarantee provisional responses are received at the originating terminal. This is the reason that the callee's IMS terminal inserts a `Require` header field with the tag `100rel` in the 183 (Session Progress) provisional response.

Figure 5.36 shows a `Privacy` header field set with the value `none`. Hence, the user does not have a requirement to maintain privacy with respect to those header fields that reveal any identity or private information. However, the reader may notice the absence of a `P-Preferred-Identity` header field. This header field is used to give a hint to the network node about the identity that the user wants to reveal. The network already knows this information, because the INVITE request was addressed to that identity. So, there is no need for the IMS terminal to indicate to which identity the INVITE request was addressed.

The SDP in the 183 (Session Progress) provisional response also includes an `a=conf:qos` line. This line requests callers to send an updated SDP when they have reserved resources in their local segment. We described earlier that the IMS uses the so-called segmented model within the precondition model. According to the local segmented quality of service, in this model each terminal is responsible for reserving resources in its own local segment. A condition for the terminating terminal to alert the user is that resources are reserved both at the originating and the terminating side. If the originating side reserves resources but does not inform the terminating side, the callee will never be alerted. To solve this problem the terminating terminal requests the originating terminal to inform it when the originating

```
SIP/2.0 183 Session Progress
Via: SIP/2.0/UDP pcscf2.visited2.net:5056;comp=sigcomp;
              branch=z9hG4bK2a2qr
Via: SIP/2.0/UDP scscf2.home2.net;branch=z9hG4bKvp2yml
Via: SIP/2.0/UDP icscf2.home2.net;branch=z9hG4bKra1ar
Via: SIP/2.0/UDP scscf1.home1.net;branch=z9hG4bKs1pp0
Via: SIP/2.0/UDP pcscf1.visited1.net;branch=z9hG4bKoh2qrz
Via: SIP/2.0/UDP [1080::8:800:200C:417A]:5059;comp=sigcomp;
              branch=z9hG4bK9h9ab
Record-Route: <sip:pcscf2.visited2.net:5056;lr;comp=sigcomp>,
              <sip:scscf2.home2.net;lr>,
              <sip:scscf1.home1.net;lr>,
              <sip:pcscf1.visited1.net;lr>
Privacy: none
P-Access-Network-Info: 3GPP-UTRAN-TDD;
                       utran-cell-id-3gpp=C359A3913B20E
From: <sip:alice@home1.net>;tag=ty20s
To: <sip:bob@home2.net>;tag=d92119
Call-ID: 3s09cs03
Cseq: 127 INVITE
Require: 100rel
Contact: <sip:[1081::5:800:200A:B2B2]:5055;comp=sigcomp>
Allow: INVITE, ACK, CANCEL, BYE, PRACK, UPDATE, REFER, MESSAGE
RSeq: 9021
Content-Type: application/sdp
Content-Length: 634

v=0
o=- 3021393216 3021393216 IN IP6 1081::5:800:200A:B2B2
s=-
c=IN IP6 1081::5:800:200A:B2B2
t=0 0
m=video 14401 RTP/AVP 98 99
b=AS:75
a=curr:qos local none
a=curr:qos remote none
a=des:qos mandatory local sendrecv
a=des:qos mandatory remote sendrecv
a=conf:qos remote sendrecv
a=rtpmap:98 H263
a=fmtp:98 profile-level-id=0
a=rtpmap:99 MP4V-ES
m=audio 6544 RTP/AVP 97 96
b=AS:25.4
a=curr:qos local none
a=curr:qos remote none
a=des:qos mandatory local sendrecv
a=des:qos mandatory remote sendrecv
a=conf:qos remote sendrecv
a=rtpmap:97 AMR
a=fmtp:97 mode-set=0,2,5,7; maxframes=2
a=rtpmap:96 telephone-event
```

Figure 5.36: (15) 183 Session Progress

terminal has its resources available, and this is indicated in the mentioned a=conf:qos line in SDP.

The IMS terminal also inserts a Contact header field whose value is a SIP URI that includes the IP address or Fully Qualified Domain Name of the IMS terminal. Additionally, the SIP URI also includes the port number bound to the security association established between the P-CSCF and the IMS terminal. Also, the IMS terminal declares the willingness to compress SIP from and to the P-CSCF in the comp=sigcomp parameter.

In addition, the callee's IMS terminal carries out the regular SIP processing to generate the 183 Session Progress toward the caller's IMS terminal.

5.7.8 Processing the 183 Response

The 183 (Session Progress) response traverses, according to SIP procedures to route responses, the same set of proxies that the corresponding INVITE request traversed.

When the P-CSCF receives the 183 response, (15) in Figure 5.19, it verifies that the IMS terminal has formulated the response correctly. For example, it verifies that the Via and Record-Route header field contain the values that the IMS terminal should have used in a response to the INVITE request. This avoids "spoof" IMS terminals that "forget" to include the address of an S-CSCF or P-CSCF in a Record-Route header field to avoid any account of the activity of the IMS terminal. In the case where the P-CSCF finds a malformed Via or Record-Route the P-CSCF discards the response (why bother for a "spoof" terminal) or rewrites that header field with the appropriate values.

When the P-CSCF received the INVITE request, it saved the value of the P-Called-Party-ID header. The value of that header indicated the Public User Identity of the receiver side. Now, when the P-CSCF receives the 183 response from the IMS terminal the P-CSCF inserts a P-Asserted-Identity header field, whose value is the same as that included in the P-Called-Party-ID header field of the INVITE request. This mechanism allows the remaining trusted network nodes to get the Public User Identity of the callee being used for this session.

Eventually, the P-CSCF forwards the 183 response, based on the contents of the value of the topmost Via header field, to the S-CSCF, (16) in Figure 5.19.

The S-CSCF receives the 183 response. The S-CSCF removes the P-Access-Network-Info header field prior to forwarding the response to the I-CSCF (17).

The I-CSCF does not undertake any special processing in the 183 response. It just forwards it to the S-CSCF in the originating network (17).

The originating S-CSCF receives the 183 response (18) and may remove the P-Asserted-Identity header field if privacy requirements indicate so. In the example we are following, as the callee set the Privacy header field to the value none the originating S-CSCF does not take any action on the P-Asserted-Identity header field.

Eventually, the S-CSCF forwards the 183 response (19) to the P-CSCF. The P-CSCF forwards the response (20) to the caller's IMS terminal.

5.7.9 The Caller's IMS Terminal Processes the 183 Response

The caller's IMS terminal receives the 183 response and focuses particularly on the SDP answer. The SDP contains the IP address of the remote terminal (i.e., the destination IP address where the media streams are sent). It also contains an indication of whether the callee accepted establishment of a session with these media streams. For instance, the callee could have accepted the establishment of an audio session without video. However, this is not the case in our example, because the callee accepted all the proposed media streams in the SDP answer.

The SDP answer also includes an indication from the callee that it wants to be notified when the caller's IMS terminal has completed its resource reservation process. This is indicated by the presence of the `a=conf:qos` line somewhere below each `m=` line.

The SDP answer indicates which of the offered codecs are supported and desired for this session at the callee. We have seen an example in Figure 5.36 in which the callee supported the two offered video codecs (H.263 and MPEG-4) and the audio codec (AMR). An IMS terminal typically tries to narrow down the number of negotiated codecs to just one codec per media stream. This allows accurate resource reservation which only requires reservation of the bandwidth of that codec. If several codecs were negotiated the terminal would reserve the maximum bandwidth required by the most demanding codec, even though during the session the actual codec used could require less bandwidth. This leads to an inefficient use of resource management and, potentially, the user being charged for bandwidth that was not actually used. For this reason, IMS terminals, once a first SDP exchange has taken place, trim down the number of codecs per media stream to just a single codec.

Hence, the caller's IMS terminal creates a new SDP offer, whose only difference is the removal of the MPEG-4 codec from the video line in SDP in favor of the H.263 codec.

As the 183 response required the response to be acknowledged, due to the presence of the `100rel` tag in the `Require` header field, the caller's IMS terminal creates a PRACK request. The new SDP offer is included in the PRACK request. Figure 5.37 shows the PRACK request that carries a piggybacked SDP offer.

In parallel with the generation of the PRACK request the IMS terminal starts the resource reservation mechanism. The actual procedure is dependent on the underlying IP Connectivity Access Network. Typically, it will require some sort of dialog exchange with the packet and radio nodes.

The PRACK request traverses all the proxies that asked to remain in the path for subsequent signaling. Typically, this path will be a subset of the proxies that the INVITE request traversed. The path is determined by the `Route` header field included in the PRACK request. The value of the `Record-Route` header field included in the 183 response sets that path.

5.7.10   The Callee's IMS Terminal Processes the PRACK request

When the PRACK request, (25) in Figure 5.19, is received at the callee the IMS terminal generates a 200 (OK) response (26) that contains an SDP answer. The 200 (OK) response is an answer to the PRACK request and should not be confused with a 200 (OK) for the INVITE that will occur later.

Because this is the second offer/answer exchange during the establishment of this particular session there is almost nothing to negotiate, just confirmation of the media streams and codecs of the session. At this time, the terminal may still be involved in its resource reservation process, and most likely at this stage the resource reservation will not be complete. Therefore, the SDP answer generated by the callee typically indicates that there are no resources available yet at the local segment (`a=curr:qos local` and `a=des:qos local`). The IMS terminal still indicates that it wants to receive an indication when the caller is ready with their resource reservation. Figure 5.38 shows the 200 (OK) response, (26) in Figure 5.19.

```
PRACK sip:[1081::5:800:200A:B2B2]:5055;comp=sigcomp SIP/2.0
Via: SIP/2.0/UDP [1080::8:800:200C:417A]:5059;comp=sigcomp;
                  branch=z9hG4bK9h9ab
Max-Forwards: 70
P-Access-Network-Info: 3GPP-UTRAN-TDD;
                          utran-cell-id-3gpp=C359A3913B20E
Route: <sip:pcscf1.visited1.net:5058;lr;comp=sigcomp>,
       <sip:scscf1.home1.net;lr>,
       <sip:scscf2.home2.net;lr>,
       <sip:pcscf2.visited2.net;lr>
From: <sip:alice@home1.net>;tag=ty20s
To: <sip:bob@home2.net>;tag=d92l19
Call-ID: 3s09cs03
Cseq: 128 PRACK
Require: precondition, sec-agree
Proxy-Require: sec-agree
Security-Verify: ipsec-3gpp; q=0.1; alg=hmac-sha-1-96;
                  spi-c=98765432; spi-s=909767;
                  port-c=5057; port-s=5058
RAck: 9021 127 INVITE
Content-Type: application/sdp
Content-Length: 553

v=0
o=- 1073055600 1073055602 IN IP6 1080::8:800:200C:417A
s=-
c=IN IP6 1080::8:800:200C:417A
t=0 0
m=video 8382 RTP/AVP 98
b=AS:75
a=curr:qos local none
a=curr:qos remote none
a=des:qos mandatory local sendrecv
a=des:qos mandatory remote sendrecv
a=rtpmap:98 H263
a=fmtp:98 profile-level-id=0
m=audio 8283 RTP/AVP 97 96
b=AS:25.4
a=curr:qos local none
a=curr:qos remote none
a=des:qos mandatory local sendrecv
a=des:qos mandatory remote sendrecv
a=rtpmap:97 AMR
a=fmtp:97 mode-set=0,2,5,7; maxframes=2
a=rtpmap:96 telephone-event
```

Figure 5.37: (21) PRACK

```
SIP/2.0 200 OK
Via: SIP/2.0/UDP pcscf2.visited2.net:5056;comp=sigcomp;
     branch=z9hG4bK2a2qr
Via: SIP/2.0/UDP scscf2.home2.net;branch=z9hG4bKvp2yml
Via: SIP/2.0/UDP scscf1.home1.net;branch=z9hG4bKs1pp0
Via: SIP/2.0/UDP pcscf1.visited1.net;branch=z9hG4bKoh2qrz
Via: SIP/2.0/UDP [1080::8:800:200C:417A]:5059;comp=sigcomp;
                 branch=z9hG4bK9h9ab
P-Access-Network-Info: 3GPP-UTRAN-TDD;
                         utran-cell-id-3gpp=C359A3913B20E
From: <sip:alice@home1.net>;tag=ty20s
To: <sip:bob@home2.net>;tag=d92l19
Call-ID: 3s09cs03
Cseq: 128 PRACK
Content-Type: application/sdp
Content-Length: 610

v=0
o=- 3021393216 3021393218 IN IP6 1081::5:800:200A:B2B2
s=-
c=IN IP6 1081::5:800:200A:B2B2
t=0 0
m=video 14401 RTP/AVP 98
b=AS:75
a=curr:qos local none
a=curr:qos remote none
a=des:qos mandatory local sendrecv
a=des:qos mandatory remote sendrecv
a=conf:qos remote sendrecv
a=rtpmap:98 H263
a=fmtp:98 profile-level-id=0
m=audio 6544 RTP/AVP 97 96
b=AS:25.4
a=curr:qos local none
a=curr:qos remote none
a=des:qos mandatory local sendrecv
a=des:qos mandatory remote sendrecv
a=conf:qos remote sendrecv
a=rtpmap:97 AMR
a=fmtp:97 mode-set=0,2,5,7; maxframes=2
a=rtpmap:96 telephone-event
```

Figure 5.38: (26) 200 OK

At the same time, the callee starts resource reservation in its own segment. Typically, this is a process that involves the terminal, the radio nodes, and the packet nodes. If the IP-CAN is a GPRS network, it involves the SGSN and GGSN on the packet network as well as the radio nodes.

The 200 (OK) response to the PRACK request traverses the same set of SIP proxies that the PRACK request traversed. Eventually, the 200 (OK) response, (30) in Figure 5.19, arrives at the caller's IMS terminal. At this stage the caller's IMS terminal is most likely still engaged in its resource reservation process.

Once the caller's IMS terminal has got the required resources from the network, it honors the request for a confirmation message, present in the form of the a=conf line in the SDP answer, (20) and (30) in Figure 5.19. The IMS terminal sends an UPDATE request that visits the same set of proxies as the PRACK request (i.e., those proxies that asked to receive subsequent SIP messages by adding their own SIP URIs to the Record-Route header field of the INVITE request), (1)–(13) in Figure 5.19.

Figure 5.39 shows an example of the contents of the UPDATE request, (31) in Figure 5.19. The UPDATE request contains another SDP offer, in which the caller's IMS terminal indicates that resources are reserved at his local segment. The presence of the a=curr:qos local sendrecv line conveys this information.

Eventually, the callee's IMS terminal receives the UPDATE request (35). The IMS terminal will generate a 200 (OK) response, (36) in Figure 5.19. According to the SDP offer/answer model a successful SDP offer has to be answered with an SDP answer. So, the callee's IMS terminal includes an SDP answer in the 200 (OK) response. At this stage the callee's IMS terminal may have already finished its resource reservation or not, depending on how much time it takes to complete the process. Therefore, the callee's IMS terminal indicates its own local quality-of-service status, which may either be complete or not. This is indicated once more in the a=curr:qos local line in SDP. Figure 5.40 shows an example of the 200 (OK) response that the callee's IMS terminal makes when resources are already available to it.

The 200 (OK) response will follow the same path as the UPDATE request. Eventually, the caller's IMS terminal will receive it and the UPDATE transaction is completed.

5.7.11 *Alerting the Callee*

On the callee's side there are two conditions that have to be met before the callee is alerted. On one side, the terminal needs to complete its local resource reservation process. On the other side, the callee's terminal has to get the information that the caller's terminal has also completed its local resource reservation process. The latter information is conveyed in the SDP of an UPDATE request. Resource reservation and reception of the UPDATE request are two independent processes that can happen in any order, but the idea is that the callee should not be alerted before resources are available at both sides of the session (e.g., caller and callee).

When the callee's IMS terminal rings, it will also generate a 180 (Ringing) provisional response, (41) in Figure 5.20. The response is sent to the caller's terminal and traverses those proxies the INVITE request traversed. Figure 5.41 shows an example of the 180 (Ringing) response. The response typically does not contain SDP, since all the session parameters, such as media streams and codecs, have already been negotiated in the previous exchanges. The response requires an acknowledgement, due to the presence of the Require header field with the tag 100rel in it.

```
UPDATE sip:[1081::5:800:200A:B2B2]:5055;comp=sigcomp SIP/2.0
Via: SIP/2.0/UDP [1080::8:800:200C:417A]:5059;comp=sigcomp;
               branch=z9hG4bK9h9ab
Max-Forwards: 70
P-Access-Network-Info: 3GPP-UTRAN-TDD;
                       utran-cell-id-3gpp=C359A3913B20E
Route: <sip:pcscf1.visited1.net:5058;lr;comp=sigcomp>,
       <sip:scscf1.home1.net;lr>,
       <sip:scscf2.home2.net;lr>,
       <sip:pcscf2.visited2.net;lr>
From: <sip:alice@home1.net>;tag=ty20s
To: <sip:bob@home2.net>;tag=d92l19
Call-ID: 3s09cs03
Cseq: 129 UPDATE
Require: sec-agree
Proxy-Require: sec-agree
Security-Verify: ipsec-3gpp; q=0.1;
                 alg=hmac-sha-1-96;
                 spi-c=98765432; spi-s=909767;
                 port-c=5057; port-s=5058
Content-Type: application/sdp
Content-Length: 561

v=0
o=- 1073055600 1073055604 IN IP6 1080::8:800:200C:417A
s=-
c=IN IP6 1080::8:800:200C:417A
t=0 0
m=video 8382 RTP/AVP 98
b=AS:75
a=curr:qos local sendrecv
a=curr:qos remote none
a=des:qos mandatory local sendrecv
a=des:qos mandatory remote sendrecv
a=rtpmap:98 H263
a=fmtp:98 profile-level-id=0
m=audio 8283 RTP/AVP 97 96
b=AS:25.4
a=curr:qos local sendrecv
a=curr:qos remote none
a=des:qos mandatory local sendrecv
a=des:qos mandatory remote sendrecv
a=rtpmap:97 AMR
a=fmtp:97 mode-set=0,2,5,7; maxframes=2
a=rtpmap:96 telephone-event
```

Figure 5.39: (31) UPDATE

```
SIP/2.0 200 OK
Via: SIP/2.0/UDP pcscf2.visited2.net:5056;comp=sigcomp;
               branch=z9hG4bK2a2qr
Via: SIP/2.0/UDP scscf2.home2.net;branch=z9hG4bKvp2yml
Via: SIP/2.0/UDP scscf1.home1.net;branch=z9hG4bKs1pp0
Via: SIP/2.0/UDP pcscf1.visited1.net;branch=z9hG4bKoh2qrz
Via: SIP/2.0/UDP [1080::8:800:200C:417A]:5059;comp=sigcomp;
               branch=z9hG4bK9h9ab
P-Access-Network-Info: 3GPP-UTRAN-TDD;
                       utran-cell-id-3gpp=C359A3913B20E
From: <sip:alice@home1.net>;tag=ty20s
To: <sip:bob@home2.net>;tag=d92ll9
Call-ID: 3s09cs03
Cseq: 129 UPDATE
Content-Type: application/sdp
Content-Length: 571

v=0
o=- 3021393216 3021393220 IN IP6 1081::5:800:200A:B2B2
s=-
c=IN IP6 1081::5:800:200A:B2B2
t=0 0
m=video 14401 RTP/AVP 98
b=AS:75
a=curr:qos local sendrecv
a=curr:qos remote sendrecv
a=des:qos mandatory local sendrecv
a=des:qos mandatory remote sendrecv
a=rtpmap:98 H263
a=fmtp:98 profile-level-id=0
m=audio 6544 RTP/AVP 97 96
b=AS:25.4
a=curr:qos local sendrecv
a=curr:qos remote sendrecv
a=des:qos mandatory local sendrecv
a=des:qos mandatory remote sendrecv
a=rtpmap:97 AMR
a=fmtp:97 mode-set=0,2,5,7; maxframes=2
a=rtpmap:96 telephone-event
```

Figure 5.40: (36) 200 OK

```
SIP/2.0 180 Ringing
Via: SIP/2.0/UDP pcscf2.visited2.net:5056;comp=sigcomp;
                 branch=z9hG4bK2a2qr
Via: SIP/2.0/UDP scscf2.home2.net;branch=z9hG4bKvp2yml
Via: SIP/2.0/UDP icscf2.home2.net;branch=z9hG4bKra1ar
Via: SIP/2.0/UDP scscf1.home1.net;branch=z9hG4bKs1pp0
Via: SIP/2.0/UDP pcscf1.visited1.net;branch=z9hG4bKoh2qrz
Via: SIP/2.0/UDP [1080::8:800:200C:417A]:5059;comp=sigcomp;
                 branch=z9hG4bK9h9ab
Record-Route: <sip:pcscf2.visited2.net:5056;lr;comp=sigcomp>,
                 <sip:scscf2.home2.net;lr>,
                 <sip:scscf1.home1.net;lr>,
                 <sip:pcscf1.visited1.net;lr>
P-Access-Network-Info: 3GPP-UTRAN-TDD;
                       utran-cell-id-3gpp=C359A3913B20E
From: <sip:alice@home1.net>;tag=ty20s
To: <sip:bob@home2.net>;tag=d92119
Call-ID: 3s09cs03
Cseq: 127 INVITE
Require: 100rel
Contact: <sip:[1081::5:800:200A:B2B2]:5055;comp=sigcomp>
Allow: INVITE, ACK, CANCEL, BYE, PRACK, UPDATE, REFER, MESSAGE
RSeq: 9022
Content-Length: 0
```

Figure 5.41: (41) 180 Ringing

When the caller's IMS terminal receives the 180 (Ringing) response, (46) in Figure 5.20, it will likely generate a locally stored ring-back tone to indicate to the caller that the peer terminal is ringing. As the 180 (Ringing) response requires acknowledgement the caller's IMS terminal generates a PRACK request, (47) in Figure 5.20, and sends it to the callee. This PRACK request does not typically contain SDP. The PRACK request visits the same proxies as the previous PRACK and UPDATE requests (i.e., those who recorded the route during the INVITE transaction). Eventually, the callee's IMS terminal will receive the PRACK request (51) and will answer it with a 200 (OK) response (52). This 200 (OK) response for the PRACK request does not typically contain SDP. The caller's IMS terminal will eventually receive this response, (56) in Figure 5.20.

When the callee finally accepts the session the IMS terminal sends a 200 (OK) response, (57) in Figure 5.20. This 200 (OK) response completes the INVITE transaction. The response does not typically contain SDP. Figure 5.42 shows an example of this response. Eventually, the caller's IMS terminal receives the response, (62) in Figure 5.20, and starts generating media-plane traffic. The caller's IMS terminal also sends an ACK request, (63) in Figure 5.20, to confirm receipt of the 200 (OK) response. The ACK request is routed back to the callee's IMS terminal. Figure 5.43 shows an example of the ACK request sent from the caller's IMS terminal.

```
SIP/2.0 200 OK
Via: SIP/2.0/UDP pcscf2.visited2.net:5056;comp=sigcomp;
                branch=z9hG4bK2a2qr
Via: SIP/2.0/UDP scscf2.home2.net;branch=z9hG4bKvp2yml
Via: SIP/2.0/UDP icscf2.home2.net;branch=z9hG4bKra1ar
Via: SIP/2.0/UDP scscf1.home1.net;branch=z9hG4bKs1pp0
Via: SIP/2.0/UDP pcscf1.visited1.net;branch=z9hG4bKoh2qrz
Via: SIP/2.0/UDP [1080::8:800:200C:417A]:5059;comp=sigcomp;
                branch=z9hG4bK9h9ab
Record-Route: <sip:pcscf2.visited2.net:5056;lr;comp=sigcomp>,
              <sip:scscf2.home2.net;lr>,
              <sip:scscf1.home1.net;lr>,
              <sip:pcscf1.visited1.net;lr>
P-Access-Network-Info: 3GPP-UTRAN-TDD;
                       utran-cell-id-3gpp=C359A3913B20E
From: <sip:alice@home1.net>;tag=ty20s
To: <sip:bob@home2.net>;tag=d92119
Call-ID: 3s09cs03
Cseq: 127 INVITE
Contact: <sip:[1081::5:800:200A:B2B2]:5055;comp=sigcomp>
Allow: INVITE, ACK, CANCEL, BYE, PRACK, UPDATE, REFER, MESSAGE
Content-Length: 0
```

Figure 5.42: (57) 200 OK

```
ACK sip:[1081::5:800:200A:B2B2]:5055;comp=sigcomp SIP/2.0
Via: SIP/2.0/UDP [1080::8:800:200C:417A]:5059;comp=sigcomp;
                branch=z9hG4bK9h9ab
Max-Forwards: 70
P-Access-Network-Info: 3GPP-UTRAN-TDD;
                       utran-cell-id-3gpp=C359A3913B20E
Route: <sip:pcscf1.visited1.net:5058;lr;comp=sigcomp>,
       <sip:scscf1.home1.net;lr>,
       <sip:scscf2.home2.net;lr>,
       <sip:pcscf2.visited2.net;lr>
From: <sip:alice@home1.net>;tag=ty20s
To: <sip:bob@home2.net>;tag=d92119
Call-ID: 3s09cs03
Cseq: 127 ACK
Content-Length: 0
```

Figure 5.43: (63) ACK

The session setup is complete, and both users can generate their respective audio and video media streams. These media streams are in general sent end to end (e.g., from the caller to the callee's IMS terminals and vice versa) via IP-CAN routers.

5.8   Application Servers: Providing Services to Users

The exploration of the basic session setup in Section 5.7 provided the reader with an overview of a basic audio/video session established between two IMS users, where none of the users had any services executed (other than the audio/video session). This goes against the spirit of IMS where the basic idea is to consider the IMS as a common platform for providing innovative services. The basic session setup example which provided no extra services was just an educational exercise created for the sole purpose of understanding the basics of the IMS, the basics of SIP routing in the IMS, and the peculiarities of SIP when used in the IMS.

A more realistic example includes providing one or more services to either the caller, the callee, or even both of them.

In this section we describe the generalities of Application Servers (ASs), the different types of AS that we can find in the IMS, the different modes of operation of an AS, how they get involved during session setup, and how they can provide a service to the user.

5.8.1   Generalities about Application Servers

In a network there is generally more than one Application Server. Typically, there will be several ASs, each specialized in providing a particular service. Some application servers will implement some technologies, such as Java technology, SIP servlets, or SIP CGI (Common Gateway Interface). All of these ASs are characterized by implementing a SIP interface toward the S-CSCF. For historical reasons the interface defined between the S-CSCF and the AS is known as the IMS Service Control (*ISC*) interface. The name was adopted prior to 3GPP agreeing to use SIP over that interface. That is the reason for its different name.

Furthermore, 3GPP specifications still define the interface as "SIP plus some possible extensions", despite the interface being pure SIP and, with the passage of time, the need not arising to define extensions to SIP on that interface, in contrast to all other interfaces arriving at the S-CSCF.

Figure 5.44 shows the possible combinations of ASs and the different interfaces. ASs can be located in the home network or in a third-party service provider network. But it is up to the S-CSCF to decide whether to involve an AS in the session setup or not.

In addition, any AS can implement other protocols such as HTTP (Hypertext Transfer Protocol, specified in RFC 2616 [101]) or WAP (Wireless Application Protocol [233]), although this option is not described in the IMS specifications (e.g., to provide a graphical interface to the user, so that the user may configure some specific details of a service over HTTP or WAP).

For instance, an AS may carry out the same task as a call-forwarding service. The user may log onto the web page of the AS to configure the SIP URI where all the sessions will be forwarded. The graphical user interface provided by a web page is an improvement to the end-user's experience when configuring services, compared with the current mechanism of configuring services in circuit-switched networks.[13]

---

[13]The mechanism to configure services in circuit-switched networks typically comprises the user dialing a collection of digits, such as *21*5551122#.

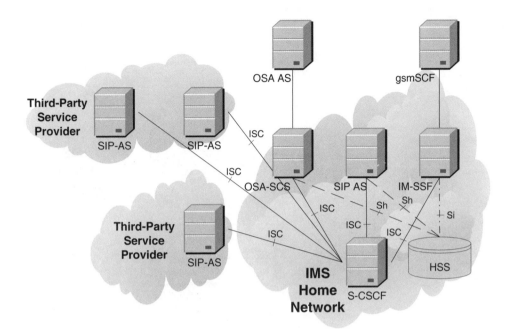

Figure 5.44: Interfaces to the Application Server

3GPP Release 6 adds yet another optional standardized interface toward an Application Server. This interface is code-named *Ut* and is used to provide the user with a protocol to configure and manage groups, policies, etc. So, this is not an interface used for live traffic. The protocol on the *Ut* interface is based on the XML Configuration Access Protocol (XCAP). At the time of writing, XCAP is being defined in the Internet-Draft "The Extensible Markup Language (XML) Configuration Access Protocol (XCAP)" [210]. XCAP defines how to use HTTP to create, modify, and delete an XML document, element, attribute, or value of a XML element. Figure 5.45 shows the *Ut* interface defined between the IMS terminal and ASs.

5.8.2 Types of Application Servers

The IMS defines three different types of Application Servers, depending on their functionality. Let's take a look at the characteristics of each of those types of AS.

5.8.2.1 The SIP Application Server

The SIP Application Server is the native AS in the IMS. New services exclusively developed for the IMS are likely to be executed by SIP ASs.

Figure 5.46 shows the relation between SIP ASs and the rest of the network. When a SIP AS is located in the home network, it can optionally implement an interface to the HSS. The implementation of the interface depends on whether the actual service logic needs to further interact with the HSS or not. The optional interface from the AS to the HSS is code-named *Sh*, and the protocol is based on Diameter (specified in RFC 3588 [60]), and an application specified in 3GPP TS 29.328 [22] and 3GPP TS 29.329 [35].

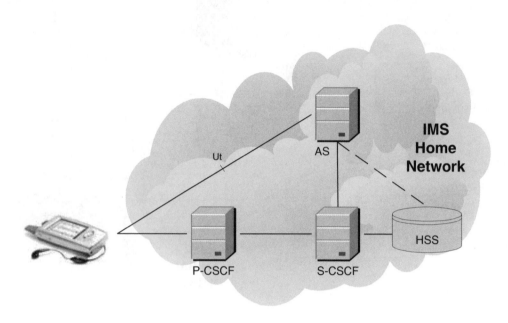

Figure 5.45: The *Ut* interface

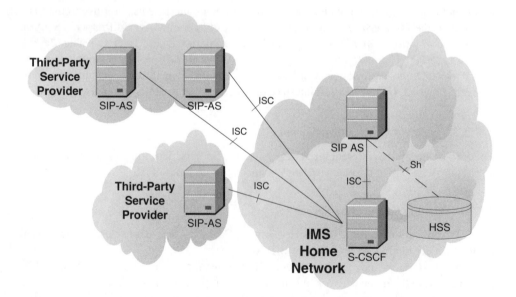

Figure 5.46: The SIP Application Server

Diameter is a highly extensible protocol designed to provide authentication, authorization, and accounting (AAA) services. We study the Diameter protocol in depth, when we describe the *AAA on the Internet* in Section 6.3. Because the *Sh* interface is based on Diameter we describe the *Sh* interface in Section 7.3.

We already mentioned that SIP ASs can be located in the home network or in a third-party service provider network. If the SIP AS is located in a third-party service provider network, it cannot implement the *Sh* interface in the HSS, as *Sh* is just an intra-operator interface.

5.8.2.2   The OSA-SCS

The Open Service Access–Service Capability Server (OSA-SCS) provides the gateway functionality to execute OSA services in the IMS. Most likely, new IMS services will not be developed in the OSA framework Application Server, but in SIP ASs instead. However, there is an existing base of services implemented in the OSA framework AS. So, in order to provide access to those services from the IMS a gateway is needed. This gateway is the OSA-SCS.

Figure 5.47 shows the OSA-SCS in the context of the IMS. The OSA-SCS provides an interface, external to the IMS, toward the OSA framework AS. This interface is based on the OSA series of specifications defined in 3GPP TS 29.198 [29]. The OSA-SCS AS also provides the *ISC* interface toward the S-CSCF based on SIP and may use the optional *Sh* interface toward the HSS. We describe the *Sh* interface in Section 7.3.

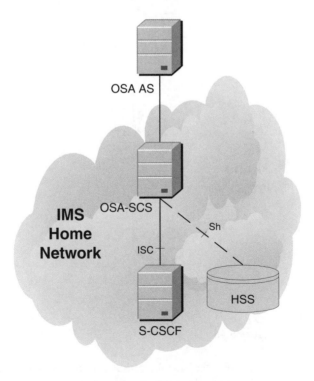

Figure 5.47: The OSA-SCS

The OSA-SCS is located in the home network, although the OSA framework allows secure third-party access; therefore, it allows the OSA framework AS to be located in a third-party service provider network.

From the perspective of the S-CSCF, the OSA-SCS appears and behaves as a SIP AS. The S-CSCF cannot differentiate an OSA-SCS from a SIP AS.

5.8.2.3   The IM-SSF Application Server

The third type of Application Server is the IP Multimedia Service Switching Function (IM-SSF). This AS provides a gateway to legacy service networks that implement CAMEL (Customized Applications for Mobile network Enhanced Logic) services, which are widely deployed in GSM networks. The IM-SSF acts as a gateway between SIP and CAMEL services, thus allowing CAMEL services to be invoked from the IMS.

Figure 5.48 shows the IM-SSF in the context of the IMS. On the IMS side the IM-SSF interfaces the S-CSCF via the ISC interface, with SIP as the protocol. The IM-SSF also interfaces the HSS with an interface code-named *Si*. The protocol on the *Si* interface is MAP (Mobile Application Part), an existing non-IMS protocol used in GSM networks. MAP is specified in 3GPP TS 29.002 [26]. 3GPP TS 23.278 [11] defines the interactions of CAMEL with the IMS.

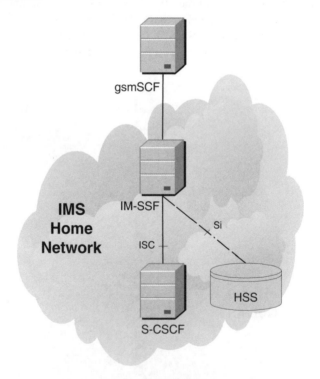

Figure 5.48: The IM-SSF Application Server

Outside the IMS the IM-SSF interfaces the gsmSCF (GSM Service Control Function), which is part of the Camel Service Environment (CSE). The protocol on this interface is CAP (CAMEL Application Part), a non-IMS protocol specified in 3GPP TS 29.278 [10].

The IM-SSF is located in the home network.

From the perspective of the S-CSCF, the IM-SSF appears and behaves as a SIP AS. The S-CSCF cannot differentiate an IM-SSF from a SIP AS.

5.8.3 The Session Setup Model through Application Servers

Although we have not yet described how the S-CSCF decides when and how to involve a particular Application Server in the signaling path (we describe it soon in Section 5.8.4), we need to investigate how ASs fit in the whole session setup path, and the different SIP behavior they can have. This description is applicable to all of the different types of AS.

From the point of view of SIP an AS can act as either an originating or a terminating SIP User Agent (i.e., an endpoint that originates or terminates SIP), a SIP proxy server, a SIP redirect server, or a SIP B2BUA (Back-to-Back User Agent). An AS may sometimes act as a SIP proxy server and at other times as a SIP User Agent, depending on the service provided to the user. If the AS decides not to provide a service, then it acts as a SIP proxy server. This guarantees that the S-CSCF receives the SIP request and continues with the process. If an AS does not want to provide a service, then it should not record the route, so that it is out of the signaling path once the SIP transaction is complete.

5.8.3.1 Application Server Acting as a SIP User Agent

Figure 5.49 shows a session setup model when the Application Server is acting as a terminating User Agent. In this example we have chosen to locate the P-CSCF in the home network, but it could be located in the visited network as well. Figure 5.49 shows an example of an AS acting as a terminating User Agent (it is acting as the destination of the session). The service is provided on the originating side (i.e., the service is provided to the caller).

We can see in Figure 5.49 an IMS terminal sending a SIP INVITE request (1) that arrives at the originating P-CSCF and the originating S-CSCF (2). The S-CSCF decides to forward the INVITE to an AS (3). The AS behaves as a SIP UA and responds with a 200 (OK) response (4) that is sent through the S-CSCF (5) and the P-CSCF (6) back to the IMS terminal.

In Figure 5.49 and subsequent figures we show examples in which the SIP request is an INVITE request. However, the SIP request could be any other initial SIP request that is subject to treatment by filter criteria (see Section 5.8.4 for a description of which requests are subject to such treatment, like SUBSCRIBE or PUBLISH requests).

For the sake of clarity, in Figure 5.49 and subsequent figures we have omitted all the signaling that may take place in between the SIP INVITE request and the 200 (OK) response. For instance, if the SIP INVITE request requires use of the precondition extension, there will be a few more messages in between the INVITE request and the 200 (OK) response.

Figure 5.50 shows a similar scenario in which an AS is also acting as a terminating SIP User Agent. But, in this case the service is provided in the terminating call leg (i.e., to the callee). It is worth noting that the AS effectively "hijacks" the session and the user never receives the SIP message (e.g., the INVITE request). As shown in Figure 5.50 a User Agent has sent an INVITE request (1) that is received at the I-CSCF. The I-CSCF forwards the INVITE request (2) to the S-CSCF. The S-CSCF decides to forward the INVITE request (3) to an AS. The AS acts as a User Agent and establishes the session; that is, it answers with a 200 (OK) response (4). The response is sent back, in (5) and (6), via the I-CSCF.

An example of a service that uses this model is any service that requires a server to handle the SIP request on behalf of the user. This is the model used in the presence service

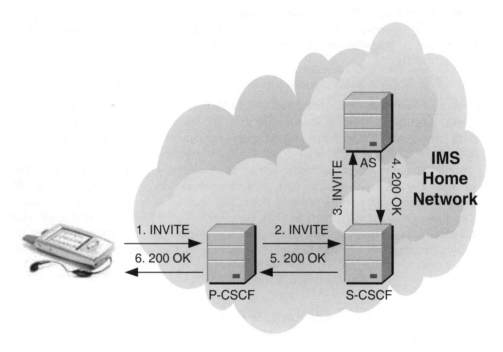

Figure 5.49: An Application Server acting as a terminating SIP UA is providing services in the originating call leg

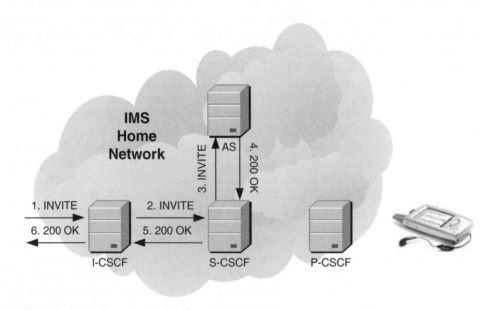

Figure 5.50: An Application Server acting as a terminating SIP UA is providing services in the terminating call leg

(e.g., when a watcher subscribes to the presence information of the presentity, or user, because in spite of the SUBSCRIBE request being addressed to the end-user the request is diverted to the presence server). The presence server (which is a specialized AS) treats the request appropriately.

Figures 5.49 and 5.50 show the case when the SIP AS acts as a terminating SIP User Agent (i.e., the AS is replacing or acting as the destination of the session). A different example is shown in Figure 5.51 where the AS is acting as a caller (i.e., the AS initiates the SIP session). In this case the service does not have an originating/terminating distinction, because the user receives the session setup.

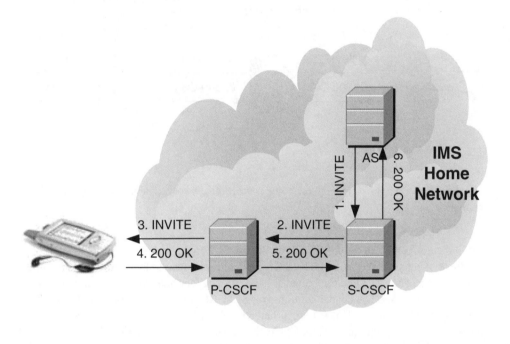

Figure 5.51: An Application Server, acting as an originating SIP UA, is providing services to the user

An example of this service is a wake-up call service, where the AS, at a certain preconfigured time, initiates a session toward the IMS terminal to provide the wake-up call.

5.8.3.2 Application Server Acting as a SIP Proxy Server

In another configuration, an Application Server may need to act as a SIP proxy server to provide the service. The configuration is shown in Figure 5.52 in an originating call leg. An IMS terminal sends a SIP INVITE request (or any other request subject to filter criteria assessment) that traverses the P-CSCF (1) and the S-CSCF (2). The S-CSCF decides to involve an AS, and forwards the INVITE request (3) to that AS. The S-CSCF inserts in the INVITE request a Route header field that points to the AS in the first place and the S-CSCF in the second place. This is to allow the AS to forward the request back to the same S-CSCF (4). Furthermore, the S-CSCF needs to insert some sort of state information in its own SIP

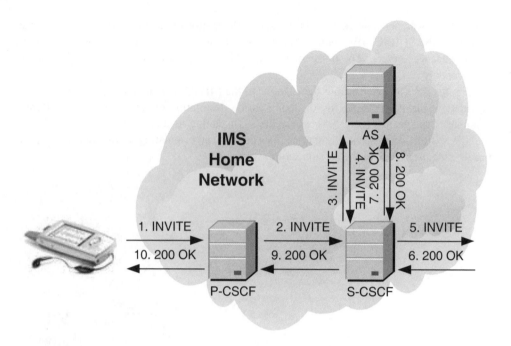

Figure 5.52: An Application Server, acting as a SIP proxy server, is providing services in the originating call leg

URI in the Route header field. This allows the S-CSCF to determine whether the SIP request has already been received from the IMS terminal and sent once to that AS.

Figure 5.53 shows an example of what the Route header field would look like in any SIP request sent to an AS. The header contains two SIP URIs, the first one points to the AS that is receiving the SIP request. This is regular behavior according to SIP loose routing rules. The second URI is pointing to the same S-CSCF, thus allowing the AS to forward the request back to the S-CSCF. There is state information contained in the "username" part of the SIP URI (state34 in our example in Figure 5.53). When the S-CSCF receives the SIP request from the AS, (4) in Figure 5.52, the S-CSCF reads the "username" part of the SIP URI in the Route header field value and is able to determine the point in the process where it needs to continue.

```
Route: <sip:as33.home1.net;lr>, <sip:state34@scscf1.home1.net;lr>
```

Figure 5.53: The Route header field in a SIP request forwarded to an AS

In our example in Figure 5.53 the S-CSCF writes the state information as part of the "username" part of the SIP URI. The S-CSCF can freely choose any other mechanism to convey this state, such as including a parameter in its own URI or choosing a different port number to receive the request from the AS. The rule is that the S-CSCF writes its own SIP URI. Since the S-CSCF itself is the only entity that will need to parse the "username" at a later point it can encode the state information it wishes to read later.

Last but not least, we need to mention that the S-CSCF is not aware of the SIP behavior of the AS. The S-CSCF, when forwarding a request to an AS, always inserts this double Route header field, no matter whether the AS is operating as a SIP User Agent, a SIP proxy server, etc. If the AS operates as a SIP User Agent the AS simply does not use the `Route` header field, because the AS answers the request.

Let's return to our examination of Figure 5.52 where we have a SIP INVITE request received at the AS (3). The AS takes different decisions and actions depending on the actual service provided to the user. For instance, the AS needs to decide whether it wants to remain in the signaling path for subsequent SIP signaling related to this session. In other words, the AS needs to decide whether it needs to insert its own SIP URI in the value of the `Record-Route` header field. The S-CSCF will not forward the remaining SIP requests that belong to the same SIP dialog if the AS does not record its own route. The behavior is different with responses, since they always follow the same path that their corresponding requests traversed. Therefore, responses to this initial request that reached the AS are always forwarded via the AS, irrespective of whether the AS recorded the route or not.

The AS may need to change a header field value in the SIP request. For instance, let's imagine that the AS is providing a speed dial service, whereby the user only dials a short string of digits, such as tel:123. The AS contains a personalized list per user that expands the dialed number with the real URI (TEL URI or SIP URI). In this case the service the AS provides consists of just rewriting the *Request-URI* and replacing tel:123 with the real destination URI. There are no other actions that the AS is requested to do. Furthermore, in the speed dial service example the AS does not even need to Record-Route, because the service is provided entirely in the INVITE request.

Eventually, the AS honors the `Route` header field contained in the received INVITE request and forwards the INVITE request, with all the needed changes, to the SIP URI indicated in the `Route` header field; that is, to the S-CSCF (4).

When the S-CSCF receives the INVITE request for the second time (4), due to the inclusion of the state information it is able to determine the next required action. In the example in Figure 5.52 the S-CSCF forwards the INVITE request toward the callee's network (5). Eventually, a 200 (OK) response is received (6). The S-CSCF, according to regular SIP routing procedures for SIP responses, forwards the 200 (OK) response (7) to the AS. The AS forwards it back to the S-CSCF (8). The S-CSCF forwards it to the P-CSCF (9) which in turns forwards it to the IMS terminal (10).

In the configuration shown in Figure 5.54 the AS is still acting as a SIP proxy server, but is providing a service to the callee. The INVITE request (1) is received at an I-CSCF. The I-CSCF forwards the INVITE request (2) to the S-CSCF that is handling the user. The S-CSCF decides to involve an AS that may provide a service; so, it forwards the INVITE request (3) to the AS. The AS executes the service, perhaps modifies some headers in the SIP request, being compliant with SIP rules which state header fields must be modified by a proxy, and then forwards the INVITE request (4) back to the S-CSCF, due to the presence of a `Route` header field pointing to the S-CSCF. The S-CSCF forwards the INVITE request to the IMS terminal via the P-CSCF, (5) and (6). The responses will traverse the same set of proxies in the reverse direction.

The call-forwarding service is an easy service that can be implemented using this configuration. The AS needs to rewrite the *Request-URI* to point to the new destination URI.

5.8.3.3 Application Server Acting as a SIP Redirect Server

An Application Server may be acting as a SIP redirect server. Figure 5.55 shows the high-level signaling flow for an AS providing services in the terminating call leg (i.e., to the callee).

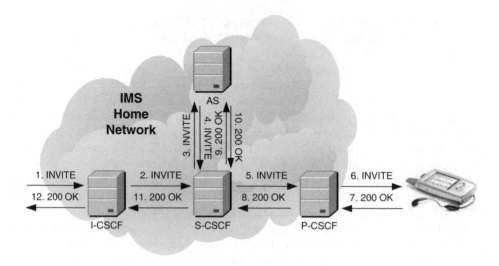

Figure 5.54: An Application Server, acting as a SIP proxy server, is providing services in the terminating call leg

According to Figure 5.55 an I-CSCF in a home network receives an INVITE request (1). The I-CSCF forwards it to the S-CSCF (2). The S-CSCF involves an AS and forwards the INVITE request to it (3). The AS acting as a SIP redirect server generates a 302 (Moved Temporarily) final response (4). The response contains a `Contact` header field that includes the new URI to contact. The response is forwarded back to the originator, (5) and (6). When the originator of the session receives the 302 (Moved Temporarily) response, it generates a new INVITE request whose *Request-URI* (the destination address) is the `Contact` header field value received in the 302 (Moved Temporarily) response. This new INVITE request may not even reach the same IMS home network (e.g., if the new destination address belongs to a different domain name).

A typical example of the applicability of SIP redirect servers is the provision of call-forwarding services. They can also be used to provide number portability services. In general, SIP redirect servers can be used whenever the session is forwarded somewhere and the operator is not interested in being part of the session.

5.8.3.4   Application Server Acting as a SIP B2BUA

The last mode of operation of an Application Server is the SIP B2BUA (Back-to-Back User Agent). A B2BUA is just two SIP User Agents connected by some application-specific logic. In general, a B2BUA performs similar actions to a SIP proxy server: it receives requests and forwards those requests somewhere else; it receives responses and relays them back to the originating entity. However, there are differences between a SIP proxy and a B2BUA. These differences are related to the type of actions that both are allowed to perform and the consequences of performing those actions.

Figure 5.56 shows the logical view of a SIP B2BUA. A SIP request A is received in one of the SIP User Agents, which will pass the request to the application-specific logic. The application-specific logic is responsible for generating a response A and creating a new SIP request B, which is partly related to request A. Requests A and B are only partly related

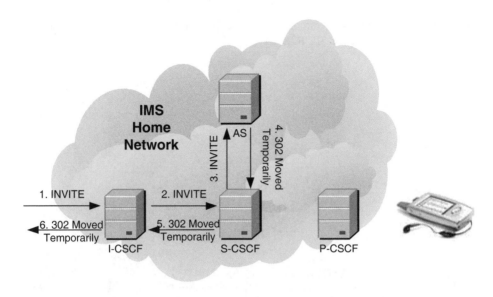

Figure 5.55: An Application Server, acting as a SIP redirect server, is providing services in the terminating call leg

because the application-specific logic may change any header, including those header fields that a SIP proxy server cannot modify, such as the From, To, Call-ID, etc. The application-specific logic may even change the method in the SIP request. The application specific logic may also change SDP as well, another action that is not allowed to be taken by SIP proxies. The B2BUA can even asynchronously generate a SIP request on one of the call legs, without having received any stimulus on the other call leg or create multiple call legs, such as a third-party call controller.

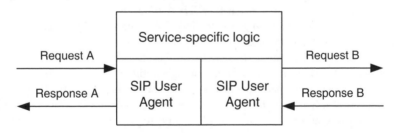

Figure 5.56: Logical view of a SIP Back-to-Back User Agent

Because a SIP B2BUA is just a collection of SIP User Agents, B2BUAs need to understand all the methods, extensions, etc. that in normal circumstances only two endpoints would need to understand. For example, in the IMS context, a SIP B2BUA needs to understand the SIP precondition extension. Proxies need not understand this extension.

Figure 5.57 shows the high-level signaling flow of an AS acting as a B2BUA and providing services to the originating party of the session. In this example "INVITE A" represents the INVITE request received by the AS, and "INVITE B" the INVITE request that the AS sends, in an effort to indicate that these INVITE requests are uncorrelated and the only point of correlation is the SIP B2BUA.

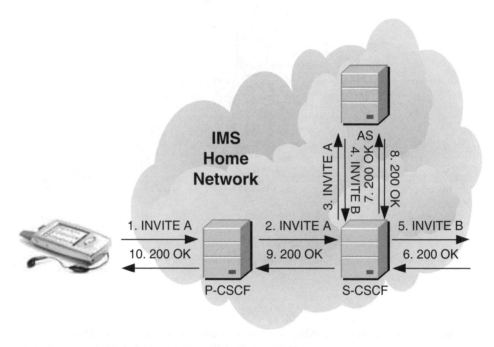

Figure 5.57: An Application Server, acting as a SIP B2BUA, is providing services in the originating call leg

The logic used by a B2BUA will also dictate how requests and responses are mapped. One possibility is that responses received on one side of the B2BUA are simply regenerated at the other side. Another is that when INVITE A is received, the B2BUA generates the corresponding response A, and then generates INVITE request B. This behavior depends on the actual service provided to the user.

An example of an AS acting as a B2BUA on the originating call leg is a prepaid AS. The prepaid AS first verifies that the user has enough credit in his account to pay for the cost generated by the session. If the user does not have enough credit the B2BUA will simply reject the call, but if it has enough credit it will proceed. Once the session is established, if the user runs out of credit, the B2BUA sends a BYE request to each of the parties to tear down the session.

Figure 5.58 shows how an AS, acting as a B2BUA, provides services to the callee. We have also distinguished the two different uncorrelated INVITE requests and named them INVITE A and INVITE B. In step (4) the AS generates a new INVITE B that is partly related to INVITE A.

A privacy AS is a good example of an AS that, acting as a SIP B2BUA, provides a service to the callee. On receiving a SIP request or response the B2BUA obfuscates those header fields that reveal information related to the user (e.g., From, Contact, etc.) and may even change the SDP to avoid peers becoming aware of the IP address of the originator of the session. So, the callee cannot see any SIP header, IP address, etc. related to the caller and vice versa.

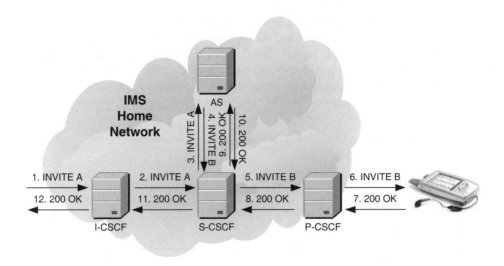

Figure 5.58: An Application Server acting as a SIP B2BUA is providing services in the terminating call leg

5.8.4 Filter Criteria

Filter criteria are among the most important pieces of user information stored in the network, because they determine the services that will be provided to each user. Filter criteria contain a collection of user-related information that helps the S-CSCF to decide when to involve (e.g., forward the SIP request to) a particular Application Server to provide the service.

3GPP TS 23.218 [20] specifies, for historical reasons, two sets of filter criteria: *initial filter criteria* and *subsequent filter criteria*. Only initial filter criteria are used. Subsequent filter criteria constituted a theoretical exercise, since the implementation of subsequent filter criteria by the S-CSCF would conflict with the SIP routing rules for proxies.

Initial filter criteria are supposed to be evaluated with those SIP initial requests that either create a dialog or are stand-alone requests (i.e., those that are not subsequent requests within a SIP dialog). For instance, the S-CSCF evaluates initial filter criteria when it receives a first SUBSCRIBE request, INVITE, OPTIONS, or any such requests that either create a dialog or are sent outside any dialog. The S-CSCF does not evaluate initial filter criteria when it receives a PRACK, NOTIFY, UPDATE, or BYE request, since they are always sent as part of an existing SIP dialog.

The subsequent filter criteria concept was that the S-CSCF would evaluate subsequent filter criteria when it receives a subsequent request within a SIP dialog. However, the result of evaluating subsequent filter criteria would lead the S-CSCF to forward a subsequent SIP request to an AS, an action that is in contradiction to the routing procedures for a subsequent request in a SIP proxy. Additionally, in the event that an AS receives this subsequent request the AS would have not (most likely) received the initial SIP request that created the SIP dialog. Therefore, the AS would reject the request and would ignore it. Consequently, the decision was made not to implement subsequent filter criteria.

The only implemented filter criteria in the specifications are the initial filter criteria. As the subsequent filter criteria do not exist the terms *initial filter criteria* and *filter criteria* are usually interchangeable.

The HSS stores all the data related to a user in a data structure named the *user profile*. Figure 5.59 shows a high-level simplified structure of the user profile. We describe the user profile in detail in Section 7.2.3, but for the time being it is enough to mention that the user profile contains the *Private User Identity* to which the user profile is applicable and one or more *service profiles*. Each service profile contains one or more *Public User Identities* to which the service profile is applicable and zero or more *filter criteria*. We focus on filter criteria in this section.

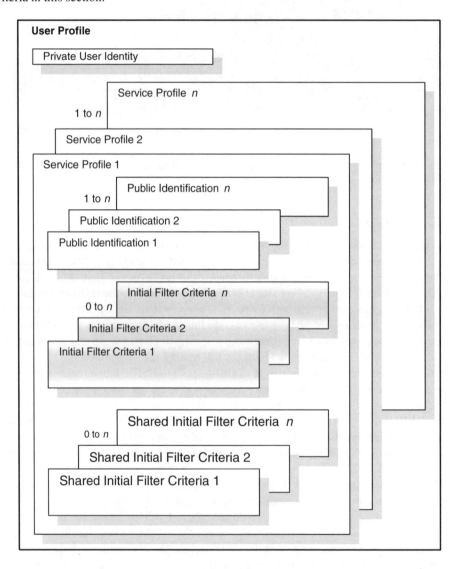

Figure 5.59: Simplified representation of the structure of the user profile

When the user registers with the S-CSCF, the S-CSCF contacts the HSS and downloads the user profile that includes the filter criteria. So, filter criteria are available in the S-CSCF at the moment the user registers.

Filter criteria determine the services that are applicable to the collection of Public User Identities listed in the Service profile. Filter criteria are further structured as sketched in Figure 5.60.

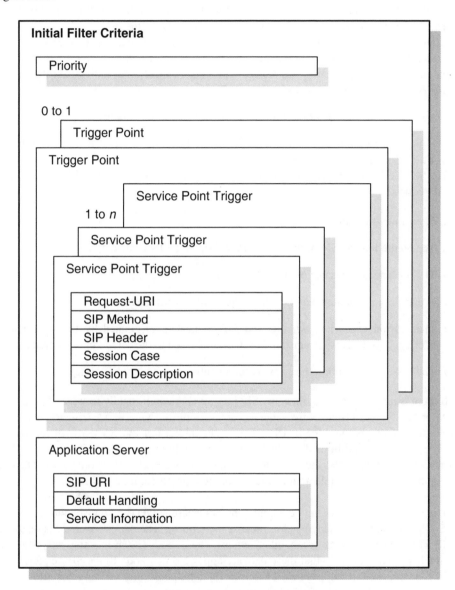

Figure 5.60: Structure of initial filter criteria

The first field in the filter criteria structure is the *Priority*. The Priority field determines the order in which these filter criteria will be assessed, compared with the remaining filter criteria that are part of the same service profile. The S-CSCF will first choose filter criteria that have higher priority, indicated by a lower figure in the Priority (i.e., priority 1 is the highest priority). After evaluating it the S-CSCF continues with the filter criteria of the next

Priority number (e.g., 2, 3, etc.). The Priority field of filter criteria is a unique number with respect to all the filter criteria that belong to the same service profile. In any case, numbers do not need to be consecutive. For instance, the highest priority of the first set of filter criteria that the S-CSCF evaluates could be Priority 100. The second could be Priority 200 and so on. This leaves room to accommodate new services in between them.

After the Priority field, there can be zero or one *Trigger Points*. A Trigger Point is an expression that needs to be evaluated in order to determine whether or not the SIP request will be forwarded to a particular AS. A trigger point is a collection of individual filters called *Service Point Triggers*. For instance, a Trigger Point can express:

```
(Method = INVITE) AND (Request-URI = sip:user@example.com)
```

In this example, one Service Point Trigger is Method = INVITE and the other is Request-URI = sip:user@example.com.

The Service Point Trigger allows us to access the information stored in different fields of the SIP request.

- The value of *Request-URI*.

- The method of the SIP request (e.g., INVITE, OPTIONS, SUBSCRIBE, etc.).

- The presence or absence of any SIP header.

- A partial or full match between the contents of any SIP header.

- The session case (i.e., whether the SIP request is originated by the served user, addressed to the served registered user, or addressed to the served unregistered user).

- The session description (i.e., any partial or full match on any SDP line).

If there is no Trigger Point, then any SIP request is unconditionally forwarded to the AS.

After the Trigger Points, which include one or more Service Point Triggers, the initial filter criteria contains the AS SIP URI. This is the address of the AS that will receive the SIP request if the conditions described in the Trigger Points are met. There is a *Default Handling* field that indicates the action to be taken if the S-CSCF cannot contact, for whatever reason, that AS. The possible actions are to continue processing the SIP request or to abort the process.

The *Service Information* field contains some transparent data (i.e., transparent to the HSS and S-CSCF) that the AS may need to process the request. The use of this field is restricted to SIP REGISTER requests or any other request where the S-CSCF is acting as a SIP User Agent Client. The reason is that these data are appended in a body to the SIP request. This action is not allowed in SIP proxies. Therefore, the only possibility of using this information is when the S-CSCF, due to initial filter criteria triggering, acting as a SIP User Agent Client generates a third-party SIP REGISTER request to the AS. Such a REGISTER request can contain the *Service Information* (if the AS needs it), whose purpose is to transfer the IMSI to the IM-SSF of the subscriber, so that the IMSI can be used by the IM-SSF.

Lastly, the user profile is encoded using the Extensible Markup Language (XML). An XML schema defining initial filter criteria is specified in 3GPP TS 29.228 [21]. Initial filter criteria are transported from the HSS to the S-CSCF over Diameter messages (Diameter is explained in Section 6.3 and the usage of Diameter in IMS in Section 7.2).

5.8.5   An Example of Service Execution

We illustrate the execution of a service, including the filter criteria that trigger that service, with an example. Let's imagine a service provided to a particular user (`sip:goodguy@ example.com`). The service automatically diverts a session setup attempt to an answering machine (e.g., MRF) when the caller is listed in a black list (e.g., `sip:badguy@ example.com`). In order to model the service, it is required that the service profile applicable to `sip:goodguy@example.com` contains initial filter criteria that have a Trigger Point that filters all the terminating INVITE requests whose caller is the bad guy. The Trigger Point is represented as:

```
(method = INVITE) AND
(P-Asserted-Identity = sip:badguy@example.com) AND
(Session Case = Terminating)
```

When the conditions are met, the S-CSCF should send the request to an Application Server identified by its SIP URI: `sip:as33.example.com`. The AS contains the logic to provide the diversion service to the MRFC, instructing the MRFC to play a pre-recorded announcement.

If we assume that the good guy in our example has only subscribed to this service, his user profile encoded in XML would look like the one represented in Figure 5.61.

Figure 5.62 shows the example in action. The bad guy sends an INVITE request (1) that is routed to the S-CSCF (2). The S-CSCF evaluates all the filter criteria that are part of the good guy's service profile. In this example we assume that there is a single filter criterion, which instructs the S-CSCF to forward to AS `sip:as33.example.com` those terminating INVITE requests that are sent by the bad guy (i.e., whose `P-Asserted-Identity` is `sip:badguy@example.com`).

The S-CSCF inserts a double `Route` header in the INVITE request, one value being the SIP URI of the AS and the other its own S-CSCF SIP URI, and then the S-CSCF forwards the INVITE request (3) to the AS. The AS receives the INVITE request and applies the service. In this case, applying the service means forwarding it to a new destination, which is the MRFC. This can be accomplished by the AS acting as a SIP proxy and replacing the *Request-URI* in the forwarded INVITE request.

Therefore, the AS replaces the *Request-URI* of the SIP INVITE request with the SIP URI of the MRFC, `sip:announcementBadGuy@mrfc.example.com`. The SIP URI contains a username part in the URI that gives a hint to the MRFC about which announcement to play. The AS honors the `Route` header in the received INVITE request (3) and forwards the INVITE request to the S-CSCF (4).

On receiving the INVITE request again, the S-CSCF detects that the *Request-URI* has changed with respect to the INVITE request initially received, so it does not assess other possible filter criteria. Instead, it forwards the request to its new destination: in this case the MRFC (5). The MRFC examines the username part contained in the *Request-URI* in order to find out what announcement to play. In this case, "announcementBadGuy" is a hint to the MRFC to play the appropriate stored announcement. The MRFC eventually sends a 200 (OK) response back to the originator. It also sends H.248 commands to the MRFP to play the stored announcement. The MRFC plays the announcement using the negotiated codec in the SDP.

In this example the tasks of the Application Server are relatively simple. In real life scenarios ASs carry out much more complex tasks. For instance, they can provide an

```xml
<?xml version="1.0" encoding="UTF-8"?>
<testDatatype xmlns:xsi="http://www.w3.org/2001/XMLSchema-instance">
    <IMSSubscription>
        <PrivateID>privategoodguy@example.com</PrivateID>
          <ServiceProfile>
            <PublicIdentity>
                <Identity>sip:goodguy@example.com</Identity>
            </PublicIdentity>
            <InitialFilterCriteria>
              <Priority>0</Priority>
              <TriggerPoint>
                <ConditionTypeCNF>1</ConditionTypeCNF>
                  <SPT>
                     <ConditionNegated>0</ConditionNegated>
                     <Group>0</Group>
                     <Method>INVITE</Method>
                  </SPT>
                  <SPT>
                     <ConditionNegated>0</ConditionNegated>
                     <Group>0</Group>
                     <SIPHeader>
                        <Header>P-Asserted-Identity</Header>
                        <Content>
                            "sip:badguy@example.com"
                        </Content>
                     </SIPHeader>
                  </SPT>
                  <SPT>
                     <ConditionNegated>0</ConditionNegated>
                     <Group>0</Group>
                     <SessionCAse>1</SessionCase>
                  </SPT>
              </TriggerPoint>
              <ApplicationServer>
                     <ServerName>sip:as33.example.com</ServerName>
                     <DefaultHandling>1</DefaultHandling>
                     </ApplicationServer>
              <InitialFilterCriteria>
          </ServiceProfile>
    </IMSSubscription>
</testDatatype>
```

Figure 5.61: An example of a user profile

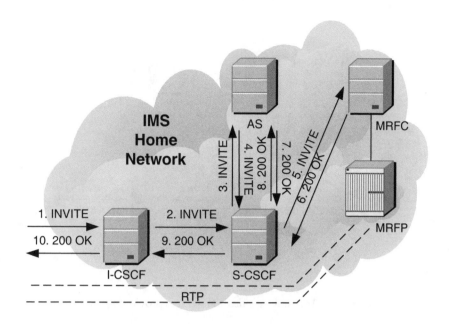

Figure 5.62: An example of a service

administrative HTTP interface that allows the user to log into the AS and manage their own black list. Every time the user adds or removes an identity from their black list the AS retrieves the filter criteria from the HSS, over the *Sh* interface, modifies them with the new addition or removal, and sends them back to the HSS. The HSS pushes the updated user profile (which includes the new updated filter criteria) to the S-CSCF, so the S-CSCF applies the new configuration of the service in real time.

So far we have shown a simple example where there is only a single AS involved in session setup. Typically, there will be several services that potentially can be provided to a user. So, in most cases there will be several ASs involved in the signaling path. Figure 5.63 shows a more complex example where the S-CSCF is evaluating all the filter criteria for the user and, as a result of this evaluation, several ASs receive the SIP INVITE request. In the example in Figure 5.63 all the ASs are acting as SIP proxies, but any other combination is possible as well.

According to Figure 5.63, the S-CSCF receives an INVITE request (2), evaluates the filter criterion that has highest priority (the one with the lowest figure in the priority field), and forwards the INVITE request (3) to AS #1. When the S-CSCF receives back the INVITE request (4) the S-CSCF evaluates the next filter criterion with the next priority and, as a result, forwards the INVITE request (4) to AS #2. The S-CSCF repeats the operation with the next filter criterion (ordered by priority) when it next receives the INVITE request (6). As a result of this last filter criteria evaluation the S-CSCF forwards the INVITE request (7) to AS #3. Once the S-CSCF receives back the INVITE request (8) and once the S-CSCF has evaluated all the different filter criteria, the S-CSCF forwards the INVITE request to the P-CSCF and the IMS terminal, (9) and (10). Of course, at any time the S-CSCF could have evaluated a filter criterion for which there was not a match in the Service Point Triggers. Consequently,

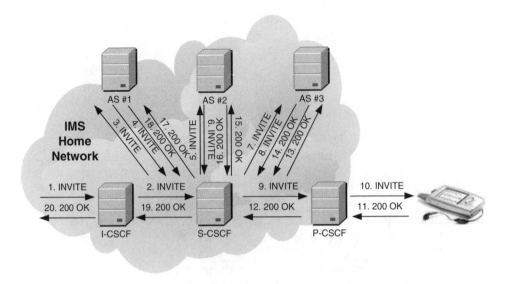

Figure 5.63: A few Application Servers providing services

the S-CSCF would have not forwarded the INVITE request to the AS indicated in that filter criterion.

Figure 5.63 indicates a few important aspects about filter criteria. First, the order in which services are executed is important. For instance, if AS #1 were an answering machine service that was automatically connected at some time of day, then AS #1 would act as a SIP User Agent and, consequently, AS #2 and AS #3 would not receive the INVITE request. A second aspect is that every time a new AS is involved there is an extra session setup delay added to the global session setup delay. If a large number of ASs were involved, then the caller may give up due to the delay in session setup time.

5.9   Interworking

The IMS will become even more attractive when its users are able to communicate with persons located in the PSTN (Public Switched Telephone Network) and on the Internet. Release 5 specifications do not define any type of interworking with other networks, but Release 6 specifications define interworking with the PSTN (3GPP TS 29.163 [19]) and with non-IMS SIP-based networks (3GPP TS 29.162 [18]).

5.9.1   SIP–PSTN Interworking

Even though the PSTN offers video services in addition to the traditional audio calls, the existing work on SIP–PSTN interworking focuses on audio-only calls. So, we will describe how to establish audio sessions between SIP user agents and PSTN terminals.

Audio calls in the PSTN are established using two types of signaling: network-to-network signaling and user-to-network signaling. Network-to-network signaling (e.g., SS7) is used between telephone switches while user-to-network signaling (e.g., DSS-1) is used between terminals and their local exchanges. We focus our description on the interworking between

SIP and network-to-network signaling protocols because this interworking is more likely to be used in the IMS in the future.

There are two levels of interworking in a call between the PSTN and a SIP network: the signaling and the media levels. They are usually handled by a distributed gateway with three components: signaling gateway, media gateway controller, and media gateway (the architecture for PSTN-to-SIP interworking is specified in RFC 3372 [232]). Figure 5.64 shows a logical gateway with these three elements.

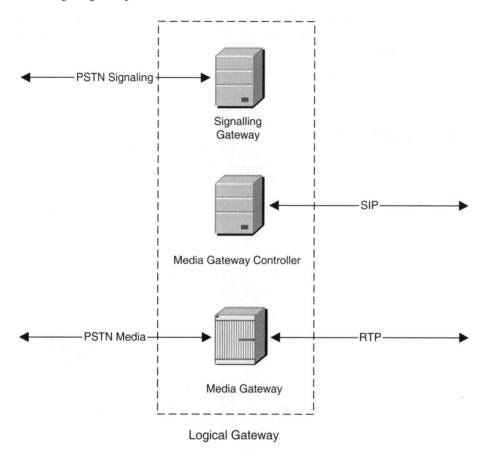

Figure 5.64: Architecture of a logical gateway

The signaling gateway receives PSTN signaling and sends it to the media gateway controller over an IP network (and vice versa). The media gateway controller performs the mapping between PSTN signaling and SIP, and controls the media gateway (ISUP-to-SIP interworking is specified in RFC 3398 [71] and RFC 3578 [72]). The media gateway performs media transcoding, if needed, and encapsulates the media received from the PSTN side in RTP packets (and vice versa).

The traffic between the signaling gateway and the media gateway controller is typically transported over SCTP (specified in RFC 2960 [230]). Keeping these two elements separate allows a single media gateway controller to handle the traffic of many signaling gateways that are usually physically distributed. Nevertheless, these two elements are sometimes co-located (i.e., they are implemented in the same physical box). This makes it unnecessary to use any transport protocol between them. Media gateway controllers use control protocols such as H.248 [143] or MGCP (Media Gateway Control Protocol, specified in RFC 2705 [47]), to control media gateways.

5.9.1.1   Gateway Architecture in the IMS

The IMS uses the gateway architecture we have just described. Figure 5.65 shows the different functions and interfaces that are involved in interworking with the PSTN (specified in 3GPP TS 29.163 [19]).

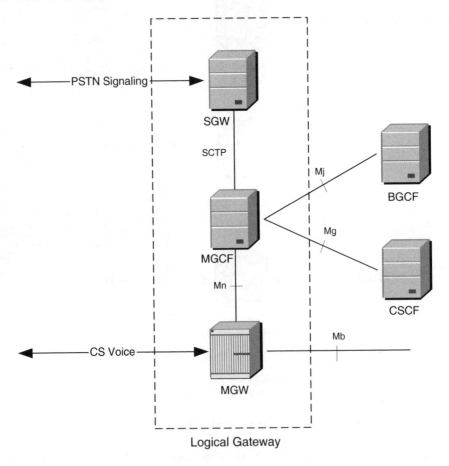

Figure 5.65: IMS-PSTN interworking architecture

The SGW (Signaling Gateway) exchanges ISUP or BICC over SCTP with the MGCF (Media Gateway Control Function). The MGCF performs BICC/ISUP-to-SIP interworking (and SIP-to-BICC/ISUP) and exchanges SIP signaling with the CSCFs using the *Mg* interface. CSCFs receiving SIP signaling from an MGCF regard the SIP signaling as coming from an S-CSCF.

The MGCF interacts with the IM-MGW (IP Multimedia Media Gateway Function) through the *Mn* interface, which is based on H.248 [143]. (The *Mn* interface is described in 3GPP TS 29.332 [24].)

The IM-MGW performs media conversions between the PSTN and the IMS. It exchanges RTP packets with the IMS over the *Mb* interface and exchanges Circuit-Switched voice data with the PSTN.

5.9.1.2 The BGCF

When a session terminates in the PSTN the BGCF (Breakout Gateway Control Function) determines which MGCF handles it. The BGCF receives a request from an S-CSCF and, based on an analysis of the destination address and on any agreements the operator may have for calls terminating in the PSTN, the BGCF decides whether the session should be handled by a local MGCF or by a remote MGCF.

If the session is to be handled by a local MGCF the BGCF relays the request to one of the MGCFs in its network, as shown in Figure 5.66. The BGCF may or may not Record-Route. If it does, it remains in the signaling path for the rest of the session. If it does not Record-Route, it steps out from the signaling path. The BGCF in Figure 5.66 does not Record-Route.

If the session is to be handled by a remote MGCF the BGCF relays the request to another BGCF in the remote network, as shown in Figure 5.67. The BGCF in the remote network chooses an MGCF and relays the request to it. None of the BGCFs in Figure 5.67 Record-Route.

Sessions originating in the PSTN and terminating in the IMS do not traverse the BGCF, because MGCF selection is performed in the circuit-switched domain instead. Figure 5.68 illustrates this point. The MGCF sends the initial INVITE request to an I-CSCF because the MGCF does not know which S-CSCF has been assigned to the destination user.

5.9.2 *Interworking with Non-IMS SIP-based Networks*

Interworking between the IMS and non-IMS SIP-based networks (specified in 3GPP TS 29.162 [18]) is still not mature. There is still a need to define the behavior of IMS entities when a SIP entity on the Internet does not support some of the extensions used in the IMS (e.g., preconditions).

In any case, one of the most important issues in this interworking is how to establish sessions between IPv6 IMS terminals and IPv4 SIP user agents on the Internet. Let's now look at how the IETF resolves this issue. The IETF IPv4/IPv6 translation mechanisms we will describe may be adopted by 3GPP and 3GPP2 in the future.

5.9.2.1 IPv4/IPv6 Interworking

Although IPv6 is the future of the Internet, IPv4 is currently the most widespread network-layer protocol. Even so, IMS terminals cannot talk directly to IPv4 Internet hosts, because the IMS is based on IPv6.

IPv6 user agents need to exchange both signaling and media with IPv4 user agents on the Internet. The signaling part is handled by SIP and the media part is typically handled using NATs (Network Address Translators), as shown in Figure 5.69.

IPv6–IPv4 interworking at the SIP layer is handled by proxy servers at the edge of every domain, which need to have dual IPv4–IPv6 stacks. Proxies within an IPv6 (or an IPv4)

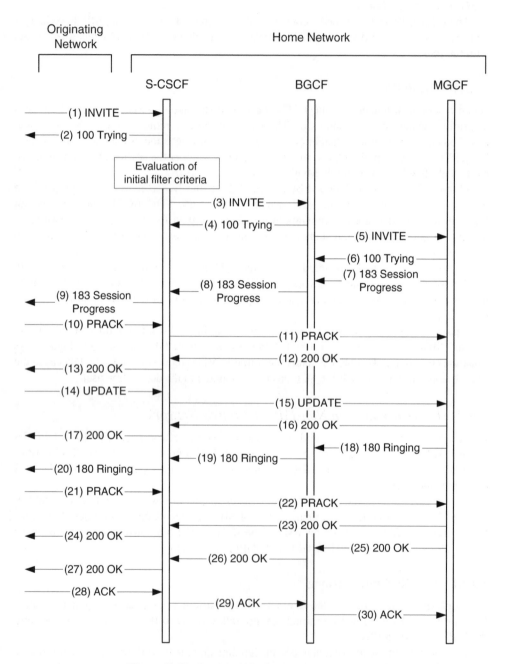

Figure 5.66: Session handled by a local MGCF

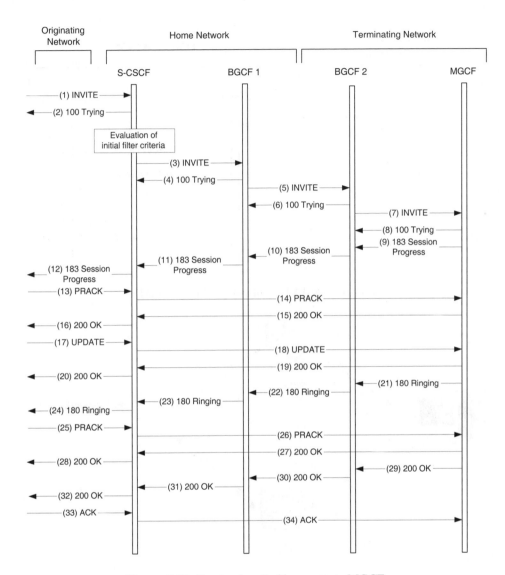

Figure 5.67: Session handled by a remote MGCF

network always receive and send SIP messages using IPv6 (or IPv4). Nevertheless, proxies at the edge of a network may receive a SIP message over IPv6 whose next hop is an IPv4 host or a SIP message over IPv4 whose next hop is an IPv6 host. In this case the proxy handling the SIP message `Record-Routes`, so that it remains in the path of all subsequent requests and responses related to that dialog. This way, it can relay messages from IPv4 to IPv6 and from IPv6 to IPv4.

Proxies at the edge of a domain need to have both IPv4 and IPv6 DNS entries in order to handle incoming sessions. IPv4 hosts that send a message to the domain use the proxy IPv4 DNS entry while IPv6 hosts use the IPv6 entry. The proxy decides whether or not it has to `Record-Route` based on the IP version of the incoming datagram.

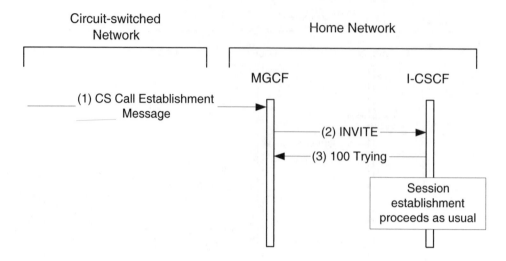

Figure 5.68: PSTN call terminating at the IMS

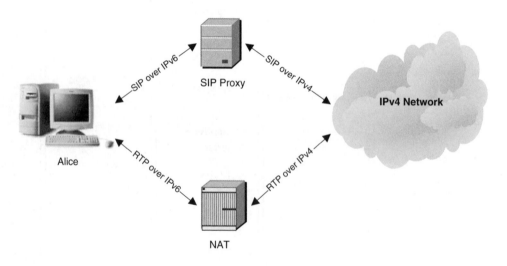

Figure 5.69: IPv4–IPv6 interworking

IPv4–IPv6 interworking at the media level is performed by media intermediaries called NATs. User agents need to somehow obtain the address of the NAT to place it in their offers and answers. In the example in Figure 5.70, Alice uses IPv6 while Bob uses IPv4. Alice sends an INVITE with an offer to Bob to establish a session. Nevertheless, Alice cannot use her IPv6 address in her offer, because Bob would not understand it; she needs to use the IPv4 address of the NAT (192.0.10.1). There are many mechanisms for obtaining the address of the NAT. ICE (specified in the Internet-Draft "Interactive Connectivity Establishment (ICE): A Methodology for Network Address Translator (NAT) Traversal for Multimedia Session

Establishment Protocols" [208]) describes a framework that uses protocols, such as STUN (specified in RFC 3489 [218]) and TURN (specified in the Internet-Draft "Traversal Using Relay NAT (TURN)" [211]), to obtain it. In addition to ICE, user agents may use the ANAT SDP grouping semantics (defined in RFC 4091 [73]) to place both IPv4 and IPv6 addresses in an offer (or an answer). This way, if the destination happens to understand IPv6, the NAT is not used and the media flows directly between both endpoints.

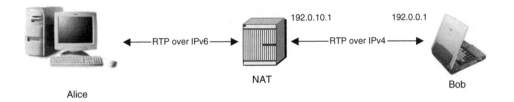

192.0.10.1   192.0.0.1

←—RTP over IPv6—→   ←—RTP over IPv4—→

NAT

Alice   Bob

Figure 5.70: IPv4–IPv6 NAT

Another way of obtaining the NAT's address is to use session policies (specified in the Internet-Draft "A Delivery Mechanism for Session-Specific Session Initiation Protocol (SIP) Session Policies" [121]). A proxy in the path informs user agents of the address of the NAT. This method could be easily adopted by the IMS to perform IPv4–IPv6 translations by getting one of the CSCFs in the signaling path to somehow control the NAT providing services to IMS terminals.

5.10 Emergency Sessions

Due to national or international regulatory requirements, most of the cellular networks in the world have to fulfill a set of requirements related to the establishment of emergency calls. The IMS is not an exception, as it is a part of a cellular network.

The requirements for support of emergency calls vary from country to country. For instance, in many European countries it is required that the user be able to make a call to the emergency center when the terminal is not even provided with a smart card. In other words, the network has to be able to route an emergency call to the emergency center even when the user is not a subscriber of the network.

In other countries the network has to provide accurate geographical location information of the user making an emergency call.

When 3GPP was specifying the first phase of the IMS (included in Release 5 of the 3GPP specifications) a number of problems were encountered. For example, as the IMS runs over different IP-CANs (IP Connectivity Access Networks) and as it might be possible to establish an emergency call when the IMS terminal is not provided with a smart card, the IP-CAN would have to provide anonymous access. However, some IP-CANs, such as GPRS, were designed with the idea of providing packet services, such as email or web browsing, which do not require anonymous access. As a consequence, in order to support emergency sessions over the IMS, 3GPP required large modifications to the GPRS to allow GPRS usage without authentication of the user.

Another interesting problem is derived from the fact that different countries have different emergency call numbers. For instance, in most European countries the emergency number is 112, in the USA it is 911, in Japan 119, etc. The problem arises when a user is roaming to a country different from the home network country. If the P-CSCF is located in the home network and the user places a local emergency call by dialing the emergency number in the visited country, the home network will not be able to recognize the number as an emergency number and, most likely, the emergency call will fail.

All of these problems can be properly addressed, but they require time. When 3GPP designed Release 5, it was felt that solving all of these problems would take some time and that Release 5 of the 3GPP specifications would be substantially delayed. However, at that time, emergency sessions were already supported over the circuit-switched domain. So, the decision was taken that Release 5 of the 3GPP specifications would not provide support for emergency calls over the IMS; instead, IMS terminals that support audio capabilities would need to support circuit-switched calls and, when users dial an emergency number, the terminal would use the circuit-switched domain to place the call. As a corollary, if an IMS terminal implements audio capabilities, the terminal would need to implement the circuit-switched part of the 3GPP specifications to provide emergency calls.

We just said that if a user dials a collection of digits that the IMS terminal recognizes as an emergency number, the terminal places the call over the circuit-switched domain. However, it may happen that the user is roaming in a country where the emergency call number is different from those known by the IMS terminal. In this case, when the user dials the emergency number, the IMS terminal may place the call over the IMS, thus sending a SIP INVITE request to the P-CSCF. The P-CSCF is configured with a list of emergency numbers in the countries of each of the roaming parties, so the P-CSCF is able to detect, even if located at home, that the user is trying to make an emergency call. But since there is no support for routing emergency calls in the IMS and as the INVITE request contains an attempt to make an emergency call, the P-CSCF returns a 380 (Alternative Service) response which includes an XML body instructing the IMS terminal to use the circuit-switched network to place the call.

3GPP is working toward providing support for emergency calls over the IMS. However, the solution has yet to be designed. It is foreseen that future versions of 3GPP (Releases 6 or 7) will include support for emergency sessions over the IMS, even when the user is not authenticated (e.g., the terminal is not provided with a smart card). Meanwhile, the IETF is also working to standardize a global emergency SIP URI that will enable users to place an emergency call.

Chapter 6

AAA on the Internet

6.1 Authentication, Authorization, and Accounting

The term *AAA* has been traditionally used to refer to *Authentication, Authorization, and Accounting* activities. All of those activities are of crucial importance for the operation of an IP network, although typically they are not so visible to the end-user.

The importance of AAA functions lies in the fact that they provide the required protection and control to accessing a network. As a consequence, the administrator of the network can bill the end-user for services used. By services we are referring to any type of services related to the access of the network, such as high bandwidth, providing routing services, gateway services, etc.

Before we proceed with this chapter, let's agree on a common terminology.

Authentication: the act of verifying the identity of an entity (subject).

Authorization: the act of determining whether a requesting entity (subject) will be allowed access to a resource (object) (e.g., network access, certain amount of bandwidth, etc.).

Accounting: the act of collecting information on resource usage for the purpose of capacity planning, auditing, billing, or cost allocation.

All of these concepts are intimately linked. For instance, it is not feasible to record the usage of a resource when the entity (subject) making usage of the resource (object) is not yet known. Therefore, in order to account for the usage of a resource the entity has to be authenticated. Once the subject is authenticated, it can be authorized to access the resource. Here, we are speaking generically. A resource could be access to a network, a radio resource, or access to a conference bridge.

The rest of this chapter describes the Internet architecture needed to provide the network functions of AAA. We will learn about the protocols that the IETF has developed to provide the mentioned functions.

6.2 AAA Framework on the Internet

At the beginning of 1997 the IETF defined the Remote Authentication Dial In User Service protocol (RADIUS, RFC 2058 [193]) as the protocol to perform AAA functions on the

Internet. The IETF revised the protocol in mid-1997 in RFC 2138 [194] and again in 2000 in RFC 2865 [195].

RADIUS offers a Network Access Server (NAS) the possibility of requesting authentication and authorization to a centralized RADIUS server. A typical example of the usage of RADIUS is shown in Figure 6.1. A user has established an agreement to access the Internet with an operator that provides a collection of dial-up access servers. A computer equipped with a modem dials up a Network Access Server. A circuit-switched connection is established between the computer (actually, the modem in the computer) and the Network Access Server. The Network Access Server does not contain a list of users who can access the network, since there may be a large collection of servers that are geographically widely spread and it would not be feasible to manage the list in all access servers. Instead, the Network Access Server is configured to request authentication and authorization from an AAA server, using an AAA protocol like RADIUS. The AAA server contains all the data needed to authenticate and then authorize the user (e.g., a password). Once the user is authenticated and authorized the user can get access to the network. The Network Access Server will be providing accounting information reports to the AAA server, so that the network operator can appropriately bill the user.

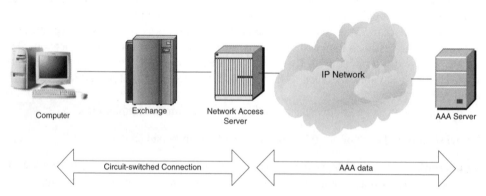

Figure 6.1: AAA functions in a dial-up scenario

The RADIUS protocol performs relatively well in small-scale configurations and for the particular application that it was designed for; that is, a user dials into a dial-up server, the dial-up server requests authentication and authorization from an AAA server. RADIUS offers problems in large environments where congestion and lost data can appear. RADIUS runs over UDP and, therefore, lacks congestion control. RADIUS lacks some functionality that is required in certain applications or networks, such as the ability of the AAA server to send an unsolicited message to the access server.

For all of these reasons the IETF has come up with an improved version of RADIUS, named Diameter (the IETF specified the Diameter base protocol in RFC 3588 [60] at the end of 2003). The IMS has selected the modern Diameter as the protocol to perform AAA functions. As the IMS relies on Diameter rather than RADIUS, we will focus in this chapter on Diameter.

6.3 The Diameter Protocol

Diameter is specified as a base protocol and a set of Diameter applications that complement the base protocol functionality. The base protocol contains the basic functionality and is implemented in all Diameter nodes, independently of any specific application. Applications are extensions to the basic functionality that are tailored for a particular usage of Diameter in a particular environment. For instance, there is an application tailored for Network Access Server configurations, another for Mobile IPv4, another for Credit Control, and even one for SIP. Applications are developed as extensions, so new applications are developed when needed. Figure 6.2 depicts the relation between the Diameter base protocol and a few applications.

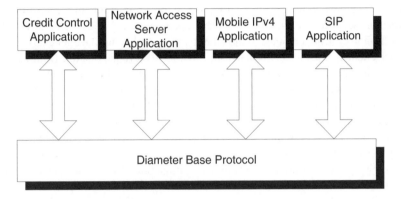

Figure 6.2: Diameter base protocol and applications

Diameter runs over a reliable transport that offers congestion control (e.g., TCP, SCTP). In particular, Diameter does not run over UDP. Unlike RADIUS, lost Diameter messages are retransmitted at each hop. Diameter provides an application-level heartbeat message that monitors the status of the connection and allows recovery in failure situations. Diameter also allows accounting messages to be routed to a different server than authentication/authorization messages (this is actually the case in the IMS).

The Diameter base protocol defines different functional entities for the purpose of performing AAA functions. These are as follows.

Diameter client: a functional entity, typically located at the edge of the network, which performs access control. Examples of Diameter clients are Network Access Servers and, in Mobile IP, mobility agents (Foreign Agents).

Diameter server: a functional entity that handles authentication, authorization, and accounting requests for a particular realm.

Proxy: a functional entity that, in addition to forwarding Diameter messages, makes policy decisions relating to resource usage and provisioning. A proxy may modify messages to implement policy decisions, such as controlling resource usage, providing admission control, and provisioning.

Relay: a functional entity that forwards Diameter messages, based on routing-related information and realm-routing table entries. A relay is typically transparent. It can modify Diameter messages only by inserting or removing routing-related data, but cannot modify other data.

Redirect agent: a functional entity that refers clients to servers and allows them to communicate directly.

Translation agent: a functional entity that performs protocol translation between Diameter and other AAA protocols, such as RADIUS.

Diameter node: a functional entity that implements the Diameter protocol and acts either as a Diameter client, Diameter server, relay, redirect agent, or translation agent.

Diameter is a peer-to-peer protocol, rather than the common client/server protocol. This means that, unlike protocols that follow the client/server model, in Diameter any of the peers can asynchronously send a request to the other peer. Note that, unlike client/server protocols, a Diameter client is *not* the functional entity that sends a request and a Diameter server is *not* the functional entity that sends an answer to the request. Instead, a Diameter client is a functional entity that performs access control, whereas a Diameter server is the functional entity that performs authentication and authorization. In Diameter, both a Diameter client and a Diameter server can send or receive requests and responses.

Diameter messages are either requests or answers. A request is answered by a single answer. Except for rare occasions, Diameter requests are always answered, so the sender of the request always gets accurate information about the fate of the request and, in the case of error, the sender can recover easily. Diameter is a binary-encoded protocol.

6.3.1   Diameter Sessions

We are used to using the term *session* in the context of SIP/SDP and with the meaning of *multimedia session*. According to SDP (RFC 2327 [115]), a *multimedia session* is:

> "a set of multimedia senders and receivers and the data streams flowing from senders to receivers. A multimedia conference is an example of a multimedia session."

Typically, a multimedia session in SIP is delimited by INVITE and BYE transactions.

Diameter also introduces the same term with a broader meaning, and care must be taken to avoid confusion between both terms. According to the Diameter base protocol (RFC 3588 [60]) a *Diameter session* is:

> "a related progression of events devoted to a particular activity."

For instance, in the context of a user dialing up a Network Access Server the session is composed of all the Diameter messages exchanged between the Network Access Server and the Diameter server from the moment the user dials until the connection is dropped. In the case of the IMS a Diameter session might be composed of all the messages exchanged between the S-CSCF (acting as a Diameter client) and the HSS (acting as a Diameter server) from the time the user registers in the IMS until the user is no longer registered.

Whenever we use the term *session* in the context of Diameter, we refer to a *Diameter session*, unless otherwise explicitly referred to as a *multimedia session*.

6.3.2 The Format of a Diameter Message

Figure 6.3 shows the format of a Diameter message. A Diameter message consists of a 20-octet header and a number of *Attribute Value Pairs (AVPs)*. The length of the header is fixed; it is always present in any Diameter message. The number of AVPs is variable, as it depends on the actual Diameter message. An AVP is a container of data (typically authentication, authorization, or accounting data).

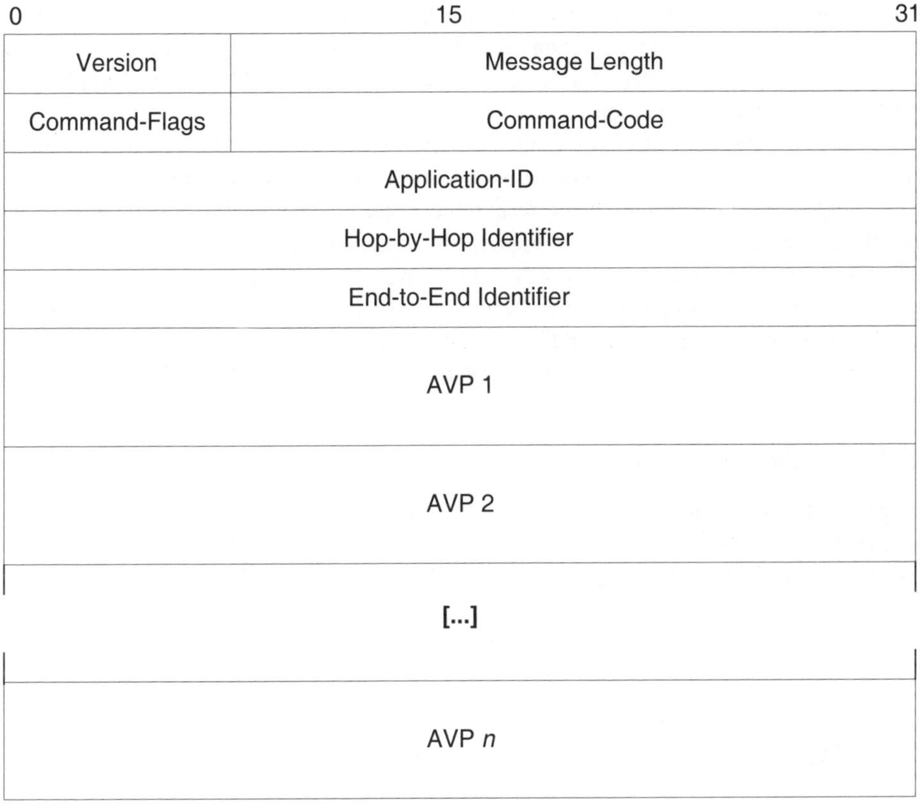

0	15	31

Version	Message Length	
Command-Flags	Command-Code	
Application-ID		
Hop-by-Hop Identifier		
End-to-End Identifier		
AVP 1		
AVP 2		
[...]		
AVP *n*		

Figure 6.3: Format of a Diameter message

The Diameter header is split into *fields*. According to Figure 6.3 a Diameter header starts with the Version field. For the time being, there is only version 1. A Message-Length field containing the length of the Diameter message including all the headers and AVPs follows in the Diameter header.

The Command-Flags field indicate:

- whether the message is a request or an answer;

- whether the message is proxiable or not;

- whether the message contains a protocol error according to the format of a Diameter message;

- whether the message is a potentially retransmitted message.

The Command-Code field identifies the actual command (i.e., the actual request or answer). Requests and answers share the same Command-Code address space. A flag present in the Command-Flags field indicates whether the message is a request or an answer.

The Application-ID field identifies the Diameter application that is sending the message. For instance, the Application-ID field can identify the application as the Diameter base protocol, a Network Access Server application, a Mobile IPv4 application, or any other Diameter application.

The Hop-by-Hop Identifier field contains a value that each hop sets when sending a request. The answer has the same Hop-by-Hop Identifier, so a Diameter node can easily correlate the answer with the corresponding request. Each Diameter node that treats a Diameter request changes the value of the Hop-by-Hop Identifier.

The sender of the request sets the value of the End-to-End Identifier field, which is a static value that does not change when Diameter nodes proxy the request. Together with the origin's host identity the End-to-End Identifier is used to detect duplicate requests. The Diameter node that generates an answer keeps the same value that was found in the request.

6.3.3 Attribute Value Pairs

Diameter messages, like RADIUS messages, transport a collection of Attribute Value Pairs (AVPs). An AVP is a container of data. Figure 6.4 depicts the structure of an AVP. Each AVP contains an AVP Code, Flags, an AVP Length, an optional Vendor-ID, and Data.

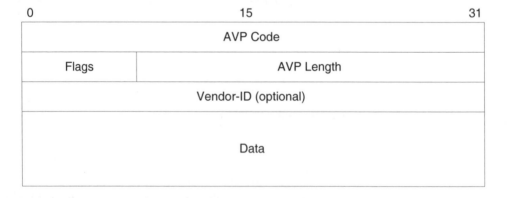

Figure 6.4: Structure of the AVP

The AVP Code in conjunction with the Vendor-ID field, if present, uniquely identify the attribute. The absence of a Vendor-ID field or a Vendor-ID field with a value set to zero indicates a standard AVP specified in an IETF specification. AVP code numbers ranging from 1 to 255 identify attributes imported from or already defined by RADIUS. AVP numbers 256 and higher identify Diameter-specific attributes.

The `Flags` field indicates:

- the need for encryption to guarantee end-to-end security;

- whether support for the AVP is mandatory or optional. If the sender indicates that support for the AVP is mandatory and the receiver does not understand the AVP the Diameter request is rejected;

- whether the optional `Vendor-ID` field is present or not.

The `AVP Length` indicates the length of the AVP, including the `AVP Code`, `AVP Length`, `Flags`, `Vendor-ID` (if present), and the `Data` field.

The `Data` field contains some data specific to the attribute. The field has a length of zero or more octets. The length of the data is derived from the `AVP Length` field.

The Diameter base protocol specifies a number of formats of the `Data` field: *OctetString*, *Integer32*, *Integer64*, *Unsigned32*, *Unsigned64*, *Float32*, *Float64*, and *Grouped*. Most of them are self-explanatory. A Grouped AVP is an AVP whose `Data` field is, in turn, a sequence of other AVPs.

Diameter allows applications to derive AVP data formats. The base protocol already defines a few derived AVPs, the most important ones being as follows.

- *Address* to convey an IPv4 or IPv6 address.

- *Time* to represent the date and time.

- *UTF8String* to represent a UTF-8 encoded string.

- *DiameterIdentity* to convey the fully qualified domain name of a Diameter node.

- *DiameterURI* to convey an AAA or AAAS URI.

- *Enumerated*, a numerical value that represents some semantics.

6.3.4   The AAA and AAAS URIs

AAA protocols are able to use an *aaa* or an *aaas* URI to identify AAA resources. The *aaas* URI indicates that transport security must be used. The syntax of these URIs is:

```
"aaa://" FQDN [ port ] [ transport ] [ protocol ]

"aaas://" FQDN [ port ] [ transport ] [ protocol ]

    port          = ":" 1*DIGIT
    transport     = ";transport=" transport-protocol
    protocol      = ";protocol=" aaa-protocol

    transport-protocol = ( "tcp" / "sctp" / "udp" )
    aaa-protocol       = ( "diameter" / "radius" /
                           "tacacs+" )
```

where FQDN is a Fully Qualified Domain Name. The URIs might be appended by an optional port number, an optional `transport` protocol, or an optional `protocol` to access the AAA resource. If the port number is not present the default Diameter port number (3868) is assumed. If the `transport` parameter is not present, then the default SCTP protocol is assumed. If the `protocol` parameter is not present, Diameter is assumed. The reader should note that the *aaa* and *aaas* URIs are able to accommodate Diameter, RADIUS, and other protocols.

Examples of *aaa* or *aaas* URIs include:

```
aaa://server.home1.net
aaas://server.home1.net
aaa://server.home1.net:8868
aaa://server.home1.net;transport=tcp;protocol=diameter
```

6.3.5 Diameter Base Protocol Commands

We have seen that Diameter messages are either requests or answers. A request and its corresponding answer are identified by a common `Command-Code` in the Diameter header. The `Command-Code` is a number that indicates the action the Diameter server needs to take. As a request and its corresponding answer share the same `Command-Code` address space, we need to refer to the `Command-Flags` to find out if the command is a request or an answer.

The Diameter base protocol (RFC 3588 [60]) specifies an initial number of command codes. An application is able to extend these basic commands and add new application-specific ones. Table 6.1 shows the initial set of requests and answers defined in the Diameter base protocol.

Table 6.1: Diameter base commands

Command-Name	Abbreviation	Command-Code
Abort-Session-Request	ASR	274
Abort-Session-Answer	ASA	274
Accounting-Request	ACR	271
Accounting-Answer	ACA	271
Capabilities-Exchange-Request	CER	275
Capabilities-Exchange-Answer	CEA	275
Device-Watchdog-Request	DWR	280
Device-Watchdog-Answer	DWA	280
Disconnect-Peer-Request	DPR	282
Disconnect-Peer-Answer	DPA	282
Re-Auth-Request	RAR	258
Re-Auth-Answer	RAA	258
Session-Termination-Request	STR	275
Session-Termination-Answer	STA	275

Typically, every message is abbreviated by its initials. For instance, the Abort-Session-Request message is typically referred to as the ASR message. Let us take a look at the semantics of the main Diameter messages.

6.3.5.1 Abort Session Request and Answer (ASR, ASA)

It might be necessary for a Diameter server to stop the service provided to the user (e.g., network access), because, say, there are new reasons that were not anticipated when the session was authorized. Among others, lack of credit, security reasons, or just an administrative order may be reasons to abort an ongoing Diameter session.

When a Diameter server decides to instruct the Diameter client to stop providing a service the Diameter server sends an Abort-Session-Request (ASR) message to the Diameter client. The Diameter client reports the execution of the command in an Abort-Session-Answer (ASA).

6.3.5.2 Accounting Request and Answer (ACR, ACA)

A Diameter node may need to report accounting events to a Diameter server that provides accounting services. Diameter provides the Accounting-Request (ACR) command, whereby a Diameter client reports usages of the service to a Diameter server. The command includes information that helps the Diameter server to record the one-time event that generated the command or the beginning or end of a service (e.g., access to a network).

6.3.5.3 Capabilities Exchange Request and Answer (CER, CEA)

The first Diameter messages that two Diameter nodes exchange, once the transport connection is established, are the Capabilities-Exchange-Request (CER) and the Capabilities-Exchange-Answer (CEA). The messages carry the node's identity and its capabilities (protocol version, the supported Diameter applications, the supported security mechanisms, etc.).

6.3.5.4 Device Watchdog Request and Answer (DWR, DWA)

It is essential for Diameter to detect transport and application-layer failures as soon as possible, in order to take corrective action. The mechanism that Diameter provides to detect these failures is based on an application-layer watchdog. During periods of traffic between two Diameter nodes, if a node sends a request and no answer is received within a certain time period, that is enough to detect a failure either at the transport or application layer. However, in the absence of regular traffic it is not possible to detect such a potential failure. Diameter solves the problem by probing the transport and application layer by means of a Diameter node sending a DWR message. The absence of the receipt of a DWA message is enough to conclude that a failure has occurred.

6.3.5.5 Disconnect Peer Request and Answer (DPR, DPA)

A Diameter node that has established a transport connection with a peer Diameter node may want to close the transport connection, (e.g., if it does not foresee more traffic toward the peer node). In this case the Diameter node sends a Disconnect-Peer-Request (DPR) to the peer node to indicate the imminent disconnection of the transport protocol. The DPR message also conveys the semantics of requesting the peer not to re-establish the connection unless it is essential (e.g., to forward a message).

6.3.5.6   Re-Authentication Request and Answer (RAR, RAA)

At any time, but especially in sessions that last a long time, the Diameter server may request a re-authentication of the user, just to confirm that there is no possible fraud. A Diameter server that wants to re-authenticate a user sends a Re-Auth-Request message to a Diameter client. The client responds with a Re-Auth-Answer message.

6.3.5.7   Session Termination Request and Answer (STR, STA)

A Diameter client reports to the Diameter server that a user is no longer making use of the service by sending a Session-Termination-Request (STR) message. The Diameter server answers with a Session-Termination-Answer (STA) message.

For instance, if the dial-up server reports that the dial-up connection has dropped, then the Diameter client sends the STR message to the Diameter server.

6.3.6   Diameter Base Protocol AVPs

Each request and answer defines which Attribute Value Pairs (AVPs) are present in the message. Some AVPs may be optional in a particular request or answer, others may be mandatory.

The presence or absence of standard defined AVPs are dependent on the actual request or response. For instance, the `Authorization-Lifetime` AVP indicates the period of time for which the authorization of a user is valid. Once the authorization lifetime is reached the Diameter client will re-authenticate the user. This AVP is only present in authorization answer messages.

The list of Diameter base protocol AVPs is quite large; we refer the reader to RFC 3588 [60] to get the complete list. We describe in the following paragraphs a few important AVPs that are very often found in Diameter messages.

The `User-Name` AVP indicates the user name under which the user is known in the realm. The `User-Name` AVP follows the format of the Network Access Identifier (NAI) specified in RFC 2486 [45]. An NAI has either the format of `username` or `username@realm`. In the IMS the `User-Name` AVP carries the Private User Identity.

Every Diameter answer message carries a `Result-Code` AVP. The value of the `Result-Code` AVP indicates whether the request was successfully completed or not and gives a list of possible values of the AVP that depend on the actual request and answer.

The `Origin-Host` AVP conveys the fully qualified domain name of the Diameter node that generates the request. The node also includes the realm of the Diameter node in the `Origin-Realm` AVP.

The `Destination-Host` AVP indicates the fully qualified domain name of the Diameter server where the user name is defined. Sometimes the user does not know the actual host name of the server, but does know the administrative domain where the user name is valid. In that case the `Destination-Realm` AVP is used.

Diameter request messages can be proxiable or non-proxiable. There is a flag in the Diameter header that indicates whether the message is proxiable or not. Proxiable messages can be routed through proxies toward the destination realm. Therefore, a proxiable request always contains a `Destination-Realm` AVP. Non-proxiable messages are just routed to the next hop and are never forwarded.

Another interesting AVP is the `Session-ID` AVP. It contains a global identifier of the session. All messages pertaining to the same session contain the same `Session-ID` AVP value. Section 6.3.1 describes the concept of session in the context of Diameter.

The Vendor-Specific-Application-Id is a grouped AVP that conveys the identity of a Diameter application that is vendor-specific (e.g., not standardized in the IETF). The Vendor-Specific-Application-Id contains either an Auth-Application-Id or an Acct-Application-id AVP, although only one of them can be present at the same time. The former identifies the authentication and authorization portion of the application, whereas the latter identifies the accounting portion of the application. The Vendor-Specific-Application-Id AVP can also contain a Vendor-Id AVP that identifies the vendor.

The Auth-Session-State AVP indicates whether the Diameter client wants to maintain a state for a particular Diameter session. The Diameter client uses this AVP as a request, and the Diameter server replies with the same AVP in the answer.

The Proxy-Info is a grouped AVP that contains a Proxy-Host and a Proxy-State AVP; it may also contain any other AVP. It allows stateless agents to include a state in a request. The corresponding answer will contain the same AVP, so the stateless agent can retrieve the state information and proceed with the Diameter session. The Proxy-Host AVP contains the host name of the proxy that inserted the information. The Proxy-State AVP contains the opaque data that are written and then read by the proxy itself.

A relay or proxy agent appends a Route-Record AVP to all the requests. The Route-Record AVP contains the identity of the Diameter node the request was received from. This allows the detection of loops. It also allows the Diameter server to verify and authorize the path a request took.

Chapter 7

AAA in the IMS

Authentication and authorization are generally linked in the IMS. In contrast, accounting is a separate function executed by different nodes. This was the reason we decided to separate the description of authentication and authorization from the description of accounting. Section 7.1 discusses authentication and authorization, and Section 7.4 discusses accounting.

7.1 Authentication and Authorization in the IMS

Figure 7.1 shows the IMS architecture for the purpose of performing authentication and authorization functions. There are three interfaces over which authentication and authorization actions are performed (namely the *Cx*, *Dx*, and *Sh* interfaces).

The *Cx* interface is specified between a Home Subscriber Server (HSS) and either an I-CSCF or an S-CSCF. When more than a single HSS is present in a home network there is a need for a Subscription Locator Function (SLF) to help the I-CSCF or S-CSCF to determine which HSS stores the data for a certain user. The *Dx* interface connects an I-CSCF or S-CSCF to an SLF running in Diameter redirect mode.

The *Sh* interface is specified between an HSS and either a SIP Application Server or an OSA Service Capability Server (for a complete description of the different types of Application Servers in the IMS see Section 5.8.2).

In all of these interfaces the protocol used between any two nodes is Diameter (specified in RFC 3588 [60]) with an IMS-specific tailored application. Such a Diameter application defines new Diameter command codes and new Attribute Value Pairs (AVPs).

7.2 The *Cx* and *Dx* Interfaces

The *Cx* interface is specified between an I-CSCF and an HSS or between an S-CSCF and an HSS. Similarly, the *Dx* interface is specified between an I-CSCF and an SLF or between an S-CSCF and an SLF. The only difference between these interfaces is that the SLF implements the functionality of a Diameter redirect agent, whereas the HSS acts as a Diameter server. In either case, both the I-CSCFs and S-CSCFs act as Diameter clients.

In networks where there is more than a single HSS, Diameter clients (S-CSCF, I-CSCF) need to contact the SLF to find out which of the several HSSs in the network stores the user information for the user identified by the Public User Identity. The Diameter command the S-CSCF or the I-CSCF sends is the same, no matter whether the message is addressed to the

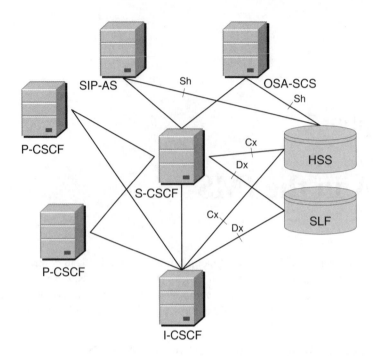

Figure 7.1: Architecture for authentication and authorization in the IMS

SLF or the HSS. The SLF acts as an enhanced Diameter redirect agent and contains a table that maps Public User Identities to the address of the HSS that contains the user information. The SLF then includes a `Redirect-Host` AVP in the answer. The `Redirect-Host` AVP contains the address of the particular HSS that the I-CSCF or S-CSCF needs to contact. The I-CSCF or S-CSCF then forwards the Diameter message addressed to that HSS.

Because Diameter messages over the *Cx* and *Dx* interfaces are the same, the *Dx* interface can be considered as transparent to describe the interactions over the *Cx* interface. In this chapter we typically refer to the *Cx* interface and the HSS, but the description applies in a similar manner to the *Dx* interface and the SLF.

Note that the P-CSCF implements neither the *Cx* nor the *Dx* interface.

For a particular user the I-CSCF and S-CSCF use the *Cx* and *Dx* interfaces to perform the following functions.

- To locate an already allocated S-CSCF to the user.

- To download the authentication vectors of the user. These vectors are stored in the HSS.

- To authorize the user to roam in a visited network.

- To record in the HSS the address of the S-CSCF allocated to the user.

- To inform the HSS about the registration state of a user's identity.

- To download from the HSS the user profile that includes the filter criteria.

- To push the user profile from the HSS to the S-CSCF when the user profile has changed.

- To provide the I-CSCF with the necessary information to select an S-CSCF.

The *Cx* and *Dx* interfaces implement a vendor-specific Diameter application called the *Diameter Application for the Cx interface*. This application is specified in 3GPP TS 29.228 [21] and 3GPP TS 29.229 [12]. The Diameter Application for the *Cx* interface is not standardized in the IETF, but the IETF has authorized, for 3GPP Release 5, this application (as specified in RFC 3589 [156]).

The Diameter Application for the *Cx* interface has formed the basis of a generic AAA application for SIP servers (documented in the Internet-Draft "Diameter SIP Application" [106]). At the time of writing, the IETF is still working on this Diameter application. It is foreseen that in future 3GPP releases the vendor-specific Diameter application for the *Cx* interface will migrate to the IETF standardized Diameter SIP Application.

7.2.1 Command Codes Defined in the Diameter Application for the Cx Interface

As previously mentioned, the I-CSCF and S-CSCF perform a number of functions over the *Cx* and *Dx* interfaces. In order to perform these functions the Diameter Application for the *Cx* interface has defined a number of new commands (requests and answers). Table 7.1 lists the new commands specified in the Diameter Application for the *Cx* interface.

Table 7.1: List of commands defined by the Diameter Application for the *Cx* interface

Command-Name	Abbreviation	Command-Code
User-Authorization-Request	UAR	300
User-Authorization-Answer	UAA	300
Server-Assignment-Request	SAR	301
Server-Assignment-Answer	SAA	301
Location-Info-Request	LIR	302
Location-Info-Answer	LIA	302
Multimedia-Auth-Request	MAR	303
Multimedia-Auth-Answer	MAA	303
Registration-Termination-Request	RTR	304
Registration-Termination-Answer	RTA	304
Push-Profile-Request	PPR	305
Push-Profile-Answer	PPA	305

7.2.1.1 User Authorization Request and Answer (UAR, UAA)

An I-CSCF sends a User-Authorization-Request (UAR) message when the I-CSCF receives a SIP REGISTER request from an IMS terminal. There are a few reasons the I-CSCF sends the Diameter UAR message to the HSS.

- The HSS first filters the Public User Identity contained in the SIP REGISTER request. For instance, the HSS verifies that the Public User Identity is allocated to a legitimate subscriber of the home network and that the user is a regular non-blocked user (e.g., due to lack of payment or any other reason).

- The HSS also verifies that the home network has a roaming agreement with the network where the P-CSCF is operating. This allows the P-CSCF network to exchange charging records with the home network.

- The I-CSCF also needs to determine whether there is an S-CSCF already allocated to the Public User Identity under registration, before the I-CSCF forwards the SIP REGISTER request to that S-CSCF. If there is not an S-CSCF allocated to the Public User Identity, the I-CSCF will receive the set of capabilities required in the S-CSCF, so that the I-CSCF is able to select an S-CSCF with those capabilities.

- The SIP REGISTER request typically carries the Private User Identity and the Public User Identity of the user. The HSS checks that the Public User Identity can use that Private User Identity for authentication purposes.

Figure 7.2 depicts a typical registration flow. When the I-CSCF receives a SIP REGISTER request (2) it sends the Diameter UAR message to the HSS (3). The HSS sends a Diameter User-Authorization-Answer (UAA) message (4) and then the I-CSCF proceeds with the registration process. The operation is also repeated in (13) and (14), since the I-CSCF does not keep a state in between these registrations. Furthermore, due to DNS load-balancing mechanisms the I-CSCF that receives the first SIP REGISTER request (2) might not be the same I-CSCF that receives the second SIP REGISTER request (12).

The HSS includes a `Result-Code` AVP in the UAA message that helps the I-CSCF to determine whether to continue with the registration or to reject it. In case the registration is authorized the UAA message contains AVPs that help the I-CSCF to determine whether there is an S-CSCF already allocated to the user or whether the I-CSCF has to select a new S-CSCF.

7.2.1.2   Multimedia Auth. Request and Answer (MAR, MAA)

Figure 7.2 is also valid for the purpose of describing the Diameter Multimedia-Auth-Request (MAR) and Multimedia-Auth-Answer (MAA) messages. When the S-CSCF receives an initial SIP REGISTER request, it has to authenticate the IMS user. However, in an initial registration the S-CSCF does not have authentication vectors to authenticate the user. These vectors are stored in the HSS. The S-CSCF sends a Diameter MAR message to the HSS with the purpose of retrieving the authentication vectors. Additionally, the S-CSCF records its own SIP URI in the user-related data stored in the HSS, so that other CSCFs (e.g., I-CSCFs) or ASs are able to get the URI of the S-CSCF allocated to that particular user by interrogating the HSS.

7.2.1.3   Server Assignment Request and Answer (SAR, SAA)

Figure 7.2 also depicts the use of the SAR and SAA messages. When the S-CSCF eventually authenticates the user (actually, the Private User Identity) the Public User Identity is registered and bound to a contact address. At that time the S-CSCF sends the SAR message to the HSS for the purpose of informing the HSS that the user is currently registered in that S-CSCF. The S-CSCF also requests the user profile associated with that user. The HSS attaches the user profile in the SAA message.

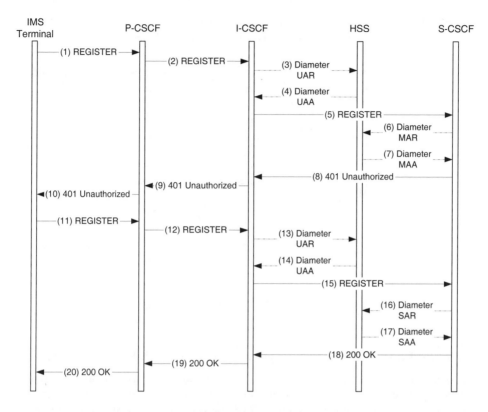

Figure 7.2: UAR/UAA, MAR/MAA, and SAR/SAA messages during registration

The S-CSCF also sends a SAR message to the HSS when the user is no longer registered in that S-CSCF, so that the HSS is aware of the user registration status. The S-CSCF might also request the HSS to continue to be the S-CSCF allocated to the user, even when the user is not registered. The final word belongs to the HSS, because it authorizes whether the S-CSCF keeps the S-CSCF allocation or not. Keeping the S-CSCF allocation allows the S-CSCF to keep the user profile information. Subsequent registrations would not require downloading of such information from the HSS.

7.2.1.4   Location Information Request and Answer (LIR, LIA)

An I-CSCF that receives a SIP request that does not contain a Route header field that points to the next SIP hop (S-CSCF) needs to find out which S-CSCF (if any) is allocated to the user. On receiving the SIP request the S-CSCF sends a Location-Info-Request (LIR) to the HSS. The HSS replies with a Location-Info-Answer (LIA). The LIA command indicates the SIP URI of the S-CSCF allocated to the user or, if there is no S-CSCF allocated to the user, then the HSS will include the set of capabilities that are required by the S-CSCF, so that the I-CSCF is able to select an S-CSCF for this user (similar to the selection that takes place during initial registration).

According to Figure 7.3 an I-CSCF that receives a SIP request that does not contain routing information sends a Diameter LIR message to the HSS. The HSS replies with a

Diameter LIA message that contains the address of the S-CSCF that is allocated to the user; therefore, the I-CSCF forwards the INVITE request to that S-CSCF.

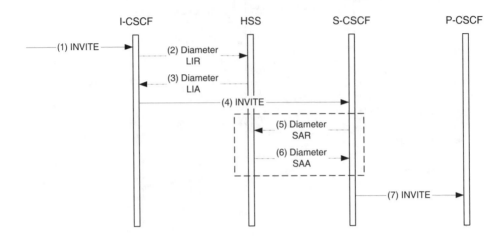

Figure 7.3: LIR/LIA and SAR/SAA messages

7.2.1.5   Registration Termination Request and Answer (RTR, RTA)

Due to administrative action the operator of the home network may wish to deregister one or more registered Public User Identities allocated to a user. When this happens the HSS sends a Registration-Termination-Request (RTR) message to the S-CSCF where the user is registered.

Figure 7.4 depicts an HSS sending an RTR message to the S-CSCF with the purpose of deregistering one or more Public User Identities. The S-CSCF notifies all the subscribers of the user's reg state; in this case, the P-CSCF and the IMS terminal. In this example both the P-CSCF and the IMS terminal have subscribed to the user's reg state, so the S-CSCF notifies the P-CSCF (3) and the IMS terminal (5) and (6).

7.2.1.6   Push Profile Request and Answer (PPR, PPA)

The HSS may change the user profile at any time, such as when a new service is available to the user; this action typically requires the addition of new filter criteria to the user profile. When the user profile is updated the HSS sends a Push-Profile-Request message to the S-CSCF allocated to the user, the message containing an updated copy of the user profile. Figure 7.5 shows an example of the PPR and PPA Diameter messages. These messages are not connected to any SIP signaling.

7.2.2   AVPs Defined in the Diameter Application for the Cx Interface

The Diameter Application for the *Cx* interface defines a number of new Attribute Value Pairs. Table 7.2 lists these new attributes together with their AVP code. The Vendor-Id field of all of these AVPs is set to the value 10415, which identifies 3GPP.

The Visited-Network-Identifier AVP conveys an identifier of the network where the P-CSCF is located. An I-CSCF maps the P-Visited-Network-ID header field included

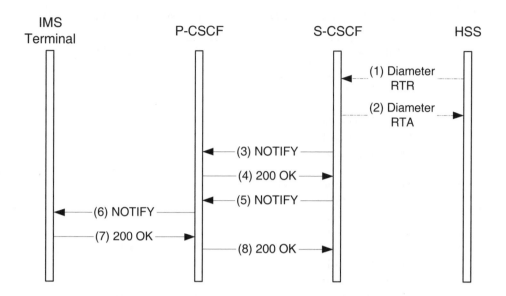

Figure 7.4: RTR/RTA messages

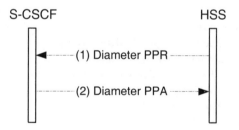

Figure 7.5: PPR/PPA messages

in a SIP REGISTER request to the `Visited-Network-Identifier` AVP. The home network is able to authorize the IMS terminal to use a P-CSCF located in that network.

The `Public-Identity` AVP carries a Public User Identity (SIP URI or TEL URI).

The `Server-Name` AVP contains the URI of the SIP server node (e.g., the S-CSCF URI).

The `Server-Capabilities` is a grouped AVP whose main purpose is to convey the required capabilities of the S-CSCF that will be serving the user. The HSS conveys these capabilities to the I-CSCF, so that the I-CSCF can select an adequate S-CSCF for that user. As this is a grouped AVP it contains, in turn, other AVPs: `Mandatory-Capability`, `Optional-Capability`, and `Server-Name`.

The `Mandatory-Capability` AVP indicates a capability that the chosen S-CSCF has to implement, whereas the `Optional-Capability` contains a capability that the S-CSCF may optionally implement. Both AVPs can be repeated a number of times to allow several capabilities to be expressed. Capabilities are represented by an integer. The home operator allocates the semantics of the capabilities to each integer, which is a valid action because the *Cx* interface is only used inside the home network. For instance, the capability of the S-CSCF

Table 7.2: AVPs defined by the Diameter Application for the *Cx* interface

Attribute name	AVP code
Visited-Network-Identifier	600
Public-Identity	601
Server-Name	602
Server-Capabilities	603
Mandatory-Capability	604
Optional-Capability	605
User-Data	606
SIP-Number-Auth-Items	607
SIP-Authentication-Scheme	608
SIP-Authenticate	609
SIP-Authorization	610
SIP-Authentication-Context	611
SIP-Auth-Data-Item	612
SIP-Item-Number	613
Server-Assignment-Type	614
Deregistration-Reason	615
Reason-Code	616
Reason-Info	617
Charging-Information	618
Primary-Event-Charging-Function-Name	619
Secondary-Event-Charging-Function-Name	620
Primary-Charging-Collection-Function-Name	621
Secondary-Charging-Collection-Function-Name	622
User-Authorization-Type	623
User-Data-Already-Available	624
Confidentiality-Key	625
Integrity-Key	626
User-Data-Request-Type	627

to execute Java code can be assigned to capability 1, the capability of the S-CSCF to run SIP CGI scripts can be assigned to capability 2, and so on.

The User-Data AVP carries the user profile. The user profile is described in detail in Section 7.2.3.

The S-CSCF indicates how many authentication vectors it wants to receive from the HSS for a particular user in the SIP-Number-Auth-Items AVP. The HSS also uses this AVP to indicate how many authentication vectors it is actually sending.

The SIP-Auth-Data-Item is a grouped AVP that contains the following AVPs: SIP-Item-Number, SIP-Authentication-Scheme, SIP-Authenticate, SIP-Authorization, SIP-Authentication-Context, Confidentiality-Key, and Integrity-Key.

When a Diameter message carries more than one SIP-Auth-Data-Item AVP and the S-CSCF has to consider the order in which to process them, then the HSS includes a sequential number in the `SIP-Item-Number` AVP that is included in each `SIP-Auth-Data-Item`.

The `SIP-Authentication-Scheme` AVP indicates the authentication scheme that is used for the authentication of SIP messages. 3GPP Release 5 only defines `Digest-AKAv1-MD5` as an authentication scheme.

The `SIP-Authenticate` AVP is used by the HSS to send the value that the S-CSCF inserts in the SIP `WWW-Authenticate` header field of a 401 Unauthorized response. When the user is authenticated the IMS terminal sends a SIP request that contains an `Authorization` header field. The value of this header is not sent to the HSS unless there is a failure of synchronization and in that case the S-CSCF copies the SIP `Authorization` header to the `SIP-Authorization` AVP.

The `SIP-Authentication-Context` is used to carry part or the complete SIP request to the S-CSCF for certain authentication mechanisms (e.g., the HTTP Digest with quality of protection set to `auth-int`, specified in RFC 2617 [102]).

`Confidentiality-Key` and `Integrity-Key` AVPs contain, respectively, the keys that the P-CSCF needs to encrypt/decrypt or protect/verify the SIP messages sent to or from the IMS terminal. The HSS sends these keys to the S-CSCF in the mentioned AVPs, the S-CSCF inserts these keys as parameters of the Digest scheme in the SIP `WWW-Authenticate` header field, and then the P-CSCF removes them.

The S-CSCF indicates in the `Server-Assignment-Type` AVP the reason for the S-CSCF contacting the HSS. Possible reasons are: the S-CSCF requires the user profile; the user has registered, re-registered, or deregistered; a timeout during registration; administrative deregistration; an authentication failure or timeout; etc.

When the HSS deregisters a user, it sends the information to the S-CSCF in a `Deregistration-Reason` AVP. The `Deregistration-Reason` is a grouped AVP that contains a `Reason-Code` AVP and, optionally, a `Reason-Info` AVP. The `Reason-Code` AVP is a numerical code that identifies the reason for network-initiated deregistration, such as a permanent termination of the IMS subscription or because a new S-CSCF has been allocated to the user. The `Reason-Info` contains a readable text string that describes the reason for deregistration.

The `Charging-Information` is a grouped AVP that conveys to the S-CSCF the AAA URIs of the Event Charging Function (ECF) and Charging Collection Function (CCF) nodes. The AAA URIs of the primary and secondary nodes are sent to the S-CSCF, and, the secondary nodes are used in case of failure of the corresponding primary nodes. The `Charging-Information` AVP contains any of the following AVPs:

- `Primary-Event-Charging-Function-Name`

- `Secondary-Event-Charging-Function-Name`

- `Primary-Charging-Collection-Function-Name`

- `Secondary-Charging-Collection-Function-Name`.

The `User-Authorization-Type` AVP indicates the type of authorization that an I-CSCF requests in a UAR message. The value can indicate an initial registration or re-registration, a deregistration, or "registration and capabilities". The I-CSCF uses the "registration and capabilities" value when the current S-CSCF allocated to the user is not reachable and the I-CSCF requests the capabilities of the S-CSCF in order to make a new S-CSCF selection.

In a SAR message the S-CSCF can also indicate to the HSS whether the S-CSCF has already got the user profile. The S-CSCF does so by including a `User-Data-Already-Available` AVP in the SAR message.

7.2.2.1 Usage of Existing AVPs

Besides the AVPs that 3GPP created to support the Diameter Application for the *Cx* interface, the requests and answers of this application also make use of existing AVPs defined in the Diameter base protocol (RFC 3588 [60]). The most important AVPs that 3GPP uses are described in Section 6.3.6. Also important is the `User-Name` AVP which in the IMS always carries the Private User Identity.

7.2.3   The User Profile

The user profile, which is stored in the HSS, contains a lot of information related to a particular user. The S-CSCF downloads the user profile when the user registers for the first time with that S-CSCF. The S-CSCF receives the user profile in a `User-Data` AVP included in a Diameter SAA message. If the user profile changes in the HSS while the user is registered to the network, then the HSS sends the updated user profile in a `User-Data` AVP included in a Diameter PPR message.

Figure 7.6 shows the structure of the user profile. A user profile is bound to a Private User Identity and to the collection of Public User Identities that are, in turn, associated with that Private User Identity.

The user profile contains a plurality of *service profiles*. Each service profile defines the service triggers that are applicable to a collection of Public User Identities. The service profile is divided into four parts: a collection of one or more *public identifications*, an optional *core network service authorization*, zero or more *initial filter criteria*, and zero or more *shared initial filter criteria*.

The public identifications included in the service profile contain the Public User Identities associated with that service profile. The service profile is applicable to all the identities listed in public identifications. Each public identification contains a tag to indicate whether the Public User Identity is barred or not. A barred Public User Identity can be used for registration purposes, but not for any other SIP traffic (such as establishing a session). A public identification contains either a SIP URI or a TEL URI.

A service profile can also contain a core network service authorization, which, in turn, contains a *subscribed media profile identifier*. The subscribed media profile identifier contains a value that identifies the set of SDP parameters that the user is authorized to request. The identifier, which is stored in the service profile, is just an integer value; the actual SDP profile is stored in the S-CSCF. The S-CSCF uses the subscribed media profile identifier to apply a particular SDP profile that helps the S-CSCF to police SDP in user-initiated requests. For instance, a user might not be authorized to use video. In this case, if the user initiates a session whose SDP contains a video stream the S-CSCF will reject the session attempt when the S-CSCF evaluates the SDP against the subscribed media profile.

The third part of the information stored in the service profile is a collection of *initial filter criteria*. These determine which SIP requests must visit a certain Application Server so that a particular service can be provided. The initial filter criteria is described in detail in Section 5.8.4.

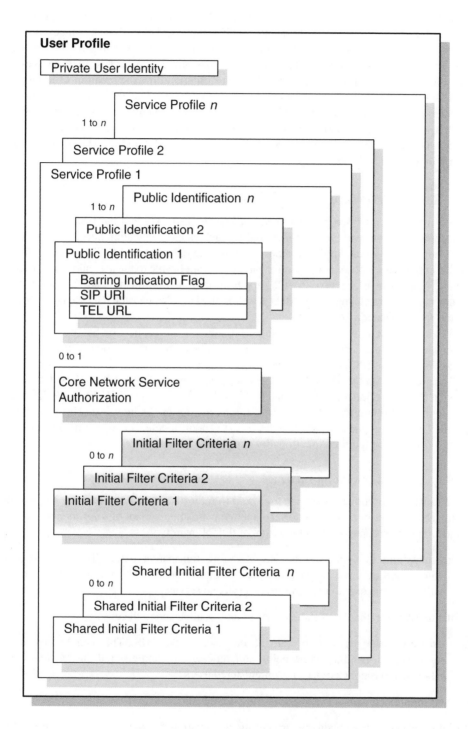

Figure 7.6: Structure of the user profile

The last part of the service profile is the *shared initial filter criteria*. This is an optional feature that requires support by both the S-CSCF and the HSS. Typically many users in a network might share a collection of initial filter criteria. It is not that optimal if every time a user registers to a S-CSCF, it downloads an initial filter criteria that has already been downloaded previously. Shared initial filter criteria allow the creation of a database of initial filter criteria that are common to a collection of users. The database is stored in both S-CSCFs and HSS. Each shared initial filter criteria is identified by a unique identifier. When a user's service profile contains one or more shared initial filter criteria, only the identifiers are downloaded to the S-CSCF; the S-CSCF has previously stored the shared initial filter criteria in its internal database.

7.3   The *Sh* Interface

The *Sh* interface is defined between a SIP AS or an OSA-SCS and the HSS. It provides a data storage and retrieval type of functionality, such as an Application Server downloading data from the HSS or an Application Server uploading data to the HSS. These data could be service execution scripts or configuration parameters applicable to the user and to a particular service. The *Sh* interface also provides a subscription and notification service, so that the AS can subscribe to changes in the data stored in the HSS. When such data change, the HSS notifies the Application Server.

The implementation of the *Sh* interface is optional in an Application Server and depends on the nature of the service provided by the Application Server: some services require interaction with the HSS, others do not.

The protocol over the *Sh* interface is Diameter (specified in RFC 3588 [60]) with the addition of a Diameter application (specified in 3GPP TS 29.328 [22] and 3GPP TS 29.329 [35]). This application is known as the *Diameter Application for the Sh interface*. This is a vendor-specific Diameter application where 3GPP is the vendor; it defines new command codes and AVPs to support the required functionality over the *Sh* interface.

7.3.0.1   User Data on the *Sh* Interface

The *Sh* interface introduces the term *user data* to refer to diverse types of data. Most of the Diameter messages over the *Sh* interface operate over some variant of *user data*. User data, in the *Sh* interface context, can refer to any of the following.

Repository data: the AS uses the HSS to store transparent data. The data are only understood by those Application Servers that implement the service. The data are different from user to user and from service to service.

Public identifiers: the list of Public User Identities allocated to the user.

IMS user state: the registration state of the user in the IMS. This can be registered, unregistered, pending while being authenticated, or unregistered but an S-CSCF is allocated to trigger services for unregistered users.

S-CSCF name: contains the address of the S-CSCF allocated to the user.

Initial filter criteria: contain the triggering information for a service. An Application Server can only get the initial filter criteria that route SIP requests to the requesting Application Server.

Location information: contains the location of the user in the circuit-switched or packet-switched domains.

User state: contains the state of the user in the circuit-switched or packet-switched domains.

Charging information: contains the addresses of the charging functions (primary and secondary event charging function or charging collection function).

7.3.1 Command Codes Defined in the Diameter Application for the Sh Interface

The *Sh* interface defines eight new Diameter messages to support the required functionality of the interface. Table 7.3 lists the new commands defined in the Diameter Application for the *Sh* interface.

Table 7.3: List of commands defined by the Diameter Application for the *Sh* interface

Command-Name	Abbreviation	Command-Code
User-Data-Request	UDR	306
User-Data-Answer	UDA	306
Profile-Update-Request	PUR	307
Profile-Update-Answer	PUA	307
Subscribe-Notifications-Request	SNR	308
Subscribe-Notifications-Answer	SNA	308
Push-Notifications-Request	PNR	309
Push-Notifications-Answer	PNA	309

7.3.1.1 User Data Request and Answer (UDR, UDA)

An Application Server sends a User-Data-Request (UDR) to the HSS to request user data for a particular user. The user data can be of a type defined over the *Sh* interface. The HSS returns the requested type of data in a Diameter User-Data-Answer (UAA) message. Figure 7.7 depicts the applicable call flow.

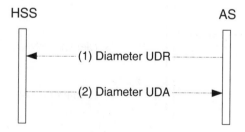

Figure 7.7: UDR/UDA messages

7.3.1.2   Profile Update Request and Answer (PUR, PUA)

An Application Server may modify repository-type user data and store them in the HSS. In doing so the Application Server sends a Profile-Update-Request (PUR) to the HSS. The HSS returns the result of the storage operation in a Diameter Profile-Update-Answer (PUA) message. Figure 7.8 depicts the applicable call flow.

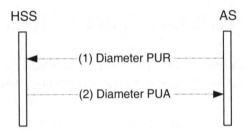

Figure 7.8: PUR/PUA messages

7.3.1.3   Subscribe Notifications Request and Answer (SNR, SNA)

An Application Server can subscribe to changes in the user data by sending a Subscribe-Notifications-Request (SNR) message to the HSS. The types of user data where notifications are allowed are: repository data, IMS user state, S-CSCF name, and initial filter criteria. The HSS informs the Application Server of the result of the subscription operation in a Subscribe-Notifications-Answer (SNA). Figure 7.9 shows the flow.

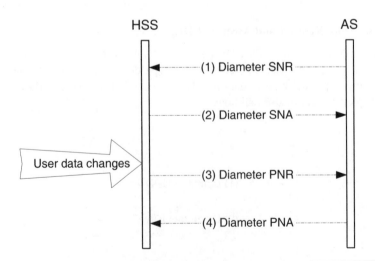

Figure 7.9: SNR/SUA and PNR/PUA messages

7.3.1.4 Push Notification Request and Answer (PNR, PNA)

When changes occur in user data stored in the HSS and an Application Server is subscribed to changes in these user data, the HSS sends a Push-Notification-Request (PNR) to the subscribed Application Server. The PNR message includes a copy of the changed data. The Application Server answers with a Push-Notification-Answer (PNA). The flow is also depicted in Figure 7.9.

7.3.2 AVPs Defined in the Diameter Application for the Sh Interface

The Diameter Application for the *Sh* interface defines a number of new Attribute Value Pairs. Table 7.4 lists these new attributes together with the AVP code.

Table 7.4: AVPs defined by the Diameter Application for the *Sh* interface

Attribute name	AVP code
User-Identity	100
MSISDN	101
User-Data	102
Data-Reference	103
Service-Indication	104
Subs-Req-Type	105
Requested-Domain	106
Current-Location	107
Server-Name	108

The User-Identity is a grouped AVP that contains the identity of the user either as a Public User Identity, in which case it contains a Public-Identity AVP (borrowed from the Diameter Application for the *Cx* interface), or as a Mobile Subscriber Integrated Services Digital Network (MSISDN) number, in which case it contains an MSISDN AVP.

The User-Data AVP contains the user data according to the definition of user data in the *Sh* interface. The type of user data is specified in a Data-Reference AVP, which can contain a value that represents any of the different types of user data.

The Service-Indication AVP contains an identifier of the repository data stored in the HSS. This allows an Application Server that implements several services to store data for each of the services in the HSS and still be able to distinguish and associate each data set with each corresponding service.

The Subs-Req-Type AVP contains an indication of whether the Application Server subscribes to the notification service in the HSS.

The Requested-Domain AVP indicates whether the Application Server is interested in accessing circuit-switched domain data or packet-switched domain data.

The Current-Location AVP indicates whether a procedure called "active location retrieval" should be initiated.

The Server-Name AVP mirrors the AVP with the same name in the Diameter Application for the *Cx* interface.

7.4   Accounting

Accounting is defined as the collection of resource consumption data for the purposes of capacity and trend analysis, cost allocation, auditing, and billing. Although we will be focusing in this section on the charging (i.e., billing) aspects of accounting, we should keep in mind that accounting records can be used for other purposes as well.

The IMS uses the Diameter protocol to transfer the accounting information that charging in the IMS is based on. The CSCFs inform the charging system about the type and the length of the sessions each user establishes. In addition, the routers (e.g., the GGSN) inform the charging system about media activity during those sessions. The charging system assembles all the accounting information related to each user in order to charge them accordingly.

Charging systems use unique identifiers used to correlate the accounting data applying to a particular session received from different entities. So, the accounting records generated by a router and by a CSCF about the same session have the same unique identifier. In this chapter, we describe how these identifiers are delivered to the relevant network elements and how these elements generate accounting information.

7.5   Charging Architecture

The IMS charging architecture (described in 3GPP TS 32.200 [42] for Release 5 and in 3GPP TS 32.240 [40] for Release 6) defines two charging models: offline charging and online charging. Offline charging is applied to users who pay for their services periodically (e.g., at the end of the month). Online charging, also known as credit-based charging, is used to charge for prepaid services.

Both the online and the offline charging models may be applied to the same session. A user may, for example, have a permanent subscription for making voice calls and a prepaid card with credit worth ten minutes of video.

Figure 7.10 shows a complete view of the 3GPP Release 5 IMS offline charging architecture. We consider all the nodes that report charging events, both in home and visited networks and in caller and callee networks. For instance, the P-CSCFs and the GGSNs may be located either in the home or in the visited network.

As we assume GPRS access, Figure 7.10 also shows the charging architecture for GPRS access. All the SIP network entities (in our figure the CSCFs, BGCF, MRFC, MGCF, and the Application Servers) involved in the session use the *Rf* interface to send accounting information to a CCF (Charging Collection Function) located in the same administrative domain. Note that the HSS and the SLF do not report charging events.

The CCF uses this information to create CDRs (whose format is specified in 3GPP TS 32.225 [41] for Release 5 and in 3GPP TS 32.260 [43] for Release 6) and sends them to the BS (Billing System) of its domain using the *Bi* interface. The entities managing GPRS access (i.e., the SGSN and the GGSN) use the *Ga* interface to report to the CGF (Charging Gateway Function), which uses the *Bp* interface to report to the BS of its domain.

BSs in different domains exchange information using nonstandard means. The *Rf* interface is based on the Diameter base protocol (specified in RFC 3588 [60]) together with a vendor-specific *Diameter Application for the Rf/Ro interfaces*. The *Bi* and *Bp* interfaces are based on a file transfer protocol (e.g., FTP), although the actual protocol is not standardized.

The 3GPP Release 6 offline charging architecture is a little different. The CGF acts as the gateway between the 3GPP network and the billing domain. Additionally, a new entity called CDF (Charging Data Function) substitutes the CCF. Consequently, the SIP IMS nodes send

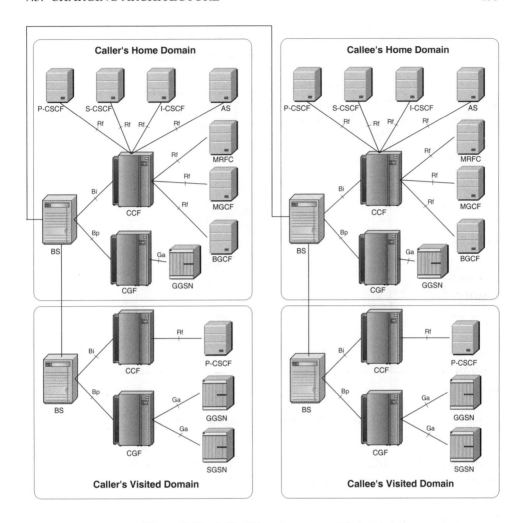

Figure 7.10: IMS offline charging architecture

accounting information to the CDF using the *Rf* interface, the CDF uses the *Ga* interface to transfer CDRs to the CGF, and the CDF transfers them to the BS.

Figure 7.11 shows the 3GPP IMS Release 5 online, or credit-based, charging architecture. The figure shows the home and visited network of the caller. The architecture on the callee's side looks the same. The external interfaces in the figure (i.e., *Ro*, IMS Service Control (*ISC*), and CAMEL Access Protocol (CAP)) were standardized either in 3GPP Release 5 or earlier releases, but the internal interfaces (i.e., *Rb*, *Rc*, *Re*) still have not been standardized.

The Session Charging Function looks like any other SIP Application Server to the S-CSCF. Consequently, SIP is used between them. The ECF implements the *Ro* interface, which is based on the Diameter base protocol (specified in RFC 3588 [60]) together with a vendor-specific *Diameter application for the Rf/Ro interfaces*. The IETF is also working in a Diameter Credit-control Application (specified in the Internet-Draft "Diameter Credit-

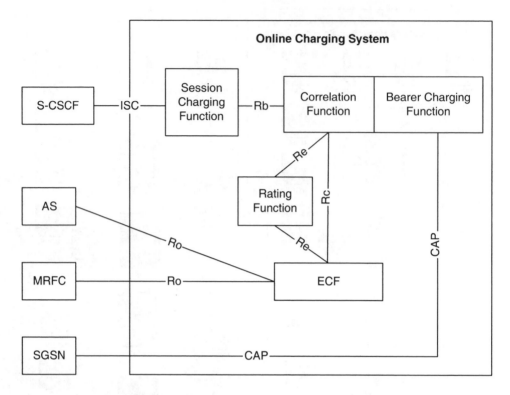

Figure 7.11: IMS online charging architecture

control Application" [162]) that will provide similar functionality. We show the SGSN's CAP interface because we assume GPRS access for this example.

7.6   Offline Charging

Let's see how all the different architectural elements interact when a session is established. In addition to the interfaces described previously, SIP entities exchange charging information among each other using SIP header fields. There are two SIP header fields (both specified in RFC 3455 [109]) that carry charging-related information in the IMS: P-Charging-Vector and P-Charging-Function-Addresses.

7.6.1   IMS Terminal in a Visited Network

Figure 7.12 shows a session establishment message flow involving a roaming user when offline charging is used. We only show the caller's home and visited domains, since the flow in the callee's home and visited domains is similar to that of the caller.

The P-CSCF receives an initial INVITE (1) and generates a globally unique identifier called ICID (IMS Charging Identity). This ICID identifies the session that is going to be established for charging purposes. The P-CSCF places the ICID in the P-Charging-Vector header field shown in Figure 7.13 and relays the INVITE to the S-CSCF (2).

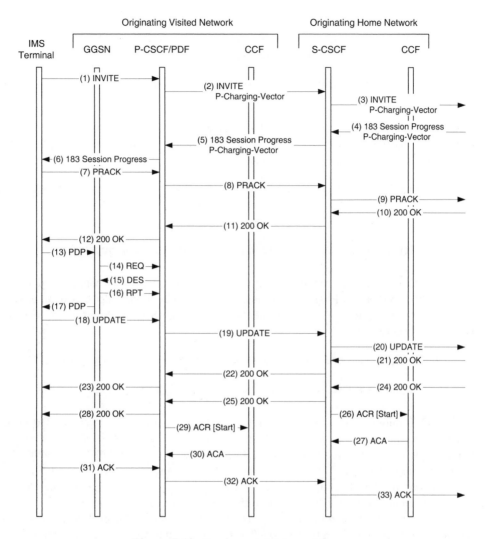

Figure 7.12: Session establishment flow

```
P-Charging-Vector: icid-value="AyretyUOdm+6O2IrT5tAFrbHLso="
```

Figure 7.13: P-Charging-Vector value in INVITE (2)

The S-CSCF inserts the caller's home network IOI (Inter-Operator Identifier) into the originating IOI parameter (orig-ioi) in the P-Charging-Vector header field. The originating IOI helps the callee's home network to identify the caller's network so that both networks can exchange charging records. The INVITE (3) sent to the callee's home network contains both the ICID and the originating IOI, as shown in Figure 7.14.

The S-CSCF in the callee's home network inserts a terminating IOI parameter (term-ioi) into the P-Charging-Vector header field of the 183 (Session Progress) (4) response.

```
P-Charging-Vector: icid-value="AyretyUOdm+6O2IrT5tAFrbHLso=";
                   orig-ioi=home1.net
```

Figure 7.14: P-Charging-Vector value in INVITE (3)

The term-ioi parameter helps the caller's home network to identify the callee's home network so that both networks can exchange charging information. An example of this P-Charging-Vector header field is shown in Figure 7.15.

```
P-Charging-Vector: icid-value="AyretyUOdm+6O2IrT5tAFrbHLso=";
                   orig-ioi=home1.net;
                   term-ioi=home2.net
```

Figure 7.15: P-Charging-Vector value in 183 (Session Progress) (4)

The S-CSCF removes the IOIs from the P-Charging-Vector header field before relaying the 183 (Session Progress) response to the P-CSCF in the caller's visited network. The resulting P-Charging-Vector header field in the 183 (Session Progress) (5) response is shown in Figure 7.16.

```
P-Charging-Vector: icid-value="AyretyUOdm+6O2IrT5tAFrbHLso="
```

Figure 7.16: P-Charging-Vector value in 183 (Session Progress) (5)

The CCF only provides an intra-operator interface. That is, only nodes located in the same administrative domain as the CCF report to it. Since the P-CSCF and the S-CSCF in our example are located in different networks, they send accounting information related to the session to different CCFs. The way the P-CSCF chooses an appropriate CCF for the session is not specified, but this choice is typically based on local configuration. The S-CSCF receives a number of CCF addresses from the HSS as part of the Diameter SAA message when the user registers with the network. The S-CSCF uses one of these addresses to send accounting information about the session. The remaining addresses are backup addresses for reliability purposes.

7.6.2   IMS Terminal in its Home Network

Figure 7.17 shows a session establishment message flow for a user in their home network when offline charging is applied. We only show the caller's home domain, since the flows in the callee's domain are similar.

The difference between Figure 7.17 and Figure 7.12 is that the P-CSCF and the S-CSCF in the former illustration are located in the same administrative domain. This implies that they share more information than the P-CSCF and the S-CSCF in Figure 7.12, which belonged to different domains. In particular, the S-CSCF provides the P-CSCF with the addresses of the CCFs (main and backup nodes) for the session in a 183 (Session Progress) response (5). In turn, the P-CSCF provides the S-CSCF with information on how the GGSN is handling the media flows in an UPDATE (19).

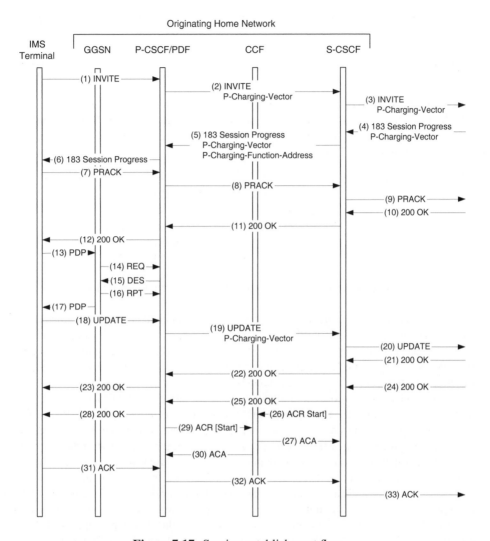

Figure 7.17: Session establishment flow

The S-CSCF inserts a P-Charging-Function-Address in the 183 (Session Progress) response (5) of Figure 7.17. Figure 7.18 shows the contents of this header field: two CCF addresses.

Basically, the S-CSCF inserts a P-Charging-Function-Address whenever the SIP request or response is addressed to a node that is located in the same network. For instance, if both caller and callee are users of the same operator and neither of them are roaming the two P-CSCFs and two S-CSCFs involved in the session report charging events to the same CCFs, whose addresses are distributed by one of the S-CSCFs.

The P-CSCF/PDF receives a COPS REQ message (14) from the GGSN. The P-CSCF/PDF responds with a COPS DES message (15) which contains the ICID for the session. The GGSN authorizes the establishment of the PDP contexts requested by the IMS terminal (recall that the terminal in this example uses a GPRS access) and sends a COPS RPT message

```
P-Charging-Function-Addresses: ccf=[5555::b99:c88:d77:e66];
                               ccf=[5555::a55:b44:c33:d22]
```

Figure 7.18: P-Charging-Function-Address value in 183 (Session Progress) (5)

(16) to the P-CSCF/PDF. This COPS RPT message (16) contains information about how the GGSN is handling the PDP contexts for this session, which the P-CSCF/PDF should pass to the S-CSCF. To do this the P-CSCF/PDF inserts the P-Charging-Vector header field in Figure 7.19 in an UPDATE request (19).

```
P-Charging-Vector: icid-value="AyretyUOdm+6O2IrT5tAFrbHLso=";
                   ggsn=[5555::4b4:3c3:2d2:1e1];
                   auth-token=4AF03C87CA;
                   pdp-info="pdp-item=1; pdp-sig=no
                             gcid=39B26CDE;
                             flow-id=({1,1},{1,2})"
```

Figure 7.19: P-Charging-Vector value in UPDATE (19)

The P-Charging-Vector header field in Figure 7.19 contains several parameters (these parameters are specified in 3GPP TS 24.229 [16], which extends the definition of the P-Charging-Vector header field in RFC 3455 [109]). The ggsn parameter contains the address of the GGSN handling the session. The auth-token parameter contains the authorization token used by the GGSN for this session. The pdp-info parameter contains a collection of information related to the PDP contexts established by the terminal. The pdp-info parameter is a quoted string that, in turn, encloses more PDP context-related parameters. The pdp-item parameter, which contains a sequential number of the PDP context information within the P-Charging-Vector header field, indicates that the following parameters pertain to a particular PDP context used in the session. There is a pdp-item parameter per PDP context established by the terminal for the session. The pdp-sig parameter indicates whether or not the PDP context is used for signaling (in our example the PDP context is not used for signaling, but for media). The gcid parameter contains the GPRS Charging Identifier (GCID), which is the charging identifier of the PDP context at the GGSN (the GGSN generates a different gcid for each new PDP context created at the GGSN).

The flow-id parameter indicates the flows that were sent or received over the PDP context. The flow-id parameter contains a collection of pairs of digits {x,y} that refer to a particular media stream in the SDP. The "x" digit indicates the order of media line (m=) in the SDP, so that a 1 refers to the first media line (m=), a 2 to the second media line (m=), etc. The "y" digit indicates the sequential order of the port number of the flow within the media line. This is so because a media stream typically is composed of more than a single flow and, therefore, each flow is sent to a different port number. For instance, if RTP (specified in RFC 3550 [225]) is the media transport protocol, RTP is sent to a port number and RTCP (also specified in RFC 3550 [225]) to the following port number. In our previous example, since the flow-id parameter contains two pairs, there are two different flows. Both are flows of the first media stream in the SDP (represented by a "1" in the "x" component of each tuple). The first pair refers to the RTP flow ("1" in the "y" component of the first tuple); the second refers to the RTCP flow of the same media stream ("2" in the "y" component of the first tuple).

7.6.3   The Rf Interface

The interface between a CCF and either a CSCF, AS, MRFC, BGCF, or MGCF (shown in Figure 7.10) is called *Rf* and is based on Diameter and the *Diameter Application for the Rf/Ro interfaces*. The Diameter messages used over this interface are Accounting-Request (ACR) and Accounting-Answer (ACA), which are part of the Diameter base protocol (specified in RFC 3588 [60]). (See Section 6.3.5 for a description of these and other Diameter base protocol commands.)

Figures 7.20 and 7.21 show the AVPs that each message can carry. These figures also indicate the specification where each AVP is specified (i.e., in the Diameter base protocol or in the 3GPP Diameter application).

Diameter Base Protocol AVPs

Session-Id
Origin-Host
Origin-Realm
Destination-Realm
Accounting-Record-Type
Accounting-Record-Number
Vendor-Specific-Application-Id
User-Name
Acct-Interim-Interval
Origin-State-Id
Event-Timestamp

3GPP Diameter Accounting AVPs

Event-Type
Role-of-Node
User-Session-Id
Calling-Party-Address
Called-Party-Address
Time-Stamps
Application-Server
Application-Provided-Called-Party-Address
Inter-Operator-Identifier
IMS-Charging-Identifier
SDP-Session-Description
SDP-Media-Component
GGSN-Address
Served-Party-IP-Address
Authorised-QoS
Server-Capabilities
Trunk-Group-Id
Bearer-Service
Service-Id
UUS-Data
Cause

Figure 7.20: AVPs in ACR messages for offline charging

The ACR and ACA messages are used to report accounting information to the CCF. ACR messages are typically triggered by the receipt of a SIP message, such as a response to an INVITE. In Figures 7.12 and 7.17 the S-CSCF exchanges an ACR (26) and an ACA (27) message with a CCF. In the same figures the P-CSCF also exchanges an ACR (29) and an ACA (30) with the same or a different CCF. Both figures show the value of the Accounting-Record-Type AVP in the ACR message. In this case the value of this AVP is START_RECORD. The value INTERIM_RECORD is used during ongoing sessions, and the value STOP_RECORD is used when a session is terminated. The value EVENT_RECORD is used for events that are not related to the management of a session, such as the reception of a SUBSCRIBE request, as shown in Figure 7.22.

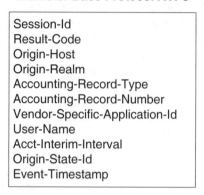

Diameter Base Protocol AVPs

Session-Id
Result-Code
Origin-Host
Origin-Realm
Accounting-Record-Type
Accounting-Record-Number
Vendor-Specific-Application-Id
User-Name
Acct-Interim-Interval
Origin-State-Id
Event-Timestamp

Figure 7.21: AVPs in ACA messages for offline charging

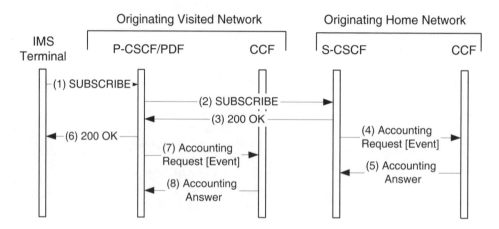

Figure 7.22: Use case for the EVENT_RECORD value in ACR messages

7.7 Online Charging

The interfaces involved in online charging in 3GPP Release 5 are shown in Figure 7.11. The S-CSCF uses the ISC interface, which is based on SIP. Application Servers (ASs) and the MRFC use the *Ro* interface, which is based on Diameter.

7.7.1 S-CSCF

Network operators configure those sessions to which online charging has to be applied by defining a filter criterion in the user profile. The filter criterion blindly sends all the SIP requests to the Application Server that is acting as a Session Charging Function (SCF).

The SCF looks like any other AS to the S-CSCF, but it does not provide services for the user in the usual sense. Instead, the SCF reports accounting information to the *Correlation Function* using the *Rb* interface (which is still not standardized). If the user runs out of credit during a session the Correlation Function informs the SCF through the *Rb* interface and the

SCF terminates the session by acting as a B2BUA and sending two BYE requests, one toward each terminal.

7.7.2 Application Servers and the MRFC

The filter criterion in the user profile may get the S-CSCF to relay a SIP message to an AS or to the MRFC that will provide the service to the user. If such a service implies online charging the AS or the MRFC uses the *Ro* interface to send charging information to the ECF. The AS or the MRFC receives the address or addresses of the ECF from the S-CSCF in the P-Charging-Function-Address header field of the SIP message, as shown in Figure 7.23.

```
P-Charging-Function-Addresses: ecf=[5555::b99:c88:d77:e66];
                                ecf=[5555::a55:b44:c33:d22]
```

Figure 7.23: P-Charging-Function-Address header field

The *Ro* interface is based on Diameter. The Diameter application used in this interface in Release 5 (specified in 3GPP TS 32.225 [41]) uses the following Diameter messages (which are all specified in RFC 3588 [60]):

- Accounting-Request (ACR)

- Accounting-Answer (ACA).

The Diameter application used in 3GPP Release 6 (specified in 3GPP TS 32.260 [43]) will use the CCR (Credit Control Request) and CCA (Credit Control Answer) messages instead. In any case, we will focus the rest of our description on Release 5, because the Release 6 specifications in this area are still not mature enough.

Figures 7.24 and 7.25 show the AVPs that each message in Release 5 can carry. These figures also indicate the specification where each AVP is defined (i.e., in the Diameter base protocol, in the 3GPP Diameter credit control application, or in the 3GPP Diameter application).

Online charging is based on credit units. Services are paid for by credit units, and users can enjoy a particular service as long as they have enough credit units in their accounts. For example, if Bob has two credit units in his account and ten minutes of streaming video costs one credit unit, Bob will be authorized to receive streaming video in his terminal for 20 minutes.

There are two types of online charging in the IMS: Immediate Event Charging (IEC) and Event Charging with Unit Reservation (ECUR). When IEC is used the ECF deducts a number of credit units from the user's account and then authorizes the MRFC or the AS to provide the service to the user.

When ECUR is used the ECF reserves a number of credit units in the user's account and authorizes the MRFC or the AS to provide the service to the user. In case a particular service costs more credit units than those originally reserved by the ECF, the MRFC or the AS can contact the ECF to request further credit unit reservations.

When the service is over the MRCF or the AS report to the ECF the number of credit units that the user spent. At this point the ECF returns to the user's account all the credit units that were reserved but not used.

Figure 7.26 shows an example of IEC. The value of the Accounting-Record-Type AVP in ACR messages used for IEC is EVENT_RECORD. The ECF receives an ACR (1) message,

Diameter Base Protocol AVPs

Session-Id
Origin-Host
Origin-Realm
Destination-Realm
Accounting-Record-Type
Accounting-Record-Number
Vendor-Specific-Application-Id
User-Name
Acct-Interim-Interval
Origin-State-Id
Event-Timestamp

3GPP Diameter Accounting AVPs

Event-Type
Role-of-Node
User-Session-Id
Calling-Party-Address
Called-Party-Address
Time-Stamps
Application-Provided-Called-Party-Address
Inter-Operator-Identifier
IMS-Charging-Identifier
SDP-Session-Description
SDP-Media-Component
GGSN-Address
Service-Id
UUS-Data
Cause

Diameter Credit Control AVPs

Subscription-Id
Requested-Action
Requested-Service-Unit
Used-Service-Unit
Tariff-Switch-Definition
Service-Parameter-Info
Abnormal-Termination-Reason
Credit-Control-Failure-Handling
Direct-Debiting-Failure-Handling

Figure 7.24: AVPs in ACR messages for online charging

Diameter Base Protocol AVPs

Session-Id
Result-Code
Origin-Host
Origin-Realm
Accounting-Record-Type
Accounting-Record-Number
Vendor-Specific-Application-Id
User-Name
Accounting-Sub-Sesssion-Id
Acct-Interim-Interval
Origin-State-Id
Event-Timestamp

Diameter Credit Control AVPs

Subscription-Id
Granted-Service-Unit
Tariff-Switch-Definition
Cost-Information
Final-Unit-Indication
Check-Balance-Result
Credit-Control-Failure-Handling

Figure 7.25: AVPs in ACA messages for online charging

deducts the appropriate number of credit units from the user's account, and returns an ACA (2) message. At this point the MRFC or the AS starts delivering the service requested by the user.

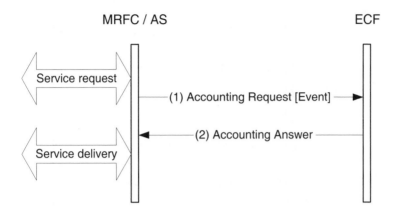

Figure 7.26: Immediate Event Charging

Figure 7.27 shows an example of ECUR. When ACR messages are used for ECUR, the values of the `Accounting-Record-Type` AVP are: START_RECORD, INTERIM_RECORD, and STOP_RECORD. The ECF receives an ACR (1) message and reserves a number of credit units in the user's account. The ECF returns an ACA (2) message and the MRFC or the AS start delivering the service.

When the user has spent most of the reserved credit units, the MRFC or the AS sends an ACR message (3) whose `Accounting-Record-Type` AVP value is INTERIM_RECORD. The ECF reserves more credit units and returns an ACA (4) message.

Once service delivery has ended the MRFC or the AS sends an ACR message (5) whose `Accounting-Record-Type` AVP value is STOP_RECORD. At this point the ECF can return to the user's account the credit units that were reserved but not used.

7.7.2.1   Unit Determination

Unit determination refers to the process of determining how many credit units need to be reserved or withdrawn from the user's account for a given service. When unit determination is performed by the EFC it is referred to as *centralized unit determination*. When unit determination is performed by the MRFC or the AS it is referred to as *decentralized unit determination*.

When decentralized unit determination is used the MRFC or the AS determines how many credit units it needs to request from the ECF. The ACR message (1) in Figures 7.26 and 7.27 contains a `Requested-Service-Unit` AVP with the number of requested credit units.

When centralized unit determination is used the ECF performs unit determination based on these AVPs related to ratings that are present in the ACR message in Figures 7.26 and 7.27.

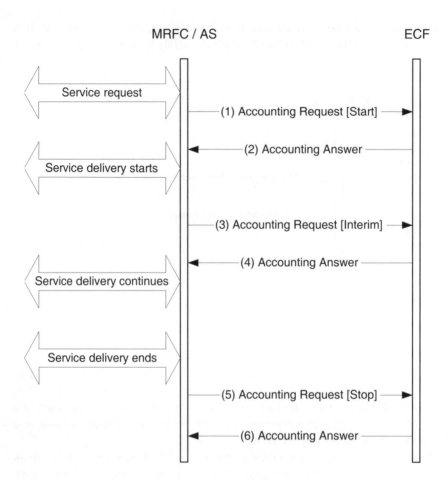

Figure 7.27: Event Charging with Unit Reservation

7.7.2.2 Rating

The conversion of credit units into monetary units is referred to as rating. That is, the price of the service is calculated based on inputs fed into the rating function. If rating is performed by the ECF it is referred to as *centralized rating*, and if rating is performed by the MRFC or the AS it is referred to as *decentralized rating*.

Centralized unit determination implies centralized rating, because it is the ECF that determines the number of credit units to be charged. Neither the MRFC nor the AS has this information.

When decentralized unit determination is used the ACA message (2) in Figures 7.26 and 7.27 contains a Unit-Value AVP and a Unit-Type AVP. The Unit-Value AVP contains the number of units that are granted and the Unit-Type AVP contains the type of units. When centralized rating is used the units are monetary units. When decentralized rating is used the units may indicate time (e.g., the service can be used for two minutes), volume (e.g., the user can download 100 kilobytes), or number of events (e.g., the user can receive two notifications).

7.7.2.3   Tariff Changes

The ECF can inform the MRFC and the AS about changes in the tariff of a given service using the `Tariff-Switch-Definition` AVP in ACA messages. The ECF can use the same ACA message to grant some credit units to be used before the tariff change and some to be used after.

Chapter 8

Security on the Internet

According to the traditional definition, network security comprises integrity, confidentiality, and availability. Message integrity ensures that if an unauthorized party modifies a message between the sender and the receiver, the receiver is able to detect this modification. In addition to message integrity, integrity mechanisms always provide some type of proof of data origin. Knowing that a message has not been modified without knowing who initially created the message would be useless.

Confidentiality mechanisms keep unauthorized parties from getting access to the contents of a message. Confidentiality is typically achieved through encryption.

Denial of Service (DoS) attacks compromise the system's availability by keeping authorized users from accessing a particular service. The most common DoS attack consists of keeping the servers busy performing an operation or sending the servers more traffic than they can handle.

SIP provides several security mechanisms to address integrity, confidentiality, and availability. Some of the security mechanisms come from the world of the web, some come from the world of email, and some of them are SIP-specific. We analyze these mechanisms in the following sections and describe how they relate to the three security properties just described.

8.1 HTTP Digest

The first problem a SIP server faces is authenticating users requesting services. SIP has inherited an authentication mechanism from HTTP called digest access authentication (specified in RFC 2617 [102]). In the SIP context the server authenticating the user (i.e., the caller) can be a proxy, a registrar, a redirect server, or a user agent (the callee's user agent). The `WWW-Authenticate` and `Authorization` header fields are used with registrars, redirect servers, and user agents, and the `Proxy-Authenticate` and `Proxy-Authorization` header fields are used with proxies.

When using digest access authentication the client and the server have a shared secret (e.g., a password), which is exchanged using an out-of-band mechanism. When a server at a given domain receives a request from a client the server challenges the client to provide valid credentials for that domain. At that point the client provides the server with a username and proves that the client knows the shared secret.

An obvious way for the client to prove that it knows the shared secret would be to send it to the server in clear text (i.e., without any encryption). In fact, this is what HTTP basic access authentication (also specified in RFC 2617 [102]) does. Nevertheless, the security risks created by sending passwords in clear text are obvious. Any attacker that manages to gain access to the message carrying the shared secret gains access to the shared secret itself. Previous SIP specifications allowed the use of basic authentication, but it has now been deprecated for some time. The use of digest access authentication is currently recommended instead.

Clients using digest can prove that they know the shared secret without sending it over the network. Digest uses hashes and nonces for this purpose. A hash algorithm is a one-way function that takes an argument of an arbitrary length and produces a fixed length result, as shown in Figure 8.1. The fact that hash algorithms are one-way functions means that it is computationally infeasible to obtain the original argument from the result. Two popular hash algorithms are MD5 (specified in RFC 1321 [196]) and SHA1 (specified in RFC 3174 [86]). A nonce is a random value that is used only once.

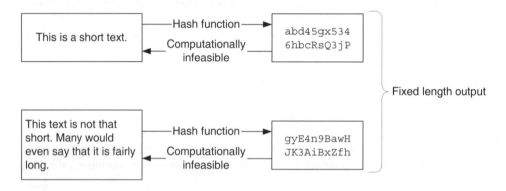

Figure 8.1: Hash function

Figure 8.2 shows how digest uses hashes and nonces. Alice sends an INVITE (1) request addressed to Bob through her outbound proxy (at domain.com). The proxy challenges the INVITE with a 407 (Proxy Authentication Required) response (2). The proxy includes a Proxy-Authenticate header field with a set of parameters. The realm parameter indicates the domain of the proxy server, so that the client knows which password to use. The qop (quality of protection) parameter indicates that the server supports integrity protection for either the request line alone (auth) or for both the request line and the message body (auth-int). The server provides the client with a random nonce in the nonce parameter. The algorithm parameter identifies the hash function (MD5, in this example).

When the client gets the response it issues a new INVITE (3) with a Proxy-Authoriza-tion header field. The Proxy-Authorization header field contains a set of parameters. The response parameter is especially interesting. It contains a hash comprising, among other things, the username, the password, the server's nonce, the client's nonce (cnonce parameter), and the request line. When the auth-int qop is chosen the message body is also fed into the hash algorithm to generate the response parameter.

When the server receives this Authorization header field it calculates another hash value using the same input as the client, but using the shared secret the server has. If the result matches the value in the response parameter of the INVITE request the server considers the

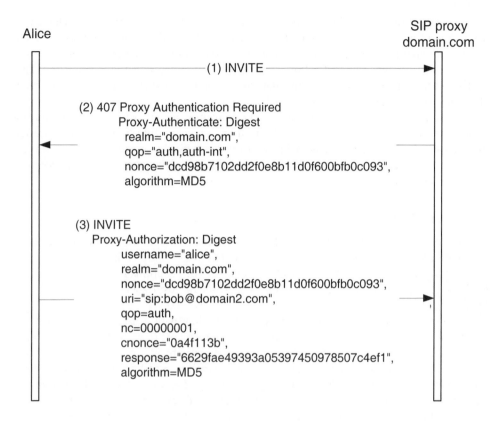

Figure 8.2: Digest operation

authentication successful and keeps on processing the INVITE. Otherwise, the server will challenge this INVITE again.

The inclusion of random nonces chosen by the server in the hash prevents replay attacks. Even if an eavesdropper manages to obtain the correct hash value for a particular nonce it will not be able to access the server, since the server will challenge it with a different random nonce.

8.1.1 Security Properties of Digest

Digest provides authentication of users and a limited degree of integrity protection (no confidentiality). Digest integrity protects the request line (i.e., method and Request-URI) and, potentially, the message body. Still, it does not integrity-protect the header fields of the message. This lack of header field protection is an important drawback, because, as we saw in previous chapters, header fields in SIP carry important information which attackers can easily manipulate if digest is used by itself. In brief, digest access authentication is vulnerable to Man-in-the-Middle (MitM) attacks.

Regarding availability, digest offers good DoS protection. Servers issuing challenges using digest can remain stateless until the new request arrives. So, an attacker issuing a large number of requests to a server will not block any resources for a long period of time.

In addition to client authentication, digest can also provide server authentication. That is, the server can prove to the client that the server knows their shared secret as well. Nevertheless, this feature is not used by SIP proxies.

Digest client authentication is common because user agents acting as clients do not typically have certificates to use more advanced and secure certificate-based authentication mechanisms. On the other hand, domain administrators typically provide their proxy servers with site certificates, which is why certificate-based proxy authentication is used instead of digest server authentication.

8.2   Certificates

Certificates are key to understanding TLS and S/MIME (which will be described in the following sections). A certificate is a statement about the truth of something signed by a trusted party. A passport is an example of a certificate. The government of a country certifies the identity of the passport holder. Everyone who trusts the government will be convinced of the identity of the holder.

Certificates used in SIP bundle together a public key with a user (e.g., `sip:Alice.Smith@domain.com`), a domain (e.g., domain.com), or a host (e.g., ws1234.domain.com). Domain certificates, also known as site certificates, are used by network elements (i.e., proxies, redirect servers, and registrars) and are usually signed by a certification authority whose public key is well known. This way, any SIP entity that knows the public key of the certification authority can check the validity of the certificates presented by another SIP entity. Within a domain, host certificates can be used to authenticate different network nodes (e.g., two proxies).

Since building and managing a certification authority hierarchy to provide signed certificates to every user has been proven to be a difficult task for a number of reasons (technical and nontechnical), SIP allows the use of self-signed certificates: that is, certificates that are not signed by any certification authority. Section 8.4 describes how self-signed certificates are used in SIP.

8.3   TLS

TLS (Transport Layer Security, specified in RFC 2246 [82]) is based on SSL (Secure Sockets Layer). SSL was designed to protect web communications and is currently implemented in most Internet browsers. TLS is more generic than SSL and can be used to protect any type of reliable connection. So, using TLS to protect SIP traffic is yet another security solution that comes from the world of the web.

TLS provides data integrity and confidentiality for reliable connections (e.g., a TCP connection). It can also provide compression, although that feature is not typically used. TLS consists of two layers: namely, the TLS handshake layer and the TLS record layer.

The TLS handshake layer handles the authentication of peers, using public keys and certificates, and the negotiation of the algorithm and the keys to encrypt the actual data transmission. Once a reliable connection at the transport layer is established (e.g., a TCP connection) a TLS handshake takes place. The handshake starts with an exchange of `hello` messages and usually takes two RTTs (Round Trip Times) to complete. During authentication the server provides the client with the server's certificates (e.g., X.509 [141] certificates) and can optionally request the client to provide its certificates as well. The server chooses the encryption algorithm to be used from among those supported by the client.

The TLS handshake also allows peers to generate a so-called premaster secret (e.g., using Diffie–Hellman) and to exchange random values. The premaster secret and the random values are used by the TLS record layer to generate the encryption key.

The TLS record layer handles the encryption of the data. It typically uses a symmetric encryption algorithm whose key is generated from the values provided by the handshake layer.

As we can see in Figure 8.3, when TLS is used over TCP peers need to wait for three RTTs (one for the TCP handshake and two for the TLS handshake) before exchanging secure data. Note that in Figure 8.3 we assume that the TLS Client Hello (3) is piggybacked on the TCP ACK that completes the TCP three-way handshake, although many TCP implementations send them in parallel rather than in a single datagram.

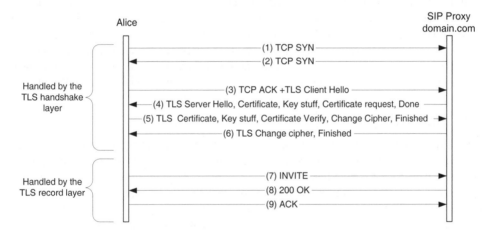

Figure 8.3: TLS connection establishment

8.3.1 SIP Usage

From the SIP point of view, TLS is a hop-by-hop security solution. That is, a TLS connection protects the messages exchanged between a SIP entity and the SIP entity one hop away (e.g., a user agent and a proxy). For example, if Alice's outbound proxy needs to send a SIP message to Bob through his proxy server at domain2.com, Alice's proxy will perform a DNS lookup for domain2.com. If the DNS records for domain2.com show that the proxy at domain2.com prefers to use TLS, they will exchange SIP traffic using a TLS connection; however, establishing this TLS connection is out of Alice's control. When Alice sent the original message to her outbound proxy she did not know anything about the domain2.com DNS records. If these DNS records had said that UDP with no security was preferred over TLS, the proxies would not have used any kind of encryption. So, Alice needs a way to tell her proxy that she wants her message to be protected all the time. That is, she wants all the proxies in the path to exchange data using TLS while handling her request. SIP provides a mechanism for user agents to request this type of treatment for a request: the SIPS URI scheme.

When a user agent sends a request to a SIPS URI it is requesting that every proxy in the path uses TLS to relay the request to the next proxy. Still, using SIPS URIs does not imply that the last proxy will use TLS to relay the request to the destination user agent. The last proxy is free to use whatever security mechanism it has agreed on with the user agent (e.g., a manually configured IPsec association). This is because users of a particular domain can agree on the security mechanisms to use within that domain fairly easily. On the other hand, doing so for inter-domain communications would be difficult to manage. It is better to use a default solution for this case (i.e., TLS). Of course, the use of SIPS URIs also implies that proxies route the responses to the request using TLS as well.   .

8.3.1.1   Client Authentication

As we said earlier, TLS supports certificate-based client and server authentication. This type of authentication is used when a proxy acts as a client sending a request to another proxy that acts as a server. In any case, while proxies typically have site certificates, at present SIP user agents do not usually have certificates. One way to provide client and server authentication without providing user agents with certificates is to use digest access authentication over TLS. The server is authenticated using certificates at the TLS handshake layer, and the user agent (the client) is authenticated using digest once the TLS connection has been established.

8.4   S/MIME

SIPS URIs ensure that messages are secured hop by hop. Still, every SIP proxy handling a message with a SIPS URI has access to *all* the contents of that message. Of course, proxies need to have access to some header fields (i.e., Route, Request-URI, Via, etc.), but user agents may want to exchange some information they do not want proxies to have access to. For example, Alice and Bob may not want the proxies to be aware of which type of session (e.g., audio or video) they are establishing, or they may want to exchange cryptographic keys to exchange media traffic in a secure way. The fact that Alice and Bob trust their proxies to route messages does not mean that they want the proxies' administrators to be able to listen to their conversation. Some type of end-to-end security is needed. SIP uses an end-to-end security solution from the email world: S/MIME.

Section 4.7 described how SIP uses the Multipurpose Internet Mail Extensions (MIME) format specified in RFC 2045 [103] to encode multipart SIP bodies. S/MIME (Secure/MIME) allows us, among other things, to sign and encrypt message parts and encode them using MIME. These message parts are usually message bodies (e.g., a session description), although sometimes header fields carrying sensitive information are protected as well.

The CMS (Cryptographic Message Syntax, specified in RFC 3369 [124]) is a binary format to encode, among other things, signed messages, encrypted messages, and the certificates related to those messages. (RFC 2633 [192] describes how to encode CMS binary objects using MIME.)

Figure 8.4 shows an example of an encrypted body using S/MIME. Message bodies are encrypted using a symmetric content encryption key, which is encrypted using the public key of the recipient (symmetric encryption is much faster than public keybased encryption), as shown in Figure 8.5. The recipient decrypts the content encryption key using their private key and then decrypts the message with the content encryption key, as shown in Figure 8.6.

```
INVITE sip:Alice.Smith@domain.com SIP/2.0
Via: SIP/2.0/UDP ws1.domain2.com:5060;branch=z9hG4bK74gh5
Max-Forwards: 70
From: Bob <sip:Bob.Brown@domain2.com>;tag=9hx34576sl
To: Alice <sip:Alice.Smith@domain.com>
Call-ID: 6328776298220188511@192.0.100.2
Cseq: 1 INVITE
Contact: <sip:bob@192.0.100.2>
Content-Length: 197
Content-Type: application/pkcs7-mime; smime-type=enveloped-data;
          name=smime.p7m
Content-Transfer-Encoding: base64
Content-Disposition: attachment; filename=smime.p7m;
          handling=required
```

```
rfvbnj756tbBghyHhHUujhJhjH77n8HHGT9HG4VQpfyF467GhIGfHfYT6
7n8HHGghyHhHUujhJh4VQpfyF467GhIGfHfYGTrfvbnjT6jH7756tbB9H
f8HHGTrfvhJhjH776tbB9HG4VQbnj7567GhIGfHfYT6ghyHhHUujpfyF4
0GhIGfHfQbnj756YT64V
```

Figure 8.4: Message body encrypted using S/MIME

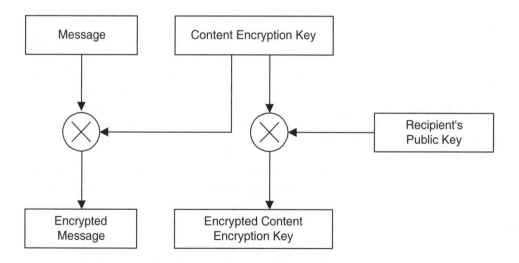

Figure 8.5: Encrypting a body using S/MIME

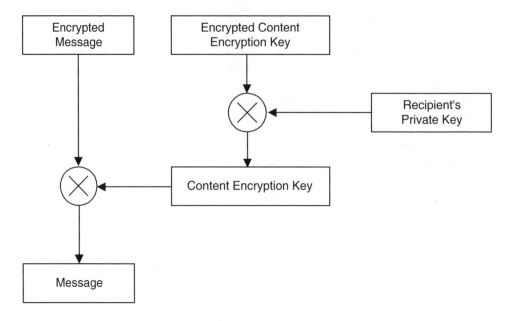

Figure 8.6: Decrypting a body using S/MIME

Figure 8.7 shows an example of a signed body using S/MIME. Message bodies are signed with the sender's private key. Signed message bodies usually carry the sender's certificate. This way the receiver obtains the sender's public key, which should be used to check the validity of the signature.

If a message body needs to be signed *and* encrypted the message bodies of Figures 8.4 and 8.7 can be nested. The encrypted message body can be signed or the signed message body can be encrypted.

The format in Figure 8.7 works well when the recipient of the message supports S/MIME. Nevertheless, at present many user agents do not support it. Such a user agent would not be able to understand Figure 8.7, since it uses the CMS binary format. There is an alternative way to sign messages that makes them understandable for S/MIME-unaware user agents (described in RFC 1847 [104]). Figure 8.8 shows an example of this format. The message body (an SDP session description) is encoded as an independent MIME part, and the signature is encoded using the S/MIME binary format. This type of signature is referred to as a CMS-detached signature, because it is not encoded together with the message body that has been signed. Even if S/MIME-unaware implementations were able to parse the session description, they of course would not be able to verify the signature. In that case the integrity of the message cannot be verified.

8.4.1   Self-signed Certificates

As explained in Section 8.2, in the absence of a certification authority S/MIME bodies can carry self-signed certificates that bundle a public key with a user. However, at first sight this seems to provide no security at all. An attacker can sign or encrypt a message with her private key and send it along with a certificate saying that her public key belongs to

```
INVITE sip:Alice.Smith@domain.com SIP/2.0
Via: SIP/2.0/UDP ws1.domain2.com:5060;branch=z9hG4bK74gh5
Max-Forwards: 70
From: Bob <sip:Bob.Brown@domain2.com>;tag=9hx34576sl
To: Alice <sip:Alice.Smith@domain.com>
Call-ID: 6328776298220188511@192.0.100.2
Cseq: 1 INVITE
Contact: <sip:bob@192.0.100.2>
Content-Length: 197
Content-Type: application/pkcs7-mime; smime-type=signed-data;
         name=smime.p7m
Content-Transfer-Encoding: base64
Content-Disposition: attachment; filename=smime.p7m;
         handling=required

567GhIGfHfYT6ghyHhHUujpfyF4f8HHGTrfvhJhjH776tbB9HG4VQbnj7
77n8HHGT9HG4VQpfyF467GhIGfHfYT6rfvbnj756tbBghyHhHUujhJhjH
HUujhJh4VQpfyF467GhIGfHfYGTrfvbnjT6jH7756tbB9H7n8HHGghyHh
6YT64VOGhIGfHfQbnj75
```

Figure 8.7: Message body signed using S/MIME

sip:Alice.Smith@domain.com. The recipient would be fooled by this certificate, since it does not have any means to check the validity of this certificate.

Self-signed certificates need a so-called *leap of faith*. That is, the first time a user exchanges SIP traffic with a second user the first user assumes that they are not under attack. When both users exchange traffic between them at a later time they assume that the keys used the first time are the valid ones. This mechanism is also used by other protocols like SSH (specified in the Internet-Draft "SSH Protocol Architecture" [237]).

In many scenarios the leap of faith provides good security properties. Two users can exchange their self-signed certificates using a cable between their user agents, or they can establish an audio session to read aloud their public keys to each other and check whether they are the same as those received in the self-signed certificate. There are many ways to obtain another user's public key in a secure enough manner. Users just need to choose the one they prefer in each situation.

8.5 Authenticated Identity Body

We have just seen how to protect message bodies using S/MIME. In addition to message bodies, S/MIME can be used to protect header fields as well. The idea is to place a copy of the header fields to be protected in a body and to protect it using S/MIME. The Authenticated Identity Body (AIB, specified in RFC 3893 [185]) is a good example of how to use S/MIME to integrity-protect header fields carrying identity information such as From, To, and Contact.

The header fields that the user wants to protect are placed in a body whose Content-Type is message/sipfrag. This type of body carries fragments of SIP messages as opposed to whole SIP messages (which are carried in message/sip bodies).

```
INVITE sip:Alice.Smith@domain.com SIP/2.0
Via: SIP/2.0/UDP ws1.domain2.com:5060;branch=z9hG4bK74gh5
Max-Forwards: 70
From: Bob <sip:Bob.Brown@domain2.com>;tag=9hx34576sl
To: Alice <sip:Alice.Smith@domain.com>
Call-ID: 6328776298220188511@192.0.100.2
Cseq: 1 INVITE
Contact: <sip:bob@192.0.100.2>
Content-Type: multipart/signed;
        protocol="application/pkcs7-signature";
        micalg=sha1; boundary="34573255067boundary"

--34573255067boundary
Content-Type: application/sdp
Content-Disposition: session

v=0
o=Alice 2790844676 2867892807 IN IP4 192.0.0.1
s=Let's talk about swimming techniques
c=IN IP4 192.0.0.1
t=0 0
m=audio 20000 RTP/AVP 0
a=sendrecv
m=video 20002 RTP/AVP 31
a=sendrecv

--34573255067boundary
Content-Type: application/pkcs7-signature; name=smime.p7s
Content-Transfer-Encoding: base64
Content-Disposition: attachment; filename=smime.p7s;

ghyHhHUujhJhjH77n8HHGTrfvbnj756tbB9HG4VQpfyF467GhIGfHfYT6
4VQpfyF467GhIGfHfYT6jH77n8HHGghyHhHUujhJh756tbB9HGTrfvbnj
n8HHGTrfvhJhjH776tbB9HG4VQbnj7567GhIGfHfYT6ghyHhHUujpfyF4
7GhIGfHfYT64VQbnj756
--34573255067boundary--
```

Figure 8.8: Message body signed using CMS-detached signatures

Figure 8.9 shows an INVITE request whose From and Contact header fields are integrity-protected. The INVITE carries three body parts: the usual SDP session description, a message/sipfrag body part with the header fields that we want to protect, and a digital signature over the previous body part.

The message/sipfrag body part carries the Call-ID, the Cseq, and the Date header fields to prevent cut-and-paste attacks. Otherwise, an attacker could cut the AIB with its signature and paste it into a different SIP message. The recipient of such a message would believe that it comes from Bob.

```
INVITE sip:Alice.Smith@domain.com SIP/2.0
Via: SIP/2.0/UDP ws1.domain2.com:5060;branch=z9hG4bK74gh5
Max-Forwards: 70
From: Bob <sip:Bob.Brown@domain2.com>;tag=9hx34576sl
To: Alice <sip:Alice.Smith@domain.com>
Call-ID: 63287762982201885110192.0.100.2
Cseq: 1 INVITE
Contact: <sip:bob@192.0.100.2>
Content-Type: multipart/mixed; boundary=--unique-boundary-1
Content-Length: 151

--unique-boundary-1
Content-Type: application/sdp

v=0
o=bob 2890844526 2890844526 IN IP4 ws1.domain2.com
s=-
c=IN IP4 192.0.100.2
t=0 0
m=audio 20000 RTP/AVP 0
a=rtpmap:0 PCMU/8000
--unique-boundary-1
Content-Type: multipart/signed;
     protocol="application/pkcs7-signature";
     micalg=sha1; boundary=--boundary2

--boundary2
Content-Type: message/sipfrag
Content-Disposition: aib; handling=optional

From: Bob <sip:Bob.Brown@domain2.com>
Call-ID: 63287762982201885110192.0.100.2
Cseq: 1 INVITE
Contact: <sip:bob@192.0.100.2>
Date: Thu, 9 Oct 2003 10:21:03 GMT
--boundary2
Content-Type: application/pkcs7-signature; name=smime.p7s
Content-Transfer-Encoding: base64
Content-Disposition: attachment; filename=smime.p7s;
     handling=required

ghyHhHUujhJhjH77n8HHGTrfvbnj756tbB9HG4VQpfyF467GhIGfHfYT6
4VQpfyF467GhIGfHfYT6jH77n8HHGghyHhHUujhJh756tbB9HGTrfvbnj
n8HHGTrfvhJhjH776tbB9HG4VQbnj7567GhIGfHfYT6ghyHhHUujpfyF4
7GhIGfHfYT64VQbnj756
--boundary2
--unique-boundary-1
```

Figure 8.9: INVITE carrying an AIB body

Proxies can also add AIB to requests and responses (as described in the Internet-Draft "Enhancements for Authenticated Identity Management in SIP" [186]). When a proxy inserts (and signs) an AIB body, in addition to providing integrity protection for some header fields it is asserting that the information in those header fields is true. For example, Bob's outbound proxy uses digest to authenticate Bob and then inserts the AIB in Figure 8.9, confirming that Bob generated the request. Other proxies that trust Bob's outbound proxy but do not share any credentials with Bob will know that this request was generated by a user that is known by at least one proxy in their circle of trust. That is usually enough for a proxy to agree to route the request further.

8.6   IPsec

IPsec (whose architecture is specified in RFC 2401 [150]) provides confidentiality and integrity protection at the network layer. Nodes that want to exchange secure IPsec-protected traffic between them set up a so-called *security association*. A security association is identified by the addresses of the nodes and by its SPI (Security Parameter Index), and it contains the security parameters (e.g., keys and algorithms) that the nodes use to protect their traffic. IKE (Internet Key Exchange, specified in RFC 2409 [118]) is the key management protocol that is typically used to set up security associations.

8.6.1   ESP and AH

IPsec provides two protocols to protect data, namely ESP (Encapsulating Security Payload) and AH (Authentication Header). ESP provides integrity and (optionally) confidentiality while AH provides integrity only. The difference between the integrity provided by ESP and AH is that ESP only protects the contents of the IP packet (excluding the IP header) while AH protects the IP header as well. Figures 8.10 and 8.11 illustrate this difference. From now on, we focus on ESP because it is more widespread than AH and is the protocol used in the IMS.

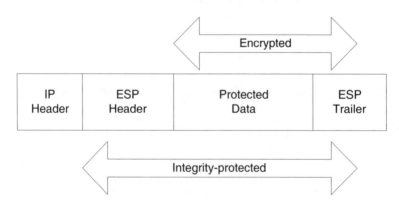

Figure 8.10: IP packet protected by ESP

ESP adds to each IP packet a header and a trailer, as shown in Figure 8.10 (some parts of the ESP trailer are encrypted and integrity-protected, while other parts are not; Figure 8.12 clarifies this point further). The ESP header contains the SPI, the sequence number of the

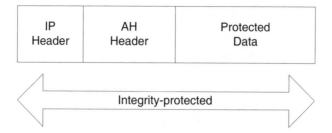

Figure 8.11: IP packet protected by AH

packet, and the initialization vector for the encryption algorithm. The ESP trailer contains optional padding in case it is required by the encryption algorithm and data related to authentication (i.e., integrity protection) of the data. Figure 8.12 shows the format of an ESP-protected packet in more detail.

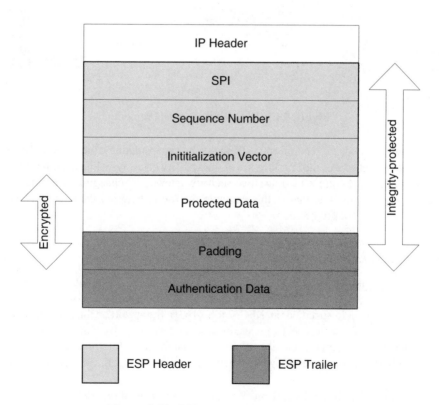

Figure 8.12: ESP header and trailer format

8.6.2   Tunnel and Transport Modes

ESP has two modes of operation: transport mode and tunnel mode. Transport mode is normally used between endpoints, while tunnel mode is typically used between security gateways to create virtual private networks.

ESP in transport mode protects the payload of an IP packet. For example, two entities exchanging TCP traffic using ESP transport mode would protect the TCP headers and the actual contents carried by TCP, as shown in Figure 8.13.

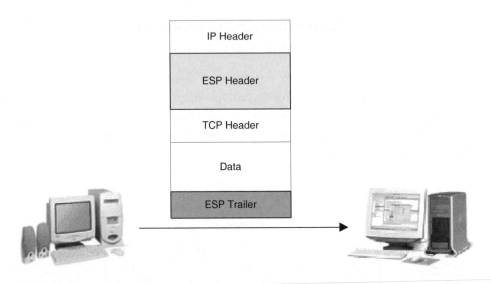

Figure 8.13: IPsec ESP in transport mode

ESP in tunnel mode protects an entire IP packet by encapsulating it into another IP packet. The outer IP packet carries the IP addresses of the security gateways while the inner IP packet remains untouched. Figure 8.14 shows two security gateways exchanging ESP tunnel-mode traffic between them. Note that in this example the traffic between the endpoints and the security gateway is not protected.

8.6.3   Internet Key Exchange

Internet Key Exchange (IKE) is the key management protocol that is typically used to set up IPsec security associations. It is based on the ISAKMP (Internet Association and Key Management Protocol, specified in RFC 2408 [163]) framework, which defines how to exchange information with a peer in a secure way to perform key exchanges. IKE defines several such key exchanges, and the IPsec DOI (Domain of Interpretation, specified in RFC 2407 [187]) document defines which attributes need to be negotiated in an IPsec security association.

The establishment of an IPsec security association consists of two steps. In the first step, peers establish a security association to run IKE on it. In the second step, peers use this security association to securely agree on the parameters of the IPsec security association that will carry the actual data traffic.

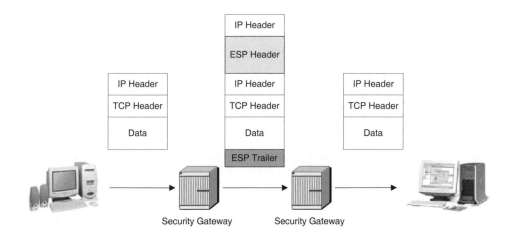

Figure 8.14: IPsec ESP in tunnel mode

8.6.3.1 IKE Security Association Establishment

There are two modes to establish the IKE security association: the *main mode* and the *aggressive mode*. The main mode consists of six messages (i.e., three RTTs) and allows peers to hide their identities from potential attackers. The aggressive mode only uses three messages, does not provide identity protection, and offers less negotiation power.

8.6.3.2 IPsec Security Association Establishment

IPsec security associations are established using the so-called *quick mode*. The quick mode uses an IKE security association and consists of three messages. After this exchange, peers are ready to exchange traffic using the IPsec security association they just established.

8.7 Privacy

Users sometimes need to establish a session without disclosing personal information, such as their identity. For example, Bob wants to remain anonymous when he calls a television program to give his opinion about a politician. To do this, Bob could send an INVITE with a From header field set to "anonymous" instead of "Bob". Nevertheless, proxies do not typically allow anonymous users to use their routing services. Bob needs to identify himself to his outbound proxy, but wants to remain anonymous to the rest of the network.

SIP user agents can request different types of privacy from a proxy by using a Privacy header field (as described in RFC 3323 [183]). The proxy typically authenticates the user and then encrypts, removes, or simply changes the contents of the header fields with personal information about the user. The proxy can also insert an AIB (as described in Section 8.5), confirming that it has authenticated the user who wishes to remain anonymous. Proxies in closed environments can insert a P-Asserted-Identity header field (specified in RFC 3325 [147]) instead of an AIB.

For example, Bob can use the following From header field in his INVITE request and then use Digest to prove to his proxy that it was he who generated the INVITE request:

```
From: "Anonymous" <sip:anonymous@anonymous.invalid>
                  ;tag=9hx34576sl
```

8.8   Encrypting Media Streams

In addition to securing SIP messages, users are often interested in securing the media they exchange. For example, they may want to have confidential conversations or to be sure that no one is modifying the contents of their instant messages.

In a single session there can be several media streams with different security requirements. One media stream may have to be encrypted while another may only need to be integrity-protected. Therefore, the natural place to introduce media security is in the session description, rather than in the SIP header fields (which would apply to the whole session).

The key management extensions for SDP (specified in the Internet-Draft "Key Management Extensions for SDP and RTSP" [48]) are encoded as SDP attributes (a= lines). An additional attribute is added to the description of a media stream that carries the name of the key management protocol to be used and the data needed by that protocol. It is possible, although not common, to offer multiple key management protocols to the answerer, who will choose one. Figure 8.15 shows an example of a media stream that uses the MIKEY key management protocol. Note that the transport protocol RTP/SAVP (Real-Time Transport Protocol/Secure Audio and Video Profile) is not the common RTP/AVP (Real-Time Transport Protocol/Audio and Video Profile) we have seen so far in session descriptions. RTP/SAVP identifies the SRTP (Secure RTP) protocol, which will be described in Section 15.2.4.

```
m=audio 20000 RTP/SAVP 0
a=key-mgmt:mikey <mikey-specific-data-in-binary-format>
```

Figure 8.15: Key management information in SDP

8.8.1   MIKEY

The key management protocol that appears in the session description in Figure 8.15 is called MIKEY (Multimedia Internet KEYing, specified in RFC 3830 [49]). MIKEY is a binary stand-alone key management protocol that uses two messages to establish keys to encrypt and/or integrity-protect a media stream. Since the offer/answer model (specified in RFC 3264 [212]) also uses two messages it is natural to embed MIKEY into the offer/answer model. That is, the offer carries the first MIKEY message in a "a=key-mgmt:mikey <data>" attribute and the answer carries the second MIKEY message in a similar attribute. Since MIKEY is a binary protocol and SDP is a textual format, MIKEY data is base64-encoded (as described in RFC 3548 [148]) before placing it in the SDP attribute.

MIKEY supports three ways of establishing keys between peers: using a pre-shared key, using public key encryption, and performing a Diffie–Hellman key exchange. In the three methods, peers establish a generation key that is used together with random nonces to generate the keys to encrypt or integrity-protect the traffic.

When peers have a shared secret (i.e., a pre-shared key), all the offerer needs to do is encrypt the generation key (along with other protocol parameters) with the pre-shared key and send it to the answerer. In the public key method the offerer encrypts the generation key (along with other protocol parameters) with a symmetric key (referred to as the envelope key) which is encrypted with the public key of the answerer. The Diffie–Hellman method consists of performing a Diffie–Hellman exchange to come up with the generation key.

Chapter 9

Security in the IMS

IMS security is divided into access security (specified in 3GPP TS 33.203 [3]) and network security (specified in 3GPP TS 33.210 [1]). Access security (which we describe in Section 9.1) includes authentication of users and the network, and protection of the traffic between the IMS terminal and the network. Network security (which we describe in Section 9.2) deals with traffic protection between network nodes, which may belong to the same or to different operators.

The IMS uses IPsec for both access and network security. However, core SIP only provides native support for TLS (as described in Section 8.3), which is the most common hop-by-hop security mechanism for SIP on the public Internet. The following sections address the extensions needed to establish and manage IPsec security associations for SIP in the IMS.

9.1 Access Security

A user accessing the IMS needs to be authenticated and authorized before they can use any IMS services. Once a user is authorized they protect their SIP traffic between their IMS terminal and the P-CSCF by using two IPsec security associations.

The authentication and authorization of the user and the establishment of their IPsec security associations is done using REGISTER transactions. The S-CSCF, armed with the authentication vectors downloaded from the HSS (Home Subscriber Server), authenticates and authorizes the user, while the P-CSCF establishes the IPsec security associations with the terminal. During the authentication process the user also authenticates the network to make sure that they are not speaking to a forged network.

9.1.1 Authentication and Authorization

Authentication and authorization in the IMS rely on the security functions present in the IMS terminal. Actually, the security functions are not implemented directly in the IMS terminal, but in a smart card that is inserted in the terminal. The functionality of the smart card is dependent on the actual network.

In 3GPP networks, the smart card is usually known as the UICC (Universal Integrated Circuit Card). The UICC contains one or more applications, as depicted in Figure 9.1. Each application stores a few configurations and parameters related to a particular usage. One of these applications is the ISIM (IP-Multimedia Services Identity Module). Other possible applications are the SIM (Subscriber Identity Module) and the USIM (UMTS Subscriber Identity Module). We describe in detail the UICC, its applications, and the parameters stored in each application in Section 3.6.

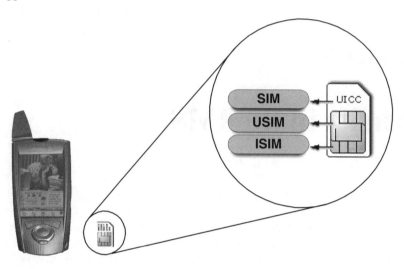

Figure 9.1: SIM, USIM, and ISIM in the UICC of 3GPP IMS terminals

3GPP networks allow access to the IMS when the UICC contains an ISIM or a USIM, although ISIM is preferred since it is tailored to the IMS. However, access with a USIM application is allowed in cases where the user has not updated their smart card to a more modern UICC that contains an ISIM. Due to the weak security functions, access to IMS with a SIM application is not allowed. Section 9.1.2 describes IMS authentication and authorization with an ISIM, whereas Section 9.1.3 describes the same procedures when the UICC contains a USIM.

The identification parameters that are stored in the ISIM in the 3GPP IMS are stored in the IMS terminal (pre-provisioned) or in the R-UIM (Removable User Identity Module) in the 3GPP2 IMS. These parameters are the same in both networks, as are the security functions. The storage can differ since 3GPP2 allows the IMS terminal or the R-UIM to store the identification parameters, but other than that there is no substantial difference. Section 9.1.2, which describes authentication and authorization with an ISIM, is also applicable for 3GPP networks.

9.1.2   Authentication and Authorization with ISIM

This section describes the authentication and authorization procedures that take place between the IMS terminal and the network when the terminal is equipped with an ISIM application.

Mutual authentication between a user and the network in the IMS is based on a long-term shared secret between the ISIM in the terminal and the HSS in the network. Every ISIM contains a secret key. The secret key of every particular ISIM is also stored in its home HSS. To achieve mutual authentication the ISIM and the HSS have to show to each other that they know the secret key. However, the terminal that contains the ISIM speaks SIP, but the

HSS does not. To resolve this issue the S-CSCF assigned to the user takes the role of the authenticator. Effectively, the HSS delegates this role to the S-CSCF.

The S-CSCF uses the Diameter protocol to obtain authentication vectors from the HSS and to challenge the user agent. These authentication vectors contain a challenge and the answer expected from the user agent to that challenge. If the user agent answers differently the S-CSCF considers the authentication to have failed. Let's see how the S-CSCF maps these challenges into a REGISTER transaction using digest access authentication.

The first thing an IMS terminal does when it logs onto an IMS network is to send a REGISTER request to its home network, as shown in Figure 9.2. The I-CSCF handling the REGISTER assigns, following the criteria obtained from the HSS in the Diameter exchange of messages (3) and (4), an S-CSCF for the user, that is tasked with authenticating and authorizing the user. To do so the S-CSCF downloads a number of authentication vectors from the HSS (7). Each vector contains a random challenge (RAND), a network authentication token (AUTN), the expected answer from the IMS terminal (XRES), a session key for integrity check (IK), and a session key for encryption (CK). The HSS creates the AUTN using the secret key that it shares with the ISIM and a sequence number (SQN) that is kept in synch between the ISIM and the HSS. Each authentication vector can be used to authenticate the ISIM only once. The S-CSCF downloads several vectors to avoid contacting the HSS every time it needs to authenticate the user again.

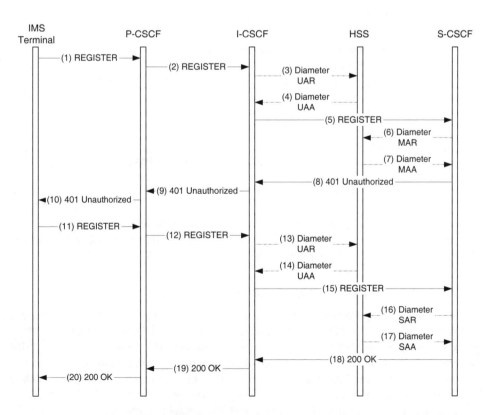

Figure 9.2: Initial REGISTER transaction

The S-CSCF uses the first authentication vector to build a digest challenge for the ISIM (as specified in RFC 3310 [168]). The S-CSCF builds a 401 (Unauthorized) response (8) that includes a `WWW-Authenticate` header field (digest operation is described in Section 8.1). The value of the `nonce` includes base64 encoding of the RAND and the AUTN. The value of the `algorithm` parameter is set to `AKAv1-MD5`. Figure 9.3 shows the contents of the `WWW-Authenticate` header field.

```
WWW-Authenticate: Digest
            realm="domain.com",
            nonce="CjPk9mRqNuT25eRkajM09uTl9nM09uTl9nMz50X25PZz==",
            qop="auth,auth-int",
            algorithm=AKAv1-MD5
```

Figure 9.3: `WWW-Authenticate` header field

When the terminal receives the 401 (Unauthorized) response (10) it deduces the RAND, the AUTN, and the CK and IK keys from the `nonce`. The IMSI calculates an AUTN using the SQN and its secret key. If it obtains the same value as the AUTN received, it considers that the network is authenticated. In this case the IMSI uses its secret key and the received RAND to generate a response value (RES), which is returned to the S-CSCF in the `Authorization` header field of a new REGISTER request (11). Figure 9.4 shows the contents of this header field.

```
Authorization: Digest
            username="Alice.Smith@domain.com",
            realm="domain.com",
            nonce="CjPk9mRqNuT25eRkajM09uTl9nM09uTl9nMz50X25PZz==",
            uri="sip:domain.com",
            qop=auth-int,
            nc=00000001,
            cnonce="0a4f113b",
            response="6629fae49393a05397450978507c4ef1",
```

Figure 9.4: `Authorization` header field

When the S-CSCF receives the REGISTER (15) it compares the RES value received with the expected value XRES in the authentication vector. If they match the S-CSCF considers that the user is authenticated and returns a 200 (OK) response (18).

9.1.3 Authentication and Authorization with USIM

An IMS terminal equipped with a UICC that contains a USIM but not an ISIM can still use the IMS. Obviously, the USIM does not contain the Private and Public User Identities, nor the long-term secret needed to authenticate the user by the IMS network. Still, the USIM contains an IMSI (equivalent to the Private User Identity in circuit- and packet-switched networks). The IMS terminal builds a temporary Private User Identity, a temporary Public User Identity, and a home network domain URI upon the contents of the IMSI. The procedure is described in detail in Section 5.5.2. The USIM also contains a long-term secret, typically used for authenticating in the circuit- and packet-switched networks. When the USIM is used to access an IMS network, both the network and the terminal use the long-term secret stored in the USIM and the HSS for authentication purposes.

In most cases the home operator would not like to disclose either the IMSI or the Private User Identity outside the home network. But, we have said that the temporary Private and Public User Identities are derived from the IMSI and, as we described in Section 5.5.2, they are visible in SIP messages. Therefore, the home operator has the ability to bar any Public User Identity, such as the temporary one, from being used in SIP messages other than the REGISTER request and its response. The IMS terminal can use any of the Public User Identities allocated to the user, as they are transferred to the terminal in the P-Associated-URI header field of the 200 (OK) response to the REGISTER. If an IMS terminal initiates a session with a barred Public User Identity, the S-CSCF will reject the session establishment.

9.1.4 Security Association Establishment

The P-CSCF and the terminal establish two IPsec security associations between them. Having two security associations, instead of one, allows terminals and P-CSCFs using UDP to receive the response to a request (port number in the Via header field of the request) on a different port than the one they use to send the request (source port of the IP packet carrying the request). Some implementors believe that implementations following this behavior are more efficient than implementations using a single port, which is why 3GPP standardized a multi-port solution. On the other hand, terminals and P-CSCFs using TCP between them send responses on the same TCP connection (i.e., using the same ports) as they received the request. Figures 9.5 and 9.6 illustrate the use of ports in UDP and TCP, respectively. In both cases, one security association is established from the terminal's client-protected port to the P-CSCF's server-protected port and the other goes from the P-CSCF's client-protected port to the terminal's server-protected port. Both security associations support traffic in both directions.

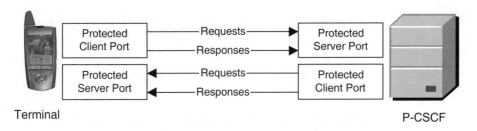

Figure 9.5: Use of ports and security associations with UDP

The P-CSCF and the terminal need to agree on a set of parameters to establish the two IPsec security associations between them (as specified in RFC 3329 [51]). The P-CSCF obtains the integrity and encryption keys (IK and CK) in a 401 (Unauthorized) response from the S-CSCF (which got them in an authentication vector from the HSS). The P-CSCF removes both keys from the response before relaying it to the IMS terminal. The P-CSCF and the IMS terminal use the same two REGISTER transactions (shown in Figure 9.2) that are used for authentication to negotiate the rest of the IPsec parameters.

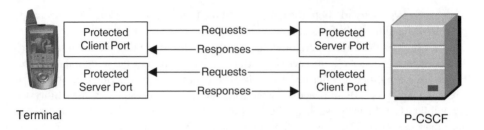

Figure 9.6: Use of ports and security associations with TCP

The terminal adds a `Security-Client` header field to the REGISTER (1) (Figure 9.2) as shown in Figure 9.7. This header field contains the mechanisms (`ipsec-3gpp`) and algorithms (`hmac-sha-1-96`) the terminal supports as well as the SPIs (identifiers for the security associations) and port numbers that it will use.

```
Security-Client: ipsec-3gpp; alg=hmac-sha-1-96;
                 spi-c=23456789; spi-s=12345678;
                 port-c=2468; port-s=1357
```

Figure 9.7: `Security-Client` header field

The P-CSCF adds a `Security-Server` header field to the 401 (Unauthorized) response (8), as shown in Figure 9.8. This header field contains the mechanisms (`ipsec-3gpp`) and algorithms (`hmac-sha-1-96`) the P-CSCF supports as well as the SPIs and port numbers that it will use. In addition to this the P-CSCF indicates its preferred security mechanism q values. Our `Security-Server` header field has a single mechanism, but, when there are more, mechanisms with higher q values are preferred.

```
Security-Server: ipsec-3gpp; q=0.1; alg=hmac-sha-1-96;
                 spi-c=98765432; spi-s=87654321;
                 port-c=8642; port-s=7531
```

Figure 9.8: `Security-Server` header field

```
Security-Verify: ipsec-3gpp; q=0.1; alg=hmac-sha-1-96;
                 spi-c=98765432; spi-s=87654321;
                 port-c=8642; port-s=7531
```

Figure 9.9: `Security-Verify` header field

The security associations are ready to be used as soon as the terminal receives the `Security-Server` header field (8). So, the terminal sends a REGISTER (9) request over one of the just established security associations. The terminal includes a `Security-Verify` header field in this REGISTER, as shown in Figure 9.9, mirroring the contents of the `Security-Server` header field received previously. This way the server will be sure that no man-in-the-middle modified the list of security mechanisms that it sent to the client.

At present, P-CSCFs only support `ipsec-3gpp`, but in the future they might support other security mechanisms. An attacker could remove the strongest security mechanisms from the `Security-Server` list to force the terminal to use weaker security. With the addition of the `Security-Verify` header field an attacker that had modified the `Security-Server` list would need to break the security mechanism chosen in real time to modify the `Security-Verify` header field as well. Otherwise, the P-CSCF would notice the attack and abort the registration. This way of selecting a security mechanism is secure as long as the weakest mechanism in the list cannot be broken in real time.

9.2 Network Security

Network security deals with securing traffic between different security domains, where a security domain is a network that is managed by a single administrative authority. For example, sessions where the P-CSCF and the S-CSCF are in different networks involve traffic exchanges between different security domains.

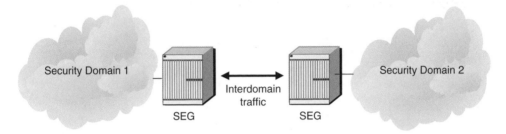

Figure 9.10: Inter-domain traffic through two security gateways

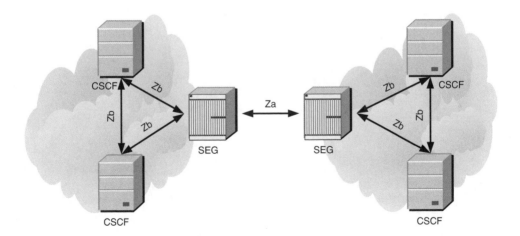

Figure 9.11: *Za* and *Zb* interfaces

All the traffic coming or leaving a security domain traverses a SEG (Security Gateway). Consequently, traffic sent from one domain to another domain traverses two SEGs, as shown in Figure 9.10.

Traffic between SEGs is protected using IPsec ESP (Encapsulated Security Payload, specified in RFC 2406 [149]) running in tunnel mode. The security associations between SEGs are established and maintained using IKE (Internet Key Exchange, specified in RFC 2409 [118]).

Within a security domain, network entities exchange traffic with the SEGs of the domain using IPsec. Network entities within a domain also use IPsec to exchange traffic between them. This way, from a security standpoint SEGs are treated as any other network entity within the domain. Figure 9.11 illustrates this point. The interface between regular network entities and between network entities and SEGs is the same: the *Zb* interface. The interface between SEGs from different domains in called *Za*.

Authentication, integrity protection, and encryption are mandatory in the *Za* interface. This offers the inter-domain IMS traffic the maximum degree of protection.

The *Zb* interface is designed to protect the IMS signaling plane. As the interface only carries intra-operator traffic, it is up to the operator to decide whether to deploy the interface and, in the case where it is deployed, which security functions to include (integrity protection is mandatory if the *Zb* interface is implemented, but encryption is optional).

Chapter 10

Policy on the Internet

The mechanisms we will describe in this chapter deal with media-level access control and QoS policies. That is, decisions about whether a user is authorized to send or receive media and, if so, with which QoS.

The media-level policy of a domain is enforced by its routers, which accept or reject access requests based on that policy. Still, routers do not usually have access to the information needed to make policy decisions, such as the profile of the user attempting to send media. On the other hand, a SIP proxy server can authenticate users and obtain their profiles before they establish a session and start sending media. So, a SIP proxy can check the user's profile and instruct the router to accept or reject the user's media. The Common Open Policy Service (COPS) protocol can be used between the SIP proxy and the routers to transmit the proxy's decision.

The COPS protocol (specified in RFC 2748 [85]) is used between a Policy Decision Point (PDP) and a Policy Enforcement Point (PEP) to transmit policy-related information. In our previous example the SIP proxy acts as the PDP and the router acts as the PEP.

COPS supports two models of policy control: outsourcing and configuration (also known as provisioning). In the outsourcing model the PEP contacts the PDP every time a policy decision has to be made. The PDP makes the decision and communicates it back to the PEP, which enforces it. In the configuration (or provisioning) model the PDP configures the PEP with the policy to be used. The PEP stores the policy received from the PDP locally and uses it to make decisions, instead of contacting the PDP every time a new event occurs.

Let's first study the fundamentals of the COPS protocol in Section 10.1. After that we will describe both the outsourcing and the configuration models in Sections 10.2 and 10.3, respectively.

10.1   The COPS Protocol

The COPS protocol is a request/response protocol used between policy servers (PDPs) and their clients (PEPs). The clients send requests to their server, which responds back. Servers can also send unsolicited information to the clients.

COPS is a framework that supports different types of PEPs, which are identified by their Client-Type values. Two examples of PEPs are an RSVP-enabled (Resource ReSerVation Protocol) router and a GGSN in a 3G network. They need to exchange different pieces of information with the PDP in order to make policy decisions, but both use COPS.

Each Client-Type defines the actual contents of the COPS messages that can be exchanged between PEPs and PDPs.

A PEP and a PDP exchange COPS traffic over a TCP connection that is always initiated by the PEP. This TCP connection can be secured using IPsec or TLS. Nonetheless, when none of these security mechanism is available, COPS messages themselves provide integrity protection.

The format of COPS messages is shown in Figure 10.1. COPS messages are binary-encoded and consist of a common header followed by a number of objects. The Op-Code contains the message type, which can take the values shown in Table 10.1. The Client-Type indicates which type of policy client the PEP is. A COPS server may support different types of clients.

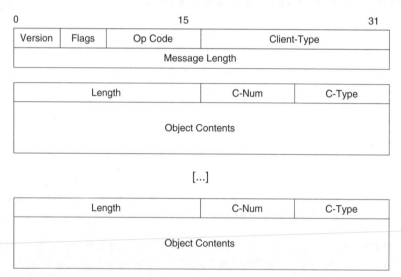

Figure 10.1: COPS Message Format

Table 10.1: Op Codes

Value	Message type	Acronym
1	Request	REQ
2	Decision	DEC
3	Report State	RPT
4	Delete Request State	DRQ
5	Synchronize State Request	SSQ
6	Client-Open	OPN
7	Client-Accept	CAT
8	Client-Close	CC
9	Keep-Alive	KA
10	Synchronize Complete	SSC

The C-Num and C-Type values identify the format of a particular object. The C-Num contains the object type and the C-Type contains its subtype. For example, objects carrying interface identifiers for incoming traffic (C-Num 3) can carry an IPv4 (C-Type 1) or an IPv6 address (C-Type 2). Table 10.2 shows the values C-Num can take.

Table 10.2: C-Num values

Value	Object type
1	Handle
2	Context
3	In Interface
4	Out Interface
5	Reason Code
6	Decision
7	LPDP Decision
8	Error
9	Client Specific Info
10	Keep-Alive Timer
11	PEP Identification
12	Report Type
13	PDP Redirect Address
14	Last PDP Address
15	Accounting Timer
16	Message Integrity

10.2 The Outsourcing Model

In the outsourcing model the PEP contacts the PDP every time a policy decision needs to be made. The COPS usage for RSVP (specified in RFC 2749 [120]) is an example of this model. Section 12.1.1 describes RSVP operations. Here, we will concentrate on the policy decisions related to those operations.

RSVP-enabled routers, which act as PEPs and use the COPS Client-Type 1, query the PDP when an RSVP message is received. The client specifies which RSVP message was received (e.g., PATH or RESV) and the expected behavior of the router (e.g., reserve resources for a particular flow). The PDP decides whether or not the requested reservation is acceptable according to the policy of the domain and communicates its decision back to the PEP. A PDP can also send unsolicited instructions to a PEP (e.g., remove the resources assigned to a particular flow).

Figure 10.2 shows an example where an RSVP-enabled router communicates with a policy server using COPS. When the RSVP-enabled router receives a PATH message (1) it informs the policy server by sending it a REQ message (2). The policy server authorizes the router to forward the PATH message (4) with a DEC message (3). When the router receives a RESV message (5) it sends a new REQ message (6) to the policy server, asking for permission to reserve resources for a flow. The policy server authorizes the router to do so with a DEC message (7). The router informs the policy server that the resources have been reserved with an RPT message (8). Finally, the router forwards the RESV (9) message to the originator of the previous PATH.

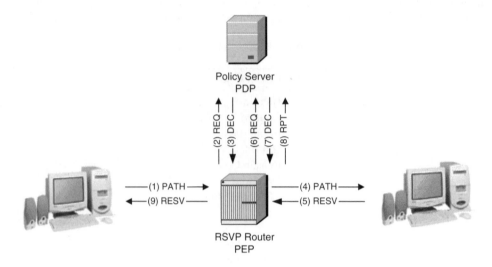

Figure 10.2: Example of COPS usage for RSVP

COPS is a stateful protocol. This means that when the router of our example receives new PATH or RESV messages (the endpoints will send them periodically following standard RSVP procedures), it does not need to contact the PDP again. It will simply refresh the resource reservations that are already in place. On the other hand, if no new RSVP messages arrive for a period of time, the router will deallocate the resources for the flow and inform the PDP about the timeout with a DRQ message.

10.3   The Configuration Model

In the configuration, or provisioning, model (known as COPS-PR and specified in RFC 3084 [78]) the PDP provides PEPs with policies and the PEP applies these policies (i.e., the PEP makes decisions locally based on the policies received from the PDP). These policies are encoded in policy documents called PIBs (Policy Information Bases).

A PIB contains a number of PRCs (Provisioning Classes), and each PRC contains a number of Provisioning Instances (PRIs). PRCs provide definitions, while PRIs are instantiations of one such definition. A PRC can be seen as a table that contains information about a particular topic (e.g., capabilities of the PEP). The PRIs inside this PRC would be the rows of the table, which would contain the actual values (e.g., each individual capability). The language used to write PIBs is called SPPI (Structure of Policy Provisioning Information, specified in RFC 3159 [164]).

When a COPS-PR client contacts a server it does not provide information about a particular event, as in the outsourcing model; rather, it provides information about the client itself. With this information the server decides which policies to send to the client. Once the client downloads this policy it does not need to contact the server to make individual decisions. This makes COPS-PR a highly scalable protocol.

COPS-PR's functionality considerably overlaps with the functionality provided by traditional network management protocols, which are used to configure network devices, such as routers. So, COPS-PR could potentially be used to transfer configuration parameters beyond policy information.

Chapter 11

Policy in the IMS

The IMS uses the COPS protocol to transfer policy-related information, but does not use a single model. A mixture of the COPS outsourcing and provisioning (COPS-PR) models (which are described in Chapter 10) is used instead. The message format and the use of PIBs (Policy Information Bases) comes from the provisioning model. The transfer of policy decisions in real time comes from the outsourcing model.

11.1 SIP Procedures

There are two types of limitations on the type of sessions a terminal can establish over an IMS network:

   (a) user-specific limitations;

   (b) general network policies.

    An example of a user-specific limitation is an audio-only subscription. A user with such a subscription is not authorized to establish video streams. General network policies are not user-specific, but apply to all the users in the network. For instance, an IMS network may not allow high-bandwidth audio codecs like G.711.

    The offer/answer exchanges that a terminal performs need to take into account these limitations. For instance, if a terminal is not allowed to establish video sessions, it should not perform offer/answer exchanges that include video media streams. The P-CSCF and the S-CSCF ensure that unauthorized offer/answer exchanges do not take place. The P-CSCF handles general network policies, while the S-CSCF deals with both network policies and user-specific limitations.

    Both the P-CSCF and the S-CSCF use the same mechanism to keep user agents from performing unauthorized offer/answer exchanges. They rely on having access to the SDP bodies that contain the offer/answer exchanges between user agents. If an offer contains information against the policy (e.g., a forbidden codec) the CSCF returns a 488 (Not Acceptable Here) response with an SDP body describing either the policy or a subset of it. For instance, the SDP body can contain a session description that would have been an acceptable offer (e.g., the forbidden codec is removed from the list of codecs). Figure 11.1 shows an S-CSCF using this mechanism. The S-CSCF receives an INVITE request (2) with the session description indicated in Figure 11.2, which contains a video stream in addition to an audio stream. The S-CSCF returns a 488 response (3), containing the session description indicated in Figure 11.3. This session description includes only an audio stream, indicating that the user cannot establish video streams.

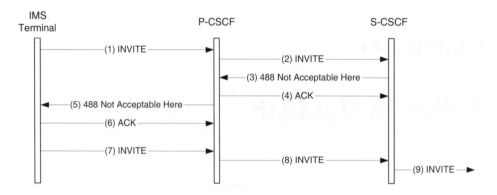

Figure 11.1: The S-CSCF does not accept the terminal's offer

```
v=0
o=- 2790844676 2867892807 IN IP6 1080::8:800:200C:417A
s=-
c=IN IP6 1080::8:800:200C:417A
t=0 0
m=audio 20000 RTP/AVP 97 96
b=AS:25.4
a=curr:qos local none
a=curr:qos remote none
a=des:qos mandatory local sendrecv
a=des:qos none remote sendrecv
a=rtpmap:97 AMR
a=fmtp:97 mode-set=0,2,5,7; maxframes=2
a=rtpmap:96 telephone-event
m=video 20002 RTP/AVP 31
```

Figure 11.2: SDP in the INVITE request

11.2   Media Authorization

When two user agents perform an offer/answer exchange that is acceptable to the P-CSCF and the S-CSCF they are allowed to exchange media. The network needs to instruct the GGSN (in 3GPP networks) to authorize the users' media traffic as long as it complies with the previously authorized offer/answer exchange. That is, two user agents that perform an offer/answer exchange to establish an audio stream are not authorized to exchange video traffic.

```
v=0
o=S-CSCF 2790844634 2867892823 IN IP6 1080::8:800:200C:2A2F
s=-
c=IN IP6 1080::8:800:200C:2A2F
t=0 0
m=audio 0 RTP/AVP 97 96
b=AS:25.4
a=rtpmap:97 AMR
a=fmtp:97 mode-set=0,2,5,7; maxframes=2
a=rtpmap:96 telephone-event
```

Figure 11.3: SDP in the 488 response

11.2.1 The Policy Decision Function

The IMS uses the COPS protocol between a network node called the PDF (Policy Decision Function) and the GGSN. The GGSN acts as the PEP and the PDF acts as the PDP. In 3GPP Release 5 the PDF is a logical entity that can either be co-located with the P-CSCF or implemented as a stand-alone unit, in which case the protocol between them is a proprietary protocol. In 3GPP Release 6 the P-CSCF and the PDF may be separate and the protocol is standardized. Figure 11.4 illustrates both architectures. The interface between the PDF and the GGSN is called the *Go* interface (which is specified in 3GPP TS 29.207 [31]) and it is based on COPS. The interface between the PDF and the P-CSCF is called the *Gq* interface (which is specified in 3GPP TS 29.209 [32]) and it is based on Diameter. We will assume the Release 5 architecture in our following discussions.

The role of the PDF consists of providing the GGSN with the characteristics of the session a user agent is authorized to establish at a particular time. For example, a user agent may be authorized to establish an audio stream whose maximum bandwidth is not greater than 20 kbit/s. The GGSN uses this information to install packet filters in its routing logic. If the user agent tries to do something which is not authorized to do, like sending media to a different destination than the one indicated in the offer/answer exchange, the packet filters discard the user agent's traffic.

The *Go* interface is optional. That is, there might be IMS networks that do not implement it. Still, we expect most commercial IMS deployments to implement this interface because it provides the operator with useful functionality.

11.2.2 Media Authorization Token

3GPP provides a mechanism for the network to authorize the establishment of media streams. The mechanism is known as the *Service-Based Local Policy (SBLP)* and is based on the network inserting a media authorization token that IMS terminals return to the network when requesting the establishment of a media stream.

The SBLP model (specified in 3GPP TS 23.207 [13]) follows the media authorization model (described in RFC 3521 [113]). Figure 11.5 shows a message flow where different entities exchange authorization tokens (3GPP TS 29.208 [14] contains additional, similar message flow examples).

The PDF generates an authentication token that the P-CSCF adds to the incoming INVITE in a P-Media-Authorization header field (specified in RFC 3313 [161]). The terminal

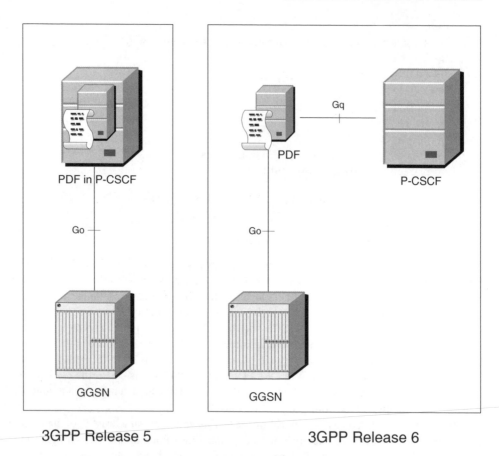

3GPP Release 5   **3GPP Release 6**

Figure 11.4: Architecture for media authorization in 3GPP Releases 5 and 6

adds this token (whose format is specified in RFC 3520 [112]) to the messages it sends to reserve network resources. In our example the terminal's access network is a GPRS network, so it performs a PDP context activation. The terminal also includes information to identify the flows it allocates to the PDP context.

The GGSN extracts the token from the received message and sends it, together with the flow identifier information, to the PDF using COPS. The PDF sends back the characteristics of the session the terminal is authorized to establish (which includes the allowed flows). This information is referred to as SBLP policy information.

Figure 11.6 shows how the authorization token is exchanged in an outgoing session. The P-CSCF adds the token to an incoming provisional response this time. Note that the authorization token is piggybacked on a SIP message that was sent by one of the user agents in both Figures 11.5 and 11.6. That is, the P-CSCF does not generate new messages to transfer tokens. It just takes advantage of SIP messages that are exchanged between the endpoints during session establishment.

The media authorization token generated by the PDF contains, in binary format, binding information that identifies the SIP dialog that the token belongs to and the address of the PDF

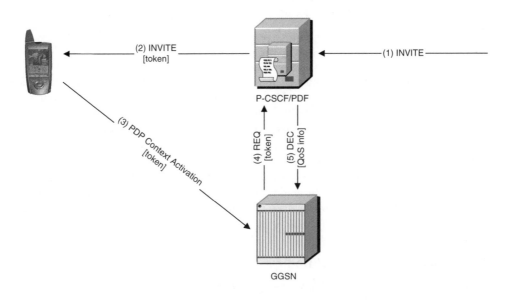

Figure 11.5: Authorization token transfer in an INVITE request

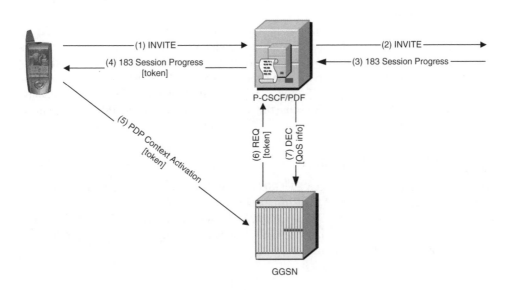

Figure 11.6: Authorization token transfer in a SIP response

that generated the token. We can see an example of an authorization token in the following P-Media-Authorization SIP header field:

```
P-Media-Authorization: 0020000100100101706466322
e76697369746564322e6e6574000c020139425633303732
```

When the GGSN receives this token, (3) in Figure 11.5 and (5) in Figure 11.6, it extracts the address of the PDF from the token and sends a COPS REQ message to that address. The GGSN attaches the token and the flow information received from the user agent to this REQ message. The PDF extracts the token from the REQ message and uses the binding information in it to identify the SIP dialog that the token belongs to. It then sends back a COPS DEC message with the authorized QoS for the session.

We describe the operations the GGSN performs with the QoS information received from the PDF in Section 13.3. For our current discussion about policy it is enough to understand that the PDF uses COPS to authorize the GGSN to route packets that belong to a specific session in a certain way (i.e., with a certain QoS).

11.3   Proxy Access to SDP Bodies

All the IMS policy mechanisms that we have described assume that P-CSCFs and S-CSCFs have access to the SDP bodies of the SIP messages. Consequently, the use of end-to-end encryption mechanisms like S/MIME (described in Section 8.4) is not allowed in the IMS. Moreover, user agents are forced to use SDP as their session description format. Otherwise, the CSCFs would not be able to understand which type of session was being established.

Furthermore, Section 13.1 describes how the P-CSCF modifies SDP bodies to convey QoS information to the terminals. So, end-to-end integrity protection mechanisms are also not allowed in the IMS.

The IETF is working on extensions to allow proxy servers to communicate different policies to the user agents without accessing end-to-end bodies such as session descriptions (see the Internet-Drafts "Session Initiation Protocol (SIP) Session Policies – Document Format and Session Independent Delivery Mechanism" [122], and "A Delivery Mechanism for Session-Specific Session Initiation Protocol (SIP) Session Policies" [121]). Unfortunately, these extensions will not be ready for some time. So, we do not expect the IMS to adopt them in the immediate future.

11.4   Initialization Procedure

The initialization procedure between a GGSN (which acts as a PEP) and a PDF (which acts as a PDP) is based on the *3GPP Go PIB* (specified in 3GPP TS 29.207 [31]). The GGSN encodes its capabilities using the *3GPP Go PIB* and sends them to the PDF in a COPS REQ message. The PDF responds with a COPS DEC message, readying both nodes to transfer policy information, as described previously.

Chapter 12

Quality of Service on the Internet

Although the Internet has been traditionally a best-effort network, the ability to provide a certain level of QoS for certain packet flows is essential for some applications. For example, while the user of a file transfer application may accept a longer transfer delay when the network is congested a multimedia user may find *trying* to maintain a conversation with a long round trip delay irritating. Such users would probably request a higher QoS for their multimedia flows than for the rest of their flows.

But, QoS is not only about requesting a better treatment for certain flows; users also want to know if the network will be able to provide them with the requested QoS. If there is a long delay or a high packet loss rate, some users may prefer to exchange instant messages instead of having a VoIP (Voice over IP) conversation.

There are two models to provide QoS on the Internet: the *Integrated Services* model and the *Differentiated Services* (DiffServ) model. We cover the former in Section 12.1 and the latter in Section 12.2.

12.1   Integrated Services

The Integrated Services architecture (specified in RFC 1633 [56]) was designed to provide end-to-end QoS. Endpoints request a certain level of QoS for their packet flows and, if the network grants it, their routers treat those flows accordingly. There are two different services available in this architecture: the controlled load service and the guaranteed service.

The controlled load service ensures that packets are treated as if the network was under moderate load. Flows using this service are not affected by network congestion when this appears. Nevertheless, the network does not guarantee a certain bandwidth or a certain delay. This service can be seen as a better-than-best-effort service.

The guaranteed service guarantees a certain bandwidth or a certain delay threshold. In practice, it is not common to see this service in use because the controlled load service is often good enough and is easier to manage.

12.1.1   RSVP

The *Integrated Services* architecture uses RSVP (Resource ReSerVation Protocol, specified in RFC 2205 [57]) as the resource reservation protocol. Endpoints send RSVP messages requesting a certain QoS (e.g., a certain bandwidth) for a flow. Routers receiving these messages obtain a description of the flow (e.g., source and destination transport addresses) as well, so that they can apply the correct treatment to all the packets that belong to it. Obviously, RSVP needs to ensure that the routers receiving resource reservation requests for a flow are the routers that will route the packets of that flow. That is, RSVP messages need to follow the same path as the packets of the flow (e.g., RTP packets carrying voice). Let's see how RSVP achieves this.

An RSVP reservation consists of a two-way handshake: a PATH message is sent and a RESV message is received, as shown in Figure 12.1. The PATH message is sent from endpoint A to endpoint B, and the network routes it as for any other IP packet. At a later time, when endpoint A sends RTP packets with voice to endpoint B they will follow the same path as the PATH message did.

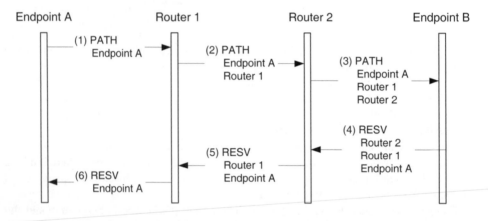

Figure 12.1: RSVP operation

PATH messages store the nodes they traverse. This allows RESV messages to be routed back to endpoint A, following the same path as the PATH message followed but in the opposite direction. In short, PATH messages leave a trail of bread crumbs so that RESV messages can find their way home. Note that SIP uses the same mechanism to route responses back to the UAC (see Section 4.6). SIP uses the `Via` header field to store the trail left by the request.

Resource reservation actually takes places when routers receive RESV messages. Therefore, resource reservation is actually performed by endpoint B; that is, the receiver of the flow (e.g., the RTP packets). PATH messages only mark the path that RESV messages have to follow.

Note, however, that the packets of the flow follow the same path as the PATH message as long as there are no routing changes in the network. If, as a consequence to a change in the network topology, packets from endpoint A to endpoint B start following a different path, a new resource reservation is needed. RSVP tackles this using soft states. Reservation soft states created by RESV messages are kept in routers only for a period of time. If they timeout before they are refreshed by a new RESV message, routers just delete them.

Endpoints periodically exchange PATH and RESV messages while the flow (e.g., the RTP packets) is active. This way, after a change in the routing logic in the network, those routers that no longer remain in the path between the endpoints do not get any new refreshes and consequently, remove their reservation state. New routers start receiving RESV messages that install the reservation state needed for the flow.

Note that RSVP is an admission control protocol, in addition to being a resource reservation protocol. A router can reject a resource reservation request either because the router does not have enough resources or because the user is not allowed to reserve them.

12.1.2 State in the Network

The main problem with the integrated services architecture is that the network needs to store a lot of state information. When a packet arrives at a router, the router needs to check all the reservations it is currently handling to see whether the packet belongs to any of them. This means that routers need to store state information about every flow and need to perform lookups before routing any packet. Even though RSVP supports aggregation of reservations for multicast sessions in order to reduce the state the network needs to keep, the general feeling is that RSVP does not scale well when implemented in the core network. On the other hand, RSVP can be used to perform admission control or to connect DiffServ clouds without these scalability problems.

12.2 Differentiated Services

The DiffServ architecture (specified in RFC 2475 [55] and RFC 3260 [111]) addresses some of the problems in the integrated services architecture. DiffServ routers need to keep minimal state, enabling them to decide which treatment a packet needs more quickly.

DiffServ routers know what treatments a packet can get. These treatments are referred to as Per-Hop Behaviors (PHBs) and are identified by 8-bit codes called Differentiated Services Codepoints (DSCPs). IP packets are marked at the edge of the network with a certain DSCP so that routers in the path apply the correct PHB to them. Two examples of standard PHBs are *expedited forwarding* (specified in RFC 3246 [79]) and *assured forwarding* (specified in RFC 2597 [119]). Packets to which the expedited forwarding PHB is applied do not see any congestion in the network. Effectively, they are treated as if they were transmitted over a TDM (Time Division Multiplexing) circuit that was exclusively reserved for them. The assured forwarding PHB provides different drop precedence levels, so that low-priority packets are discarded before high-priority ones under congestion conditions.

0		15			31
Version	Header Length	Type of Service	Total Length		
Identification			Flags	Fragment Offset	
Time to Live		Protocol	Header Checksum		
Source Address					
Destination Address					

Figure 12.2: DSCP is encoded in the *Type of Service* field in IPv4

IP packets carry their DSCPs in their IP header. The format of IPv4 and IPv6 headers is shown in Figures 12.2 and 12.3 respectively. The DSCP for a packet is placed in the *Type of Service* field in IPv4 and in the *Traffic Class* field in IPv6.

0		15		31
Version	Traffic Class	Flow Label		
Payload Length		Next Header		Hop Limit
Source Address				
Destination Address				

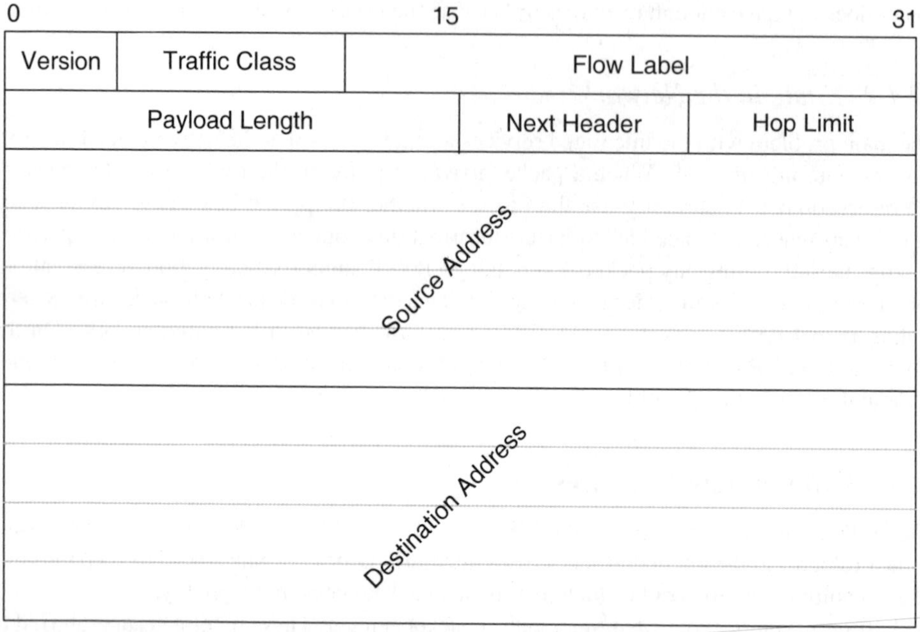

Figure 12.3: DSCP is encoded in the *Traffic Class* field in IPv6

Chapter 13

Quality of Service in the IMS

The IMS supports several end-to-end QoS models (described in 3GPP TS 23.207 [13]): terminals may use link-layer resource reservation protocols (e.g., PDP context activation), RSVP, or DiffServ codes directly; while networks use DiffServ and may use RSVP. In any case the most common model is to have terminals use link-layer protocols and to have the GGSN map link-layer resource reservation flows to DiffServ codes in the network. In this chapter we will focus on GPRS access networks. Consequently, we will describe the procedures to create and manage PDP contexts.

13.1 Instructions to Perform Resource Reservations

Terminals need to be able to map the media streams of a session into resource reservation flows. A terminal that establishes an audio and a video stream may choose to request a single reservation flow for both streams or to request two reservation flows, one for video and one for audio. Requesting a reservation flow may consist of creating a secondary PDP context or sending RSVP PATH messages, for instance.

The P-CSCF instructs the terminal to perform resource reservation. To do so the P-CSCF uses the SRF (Single Reservation Flow) semantics (specified in RFC 3524 [68]) of the SDP grouping framework.

The SDP grouping framework (specified in RFC 3388 [65]) allows us to group media streams and to describe the semantics of the group. For example, LS (Lip Synchronization) semantics indicate that the play-out of media streams in the group need to be synchronized. LS semantics are typically used to group an audio and a video stream, as shown in Figure 13.1. The a=group line carries the semantics of the group (LS in this case) and the identifiers of the streams (the a=mid line in the streams).

SRF semantics indicate that all the streams in the group should use the same resource reservation flow. Consequently, the two audio streams of the session description in Figure 13.2 would use the same PDP context (assuming a GPRS access), while the video stream would use its own PDP context.

The P-CSCF adds a=mid and a=group:SRF lines to the session descriptions before relaying them to the terminals (as described in 3GPP TS 24.229 [16]). The terminals use this information to perform resource reservation. Figures 13.3 and 13.4 illustrate this point.

The 3G IP Multimedia Subsystem (IMS) Second Edition   Gonzalo Camarillo and Miguel A. García-Martín
© 2006 John Wiley & Sons, Ltd

```
v=0
o=- 289083124 289083124 IN IP6 1080::8:800:200C:417A
t=0 0
c=IN IP6 1080::8:800:200C:417A
a=group:LS 1 2
m=audio 20000 RTP/AVP 0
a=mid:1
m=video 20002 RTP/AVP 31
a=mid:2
```

Figure 13.1: Grouping streams using LS semantics

```
v=0
o=- 289083124 289083124 IN IP6 1080::8:800:200C:417A
t=0 0
c=IN IP6 1080::8:800:200C:417A
a=group:SRF 1 2
a=group:SRF 3
m=audio 20000 RTP/AVP 0
a=mid:1
m=audio 20002 RTP/AVP 0
a=mid:2
m=video 20004 RTP/AVP 31
a=mid:3
```

Figure 13.2: Grouping streams using SRF semantics

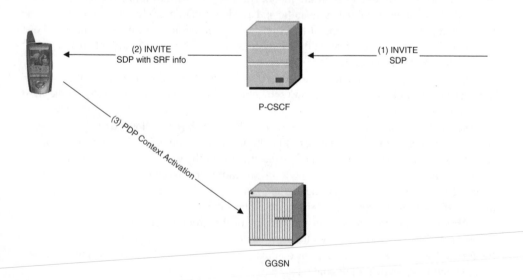

Figure 13.3: P-CSCF adds SRF info to an incoming INVITE request

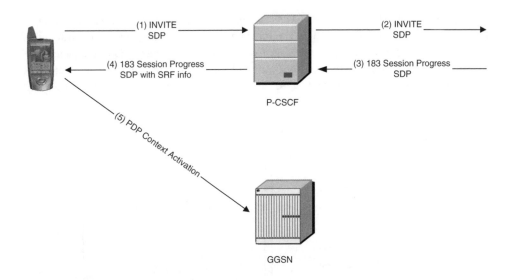

Figure 13.4: P-CSCF adds SRF info to an incoming 183 response

The P-CSCF may or may not use the mechanism we have just described in this section. It may happen, therefore, that the SDP that the IMS terminal receives does not contain any instructions to perform resource reservation. In that case the IMS terminal is free to decide how to group media streams into reservation flows.

13.1.1 Proxy Modifying Bodies

The mechanism just described assumes that the P-CSCF can both understand and modify the session descriptions exchanged by the terminals. This implies that terminals can only use SDP as their session description format and that end-to-end integrity protection or encryption mechanisms like S/MIME (described in Section 8.4) are not allowed in the IMS. Section 11.3 provides further reasons for these restrictions.

The IETF is working on extensions to allow proxy servers to communicate information, such as resource reservation instructions, to user agents without accessing end-to-end bodies, such as session descriptions (see the Internet-Drafts "Session Initiation Protocol (SIP) Session Policies – Document Format and Session Independent Delivery Mechanism" [122], and "A Delivery Mechanism for Session-Specific Session Initiation Protocol (SIP) Session Policies" [121]). As mentioned in Section 11.3, these extensions will not be ready for some time. So, we do not expect the IMS to adopt them in the immediate future.

13.2 Reservations by the Terminals

Terminals use the SRF information received in session descriptions to determine the number of resource reservation flows that need to be established. When the access network is GPRS a resource reservation flow is a PDP context. Let's see how the terminals establish PDP contexts (this is described in 3GPP TS 24.008 [27]).

A terminal establishes a PDP context to exchange SIP signaling right after performing a GPRS attach, as shown in Figure 13.5. The information stored by the network regarding this PDP context includes the terminal's IP address and the PDP context's QoS characteristics including its traffic class. There exist four traffic classes: best effort, interactive, streaming, and conversational. The PDP contexts used for SIP signaling are always conversational.

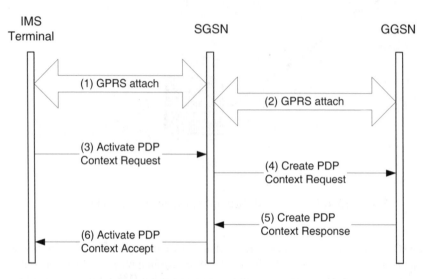

Figure 13.5: PDP context activation

IMS terminals establish additional PDP contexts to send and receive media, as shown in Figure 13.6. The number of additional PDP contexts, referred to as *secondary PDP contexts*, depends on the instructions received from the P-CSCF in the form of a=group:SRF lines. Secondary PDP contexts use the same IP address as the primary PDP context, but may have different QoS characteristics.

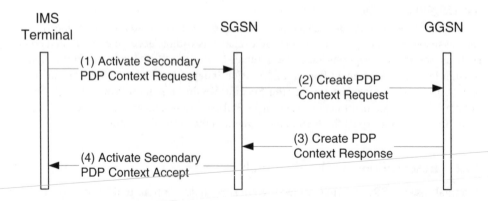

Figure 13.6: Secondary PDP context activation

13.3 Network Authorization

Terminals place the authorization token they receive from the PDF (see Section 11.2.2) in their *Activate Secondary PDP Context Request* messages. The SGSN receives this message and checks the user's subscription information stored in the HSS using MAP (Mobile Application Part, described in 3GPP TS 29.002 [26]). If the terminal requests more bandwidth than the user's subscription allows, the SGSN downgrades the requested bandwidth to an acceptable figure. The SGSN then sends a *Create PDP Context Request* to the GGSN. This message includes the authorization token received in message (1). The GGSN receives this token in message (2) of Figure 13.7 and sends it to the PDF in a COPS REQ message (3). The PDF responds with a DEC message (4), which contains the authorized QoS characteristics, and identifies the packet flows that can use the PDP context. This mechanism is referred to as SBLP (Service-Based Local Policy).

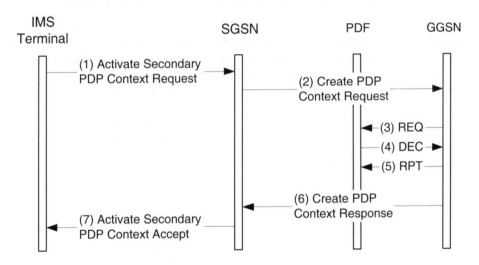

Figure 13.7: QoS authorization by the network

The GGSN uses packet filters to ensure that only authorized packet flows are sent over a given PDP context. Packet flows identifiers consist of 5-tuples that contain the source and destination addresses, source and destination ports, and the transport protocol. For example, a PDP context may only carry traffic sent to the IP address that the remote user agent provided in its answer (or offer). If the GGSN did not implement this type of packet filter a rogue user could use a PDP context authorized by the PDF for a different purpose than the one the authorization was issued for.

The GGSN, after receiving the DEC message (4), returns an RPT message (5) to the PDF indicating that it will comply with the PDF's policy. Then, it authorizes the requested PDP context (6) using the QoS parameters that were received from the PDF.

13.4 QoS in the Network

The GGSN receives traffic from a given terminal over a PDP context, assigns it an appropriate DSCP (Differentiated Services Codepoint), and sends it out into a DiffServ-enabled network, as shown in Figure 13.8. In short, the GGSN implements the DiffServ edge function.

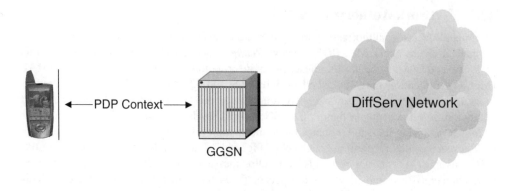

Figure 13.8: Mapping PDP context information to DSCPs by the GGSN

The DSCP that corresponds to a particular PDP context is typically assigned based on statically configured rules in the GGSN. Still, when RSVP is used the GGSN may use the information carried in RSVP signaling to decide which DSCP to use.

Part III

The Media Plane in the IMS

Part III deals with the IMS media plane. We tackle several media types, although our focus is set on audio and video, especially in Chapter 14. In Chapter 14 we describe how to represent audio and video signals in digital format using audio and video codecs. Chapter 15 describes how to transport encoded media in an IP network.

We follow a similar approach to the one we chose for Part II: we first introduce the technology and then explain how it is used in the IMS.

Chapter 14

Media Encoding

Media encoding is not an Internet technology. Hence, it is different from most of the technologies described so far. While many of the technologies we discussed earlier were originally developed for the Internet and later adapted to work in the IMS, media-encoding technologies were developed for other environments. For example, speech coding was developed with circuit-switched telephony in mind.

Because of this we will not follow the same structure as in previous chapters, where we described how each technology is used on both the Internet and in the IMS. In this chapter we will describe how speech-encoding technologies evolved from the simple codecs used in the fixed PSTN to the advanced codecs used in today's cellular networks. In addition, we will look at video-encoding technologies.

14.1 Speech Encoding

Audio encoding consists of converting an analog audio signal into a digital signal. This digital signal is transmitted over the network and decoded at the receiver's side, as shown in Figure 14.1. If the audio signal consists of human speech, the process just described is referred to as speech encoding. We will focus on speech encoding in this section.

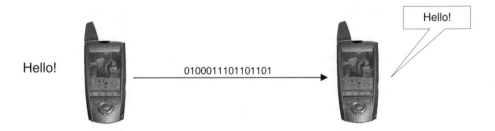

Figure 14.1: Digital transport of analog signals

The algorithm used to encode and decode the digital signal is referred to as a codec. Two important characteristics of any codec are its speech quality and its bandwidth. In general, codecs with higher bandwidth can achieve better quality. However, modern speech codecs achieve good speech quality with bandwidths below 10 kbit/s.

14.1.1   Pulse Code Modulation

The G.711 (ITU-T Recommendation G.711 [131]) codec, better known as PCM (Pulse Code Modulation), is the codec used in the fixed PSTN. In addition, nearly all the SIP phones on the Internet support this codec.

PCM uses a sampling rate of 8000 Hz. That is, PCM takes 8000 samples per second and encodes each of them separately. This way, each sample contains 0.125 ms of speech, as shown in Figure 14.2.

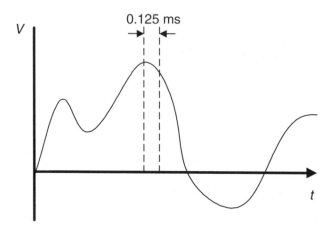

Figure 14.2: Sampling of an analog signal

For each sample it is necessary to encode the amplitude of the audio wave at that point. Still, even if the amplitude of the audio wave varies within the sampling interval (as we can see in Figure 14.2), a single value of the amplitude is provided per sample. Approximating the value of the amplitude in an interval by a single number creates some distortions, which could be reduced by making the sampling interval smaller. The sampling interval chosen by PCM (0.125 ms) provides enough quality to encode human speech.

Compact Discs (CDs) storing music use a PCM codec at 44.1 kHz. This sampling frequency yields a sampling interval that is small enough to encode music with high fidelity, but requires a much larger bandwidth than the PCM codec used for telephony which we are describing.

Digital codecs like PCM work with discrete values. So, PCM needs to provide a discrete value for the amplitude of the audio wave for each sample. PCM encodes this value using 8 bits, which give 256 different values. The mapping between the measured value of the amplitude and these 256 values is done by dividing the amplitude range into intervals and assigning a value to each interval, as shown in Figure 14.3. This process is known as quantization.

In quantization, the narrower the intervals the greater the fidelity and the higher the bandwidth (there are more values to encode). So, quantization offers the same tradeoff as sampling. In any case the distortion caused by quantization can be reduced if we know what our audio input looks like. In our case we know what the audio input looks like because human speech has been extensively studied.

PCM assigns narrower quantization intervals to amplitude values that are more common in human speech. This way we obtain a higher quality in most cases and lower quality only

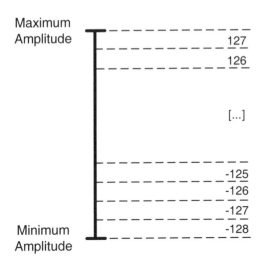

Figure 14.3: Quantization of an amplitude range

in corner cases (e.g., people screaming on the phone). PCM comes in two flavours called A-law and μ-law, depending on how the quantization is performed (both use logarithmic scales). A-law is used in Europe, and μ-law is used in the USA. Other codecs, such as G.726 (ITU-T Recommendation G.726 [132]), use adaptive quantization. That is, the quantization intervals are made larger or smaller depending on the characteristics of the previous samples (sometimes, even later samples are taken into consideration).

The PCM codec we have just described produces 8000 8-bit samples per second. This yields a total bandwidth of 64 kbit/s. Now, let's look at how other codecs achieve an acceptable voice quality using a lower bandwidth than PCM.

14.1.2 Linear Prediction

Modern codecs can encode human speech using bandwidths well below the 64 kbit/s of PCM. One of the tools that allows us to do that is linear prediction. Codecs using linear prediction take advantage of the correlation between different voice samples to encode speech more efficiently. Linear prediction works on audio frames rather than on individual samples. An audio frame consists of a number of consecutive samples; a typical choice is to use 20 ms frames.

Human speech is highly correlated. That is, a particular voice frame can be expressed fairly accurately as a linear combination of previous frames. This property is used by linear prediction codecs. These codecs send the linear coefficients that describe the frame at hand; the decoder reconstructs the voice frame using these coefficients and the voice frames that it previously decoded. We refer to these coefficients as Linear Predictive Coding (LPC) coefficients.

However, LPC coefficients only describe a voice frame approximately. The difference between the original signal and the signal described by the LPC coefficients is the residual signal, as shown in Figure 14.4.

This residual signal does not show any correlation between nearby samples (also called short-term correlation) because the LPC coefficients removed this correlation. Nevertheless,

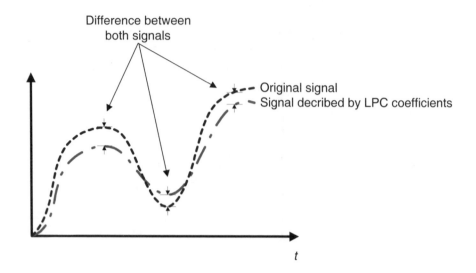

Figure 14.4: Residual signal after calculating the LPC coefficients

there is still a long-term correlation in this residual signal, caused mainly by the pitch of the voice. It is possible to apply linear prediction again, this time to this residual signal, to remove this long-term correlation. The parameters that approximate this residual signal are called Long-Term Predictors (LTPs).

If we add the signals described by the LPC coefficients and by the LTPs, we have a signal which is fairly close to the original signal. The difference between the original signal and this signal is another residual signal. This residual signal is typically encoded by the codec (different codecs use different mechanisms to encode this signal).

When the coder finishes the process that we have just described, it sends the LPC coefficients, the LTPs, and the encoded residual signal to the decoder. This information enables the decoder to produce a signal that is close enough to the original. Figure 14.5 shows the different contributors to the decoded signal (we do not show the original signal in this figure).

14.1.3 GSM-FR

Let's look at how a real codec uses the linear prediction mechanisms we have just described to encode human speech. We have chosen the GSM-FR (GSM Full Rate, specified in ETSI GSM 06.10 [88]) codec, a codec that has been implemented in millions of GSM terminals around the world. GSM-FR encodes speech at 13 kbit/s.

GSM-FR uses 20 ms frames to calculate eight LPC coefficients: two 6-bit coefficients, two 5-bit coefficients, two 4-bit coefficients, and two 3-bit coefficients. This yields a total size of 36 bits.

The residual 20 ms signal (the original signal minus the signal described by the LPC coefficients) is divided into four 5 ms subframes in order to calculate the LTPs. GSM-FR uses two LTP parameters to approximate this residual signal: the 7-bit LTP lag (the lag value indicates the period of the pitch) and the 2-bit LTP gain.

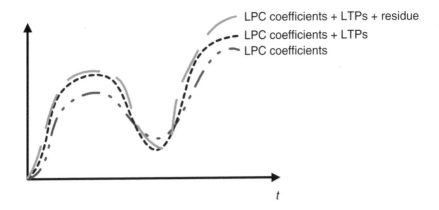

Figure 14.5: Contributors to the decoded signal

The next step consists of encoding the residual signal remaining after having calculated the LTPs. This signal is encoded using a procedure called RPE (Regular Pulse Excitation).

The residual signal after having calculated the LTPs consists of a 5 ms frame that contains 40 samples (each sample is 0.125 ms long). Instead of encoding all 40 samples, GSM-FR only encodes one-third of them in order to save bandwidth. The codec defines the following four sequences of samples.

(1) Sample 0, sample 3, sample 6, ..., sample 36.

(2) Sample 1, sample 4, sample 7, ..., sample 37.

(3) Sample 2, sample 5, sample 8, ..., sample 38.

(4) Sample 3, sample 6, sample 9, ..., sample 39.

The codec chooses the sequence that best approximates the residual signal (sequence selection is encoded using 2 bits, since there are four sequences) and encodes the samples of the chosen sequence using 3 bits per sample plus 6 bits to encode the maximum energy of all the samples in the sequence. Figure 14.6 shows how each parameter contributes to the total final bandwidth of 13 kbit/s.

14.1.4 AMR

Let's now describe the AMR (Adaptive Multi-Rate) speech codec (defined in 3GPP TS 26.090 [6]), which is the mandatory speech codec for 3GPP IMS terminals. That is, every 3GPP IMS terminal supports AMR and may optionally support other speech codecs, such as those described in previous sections.

14.1.4.1 AMR Modes

AMR consists of eight different codecs, each with a different bandwidth. Their bandwidths are 12.2, 10.2, 7.95, 7.40, 6.70, 5.90, 5.15, and 4.75 kbit/s. These codecs are referred to as AMR modes. The 12.2, 7.40, and 6.70 kbit/s AMR modes are also known as GSM-EFR (Enhanced Full Rate), TDMA-EFR, and PDC-EFR, respectively.

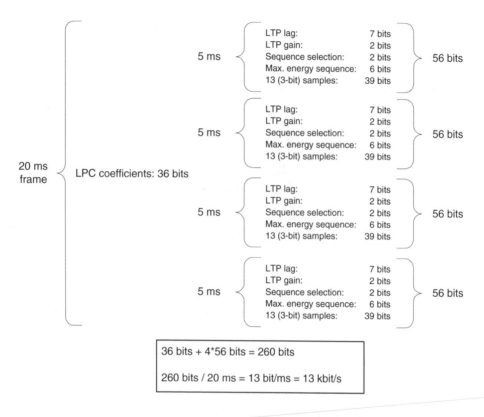

Figure 14.6: Contributors to the total bandwidth of GSM-FR

The AMR modes were initially designed to be used in GSM networks that provide a fixed bit rate for circuit-switched voice calls. This rate is split into channel encoding and data encoding. When the signal-to-noise ratio is low (poor radio quality) the terminals use low-bandwidth modes (e.g., 4.75 kbit/s). When the signal-to-noise ratio is high (good radio quality) the terminals use high-bandwidth modes (e.g., 12.2 kbit/s).

The AMR codec itself is capable of switching mode on a frame-by-frame basis. That is, one 20 ms speech frame may be encoded using an AMR mode and the next speech frame may be encoded using a different mode. Still, some circuit-switched transports impose tougher limitations on how often mode switching can be performed.

Third-generation (3G) networks using WCDMA access do not use AMR modes in the same way as GSM networks. WCDMA uses fast power control and does not perform mode adaptation at the channel-encoding level. WCDMA networks use low-bandwidth modes to gain capacity when many users make voice calls at the same time.

When AMR is transported over RTP/UDP/IP (see Section 15.2.2) the overhead introduced by the RTP, UDP, and IP headers is fairly large. So, unless header compression is used the network does not gain much extra capacity by using the AMR low-bandwidth modes instead of high-bandwidth modes. Still, all the modes need to be supported in order to make it easier to interoperate with circuit-switched networks.

14.1.4.2 LPC Coefficients Calculation

Let's now see how AMR encodes the speech. AMR uses linear prediction; therefore it uses LPC coefficients to approximate speech signals. These LPC coefficients are calculated differently in the 12.2 kbit/s mode from other modes. Yet, speech is divided into 20 ms frames in all modes. Each of these frames contains 160 0.125 ms samples.

In the 12.2 kbit/s mode, linear prediction is performed twice on a 30 ms window that contains the last 10 ms of the previous speech frame and the whole 20 ms of the speech frame at hand, as shown in Figure 14.7. Linear prediction is performed twice over the same window, but using different functions. In the remaining modes, linear prediction is performed only once over the 30 ms window.

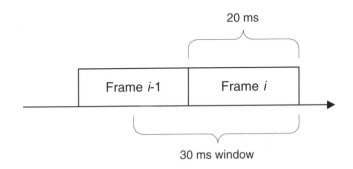

Figure 14.7: Window selection to perform linear prediction

Once the LPC coefficients are calculated they are transformed into LSPs (Line Spectral Pair). LSPs are the root of two polynomials that are formed using LPCs. Encoding LSPs instead of LPCs allows AMR to achieve a more compressed representation of the signal.

14.1.4.3 Codebooks

AMR uses two codebooks to encode the residual signal that we get after calculating the LSPs. The codebooks contain *excitation vectors* that are used by the decoder to synthesize a signal that approximates the residual signal being encoded. This way of encoding the residual signal is referred to as CELP (Code Excited Linear Prediction). AMR uses an adaptive codebook and a fixed codebook.

The contents of the adaptive codebook vary as subsequent speech frames are encoded. The adaptive codebook is used to encode the pitch of the speech. The parameters for the adaptive codebook are the *lag* and the *gain*.

The fixed codebook is implemented using an algebraic codebook. That is, an algebraic code is used to generate excitation vectors, also known as *innovation vectors*.

14.1.4.4 Adaptive Codebook

AMR uses both open-loop and closed-loop analysis to estimate the pitch of the speech. Open-loop analysis is performed first and provides near-optimal lag values. Closed-loop analysis is performed around lag values that were obtained previously and provides the final lag values to be encoded. Performing open-loop analysis before close-loop analysis reduces the difficulty of searching for the lag values.

The 12.2, 10.2, 7.95, 7.40, 6.70, and 5.90 kbit/s AMR modes perform open-loop pitch analysis on 10 ms speech windows, which gives two estimates of the pitch lag for each 20 ms frame. The 5.15 and 4.75 kbit/s AMR modes perform open-loop pitch analysis on 20 ms speech windows, which gives a single estimate of the pitch lag for each 20 ms frame.

Once the estimates of the pitch lag are obtained, AMR performs closed-loop pitch analysis around these estimates. This process results in the final value for the pitch lag. Then, AMR calculates the adaptive codebook gain value.

14.1.4.5   Fixed Codebook

The residual signal after calculating the parameters of the adaptive codebook (lag and gain) is approximated using a fixed algebraic codebook. This codebook contains a set of fixed pulses. The coder performs a search in this codebook to decide which pulse approximates the residual signal best.

14.1.4.6   Gains

The adaptive codebook gain was calculated after calculating the pitch lag. Let's now see how AMR calculates the fixed codebook gain.

Given the algebraic code, it is possible to predict the fixed codebook gain value. AMR takes advantage of this and only sends a correction factor to the decoder. The decoder obtains the real value for the fixed codebook gain using the predicted value and the correction factor.

The 12.2 and 7.95 kbit/s AMR modes encode the adaptive codebook gain and the correction factor separately, using 4 bits and 5 bits, respectively. Both values are provided for 5 ms speech windows.

The remaining modes encode the adaptive codebook gain and the correction factor jointly. The 10.2, 7.40, and 6.70 kbit/s modes use 7 bits per 5 ms speech window. The 5.90 and 5.15 kbit/s modes use 6 bits per 5 ms window. The 4.75 kbit/s mode uses 8 bits per 10 ms window.

Figures 14.8–14.15 show the different contributors to the final bandwidth of all the AMR modes.

14.1.5   AMR-WB

The AMR-WB (AMR-WideBand) codec (defined in 3GPP TS 26.190 [39]) encodes voice using 16 000 samples per second, as opposed to the 8000 samples per second used by AMR. This higher sampling frequency allows AMR-WB to encode a wider range of frequencies. Consequently, AMR-WB encodes speech with higher quality than the codecs we described earlier.

AMR-WB consists of a family of codecs which encode audio using the following bandwidths: 23.85, 23.05, 19.85, 18.25, 15.85, 14.25, 12.65, 8.85, and 6.60 kbit/s. AMR-WB is the mandatory codec for 3GPP IMS terminals that provide wideband services.

14.1.6   SMV

3GPP2 IMS terminals support the EVRC (Enhanced Variable Rate Codec) and the SMV (Selectable Mode Vocoder) codecs. First- and second-generation CDMA terminals already implement EVRC, while SMV is the preferred speech codec for CDMA2000® access. Let's describe how SMV works.

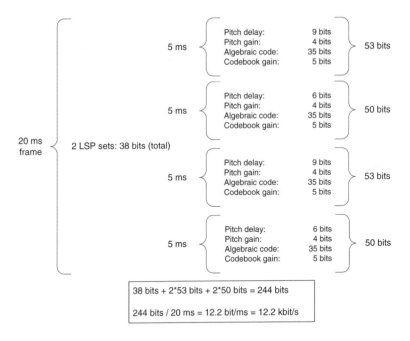

Figure 14.8: Contributors to the total bandwidth of 12.2 kbit/s AMR

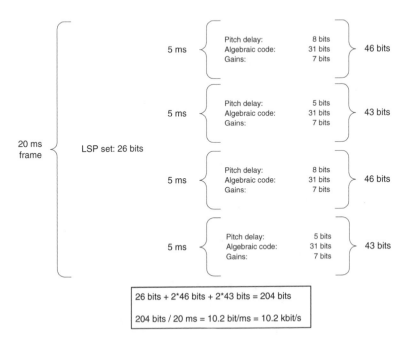

Figure 14.9: Contributors to the total bandwidth of 10.2 kbit/s AMR

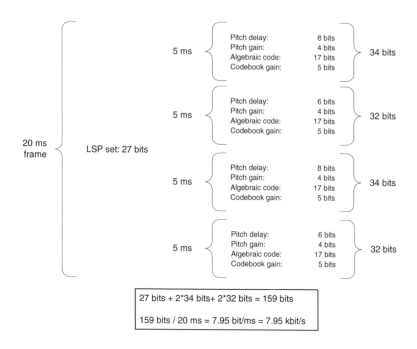

Figure 14.10: Contributors to the total bandwidth of 7.95 kbit/s AMR

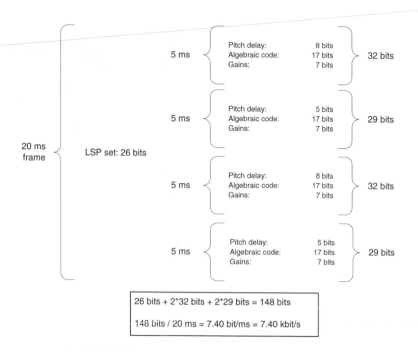

Figure 14.11: Contributors to the total bandwidth of 7.40 kbit/s AMR

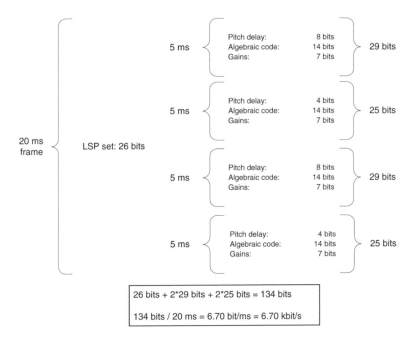

Figure 14.12: Contributors to the total bandwidth of 6.70 kbit/s AMR

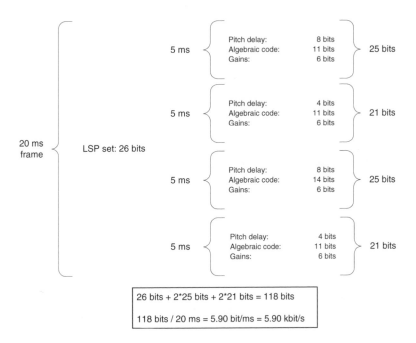

Figure 14.13: Contributors to the total bandwidth of 5.90 kbit/s AMR

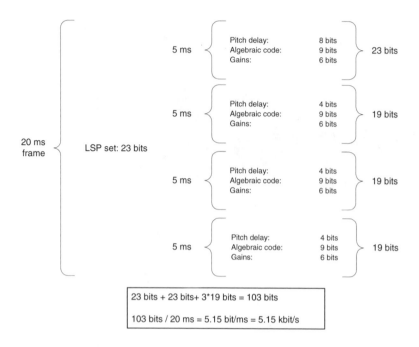

Figure 14.14: Contributors to the total bandwidth of 5.15 kbit/s AMR

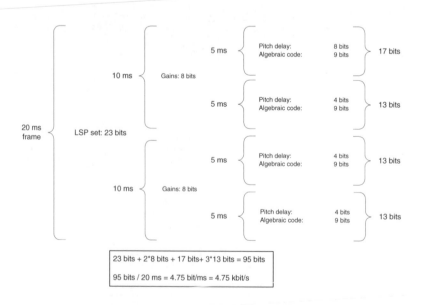

Figure 14.15: Contributors to the total bandwidth of 4.75 kbit/s AMR

SMV is a family of four codecs which are called Rate 1, Rate 1/2, Rate 1/4, and Rate 1/8. All of them work on 20 ms audio frames. We can find four modes in SMV: Mode 0 (premium mode), Mode 1 (standard mode), Mode 2 (economy mode), and Mode 3 (capacity-saving mode).

Every 20 ms audio frame is processed by one of the four codecs. The codec to be used is chosen based on the mode and on the characteristics of the audio frame at hand.

The Rate 1 and Rate 1/2 codecs use CELP (Code-Excited Linear Prediction) to encode the audio frames (we saw in Section 14.1.4.3 that AMR also uses CELP). These two codecs use two codebooks, one adaptive and one fixed, and their respective gains to encode the audio signal.

The Rate 1/4 and Rate 1/8 codecs use random number generators to generate the LPC excitation. The Rate 1/4 codec provides a gain factor per 2 ms subframe while the Rate 1/8 codec provides a single gain factor per 20 ms frame. Table 14.1 shows how many bits each of the codecs use to encode a 20 ms frame and the resulting final bandwidth of the codec.

Table 14.1: SMV bandwidths

Codec	Bits per 20 ms frame	Resulting bandwidth (kbit/s)
Rate 1	171	8.55
Rate 1/2	80	4
Rate 1/4	40	2
Rate 1/8	16	0.8

14.2 Video Encoding

The building block of video encoding is image encoding. Given a set of encoded still images taken with a short interval between them, we can play them to create a video. We just need to show these images one after another quickly enough so that the human eye perceives this succession of images as a moving image.

Still image encoders divide the image to be encoded into colored dots called *pixels* (combination of the words "picture" and "element"). For a display working at its maximum resolution a pixel corresponds to the smallest element of the display which can be assigned a color. The resolution of a picture is given by the number of pixels used (horizontally and vertically) to encode it; a typical value is 640×480.

Figure 14.16 shows a black-and-white picture encoded using a resolution of 10×10 pixels. Each of the pixels of this picture takes one of the following two values: black or white. So, we need one bit to encode the color of each pixel. Given that the picture has 10×10 pixels the size of the encoded picture will be 100 bits. Following the procedure we have just described, to encode a 640×480 picture with a color depth of 8 bits would yield 300 kilobytes.

If we want to have 25 frames per second (a value used by many television systems) using 300-kilobyte pictures we need a bandwidth of over 60 megabit/s. This is a very high figure that needs to be reduced somehow. There are a number of video compression techniques that make it possible to encode video using much lower bandwidth while still achieving acceptable quality. We will describe the techniques used by H.263 (the mandatory video codec in the

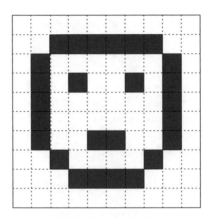

Figure 14.16: 10 × 10 black-and-white picture

3GPP IMS) in Section 14.2.2. Still, before dealing with H.263 we will mention other video codecs that are commonly used in different environments in Section 14.2.1.

14.2.1   Common Video Codecs

In this section we list a set of common video codecs and explain their relation with well-known video storage formats, such as DVD. In particular, we deal with the MPEG (Motion Picture Experts Group) and the H.261 video standards.

The MPEG video standards are used for both media storage and for videoconferencing. MPEG-1 (specified in ISO/IEC 11172 [126]) was developed to encode audio and video at rates of about 1.5 megabit/s. MPEG-1 defines an elementary stream as an audio or a video stream encoded following certain procedures. A videoconference with a single audio channel would consist of two elementary streams: audio and video. MPEG-1 also defines an encapsulation format that contains elementary streams and information on how to play these streams together (e.g., time stamps to perform lip synchronization).

Two user agents can exchange elementary streams and use external means (e.g., RTCP timestamps, described in Section 15.2.3) to do things like media synchronization or exchange a single stream (using the encapsulation format) with all the information needed to play all the elementary streams in it (as described in [123]). Streams following the encapsulation format are referred to as system streams.

MPEG-2 (specified in ISO/IEC 13818 [127]) was developed to encode audio and video at rates higher than MPEG-1, providing higher quality as well. The VCD (Video CD) format is based on MPEG-1, while the SVCD (Super Video CD) format is based on MPEG-2. The format used in DVDs is based on MPEG-2.

MPEG-4 (specified in ISO/IEC 14496-1 [128]) uses audio-visual objects to describe audio-visual scenes, and it uses low bit rates. The DivX and XviD formats are based on MPEG-4.

MPEG defines three layers to encode audio: the higher the layer the higher the audio codec complexity. The third layer of MPEG is the well-known MP3 format, which is used to store music in different devices.

Another common video-encoding format is H.261 (ITU-T Recommendation H.261 [135]). H.261 is the video codec used by the H.320 (ITU-T Recommendation H.320 [138]) video-teleconferencing framework. This codec was designed to be used over ISDN lines and therefore, produces bandwidths that are multiples of 64 kbit/s (which is the bandwidth provided by an ISDN channel), ranging from 64 to 1984 kbit/s.

14.2.2   H.263

The video codec H.263 (ITU-T Recommendation H.263 [137]) was developed for the H.324 (ITU-T Recommendation H.324 [144]) multimedia framework. H.263 was designed for low bit rate communications.

14.2.3   Image Encoding

H.263 supports five formats to encode images: sub-QCIF (Quarter Common Intermediate Format), QCIF, CIF, 4CIF, and 16CIF. All of these formats encode the color of the pixels using a luminance component and two chrominance components, but support different resolutions.

Images can be encoded using different color spaces. A common color space is RGB (red green blue), where the color of a pixel is represented by its red, green, and blue components. The color space used in H.263 consists of providing the luminance component and two chrominance components.

The luminance component is the black-and-white component of the image. That is, how dark (or light) each pixel is. The chrominance components are color difference components. They provide the color of a pixel in relation to the colors red and blue.

The image formats used by H.263 take advantage of the fact that the human eye is more sensitive to luminance information than to chrominance information and, therefore, only provide half-horizontal and half-vertical resolutions for chrominance. Table 14.2 shows the resolutions used for both values in each format.

Table 14.2: Luminance and chrominance resolution in CIF formats

Format	Luminance resolution	Chrominance resolution
sub-QCIF	128×96	64×48
QCIF	176×144	88×72
CIF	352×288	176×144
4CIF	704×576	352×288
16CIF	1408×1152	704×576

As a result of providing less chrominance information than luminance information, blocks of 2×2 pixels share the same chrominance information. Figure 14.17 illustrates this point.

14.2.4   Temporal Correlation

H.263 supports two modes: *inter-coding* and *intra-coding*. In inter-coding mode the codec takes advantage of the temporal redundancy of the pictures. That is, it encodes a picture referencing those pictures that were encoded previously. In intra-coding mode, no references to other pictures are used to encode a particular picture.

L1	L2	L3	L4
C1	C1	C2	C2
L5	L6	L7	L8
C1	C1	C2	C2
L9	L10	L11	L12
C3	C3	C4	C4
L13	L14	L15	L16
C3	C3	C4	C4

Figure 14.17: Luminance and chrominance values for a block of pixels

Interpicture prediction is implemented in H.263 using motion estimation and compensation. The encoder finds corresponding blocks of pixels between frames and represents their relationship using motion vectors. These motion vectors are approximated using those motion vectors used in previous frames. The difference between this approximation (estimation) and the real motion vector is encoded separately (compensation).

14.2.5   Spatial Correlation

H.263 uses the DCT (Discrete Cosine Transform) to exploit spatial redundancy in the images and to achieve a compact representation of the coefficients that describe them. These coefficients are entropy-coded before being transmitted.

Entropy coding consists of performing lossless compression on the DCT coefficients. That is, using a reversible compression algorithm similar to those used to compress files in any personal computer.

14.3   Text Encoding

Text is already digital information, so there is no need to perform the analog-to-digital and digital-to-analog conversions that are necessary for audio and video. There are two types of text communications: instant messages and real-time text.

Instant messages convey a whole message, such as: "How are you?" That is, the sender types the message, edits it if necessary, and sends it. This way the receiver only gets the final version of the message. Chapter 18 looks at several ways of transporting instant messages.

Real-time text consists of transferring keystrokes instead of text. If the sender writes something and then deletes it to write something else, the receiver will see how these changes are performed. That is, the receiver gets letters and commands (e.g., carriage returns or delete characters) one by one as they are typed. The most common format for real-time text is T.140 (ITU-T Recommendation T.140 [136]).

14.4 Mandatory Codecs in the IMS

It would be desirable for all IMS terminals to support a common codec so that they could communicate directly with each other without the need for transcoders. Nevertheless, 3GPP and 3GPP2 could not agree on a common mandatory codec for their terminals. So, audio and video sessions between 3GPP and 3GPP2 IMS terminals usually involve transcoders. Still, both 3GPP and 3GPP2 specify which codecs must be supported by their IMS terminals.

All 3GPP IMS terminals support the AMR speech codec and the H.263 video codec (as specified in 3GPP TS 26.235 [30]). 3GPP IMS terminals providing wideband services support the AMR-WB audio codec and those terminals providing real-time text conversational services support T.140 (ITU-T Recommendation T.140 [136]).

Chapter 15

Media Transport

There are two types of media when it comes to media transport: media that tolerates a certain degree of packet loss and media that does not. Examples of the first type are audio and video and examples of the second type are web pages and instant messages. If a few of the packets transporting an audio stream get lost the recipient may notice a decrease in the quality of the sound, but will probably be able to understand anyway. On the other hand, an instant message under packet loss may change from "I will *not* come tomorrow" into "I will come tomorrow".

The transport protocol for a particular media is chosen based on the type of media. Traditionally, TCP has been used to transport media reliably and UDP to transport media unreliably. Nevertheless, UDP is not suitable for transporting large amounts of data traffic because it lacks congestion control mechanisms (i.e., congestion control would need to be implemented at the application layer, but it is seldom implemented at all). To address this issue the IETF is developing the Datagram Congestion Control Protocol (DCCP).

15.1 Reliable Media Transport

There are two transport protocols that provide reliable delivery of user data: TCP and SCTP (Stream Control Transmission Protocol, specified in RFC 2960 [230]). TCP delivers a stream of bytes, while SCTP delivers messages.

TCP works best when the recipient of the data does not need to wait until all the data have arrived in order to start processing it. An Internet browser is a good example of such a case. When the user requests a web page using HTTP the browser receives the contents of that web page over a TCP connection (or several TCP connections). The browser displays the information received so far to the user at every moment; it starts to draw pictures and display text. This way the user does not need to wait until the whole web page is downloaded to start reading it.

On the other hand, when the application is interested in receiving all the data at once, SCTP is a better choice. SCTP delivers messages rather than a stream of bytes to the application. A message may, for instance, contain an entire instant message (e.g., "Hello, Bob!") or a file transferred between two users. Furthermore, SCTP provides features that are not present in TCP, such as better protection against DoS attacks, multi-homing, and multiple streams per SCTP association (connections are referred to as *associations* in SCTP).

Although SCTP provides some advantages over TCP for some applications, it still is not widely used. At present, TCP is by far the most widespread reliable transport protocol.

The 3G IP Multimedia Subsystem (IMS) Second Edition   Gonzalo Camarillo and Miguel A. García-Martín
© 2006 John Wiley & Sons, Ltd

Many protocols establish TCP connections to carry control messages and user data. This is the case of HTTP, SMTP, and Telnet, to name a few. In these protocols the client sends a request to a server over a TCP connection and the server sends back the data requested by the client over the same connection. Clients establish this type of connection to well-known ports that identify the protocol in use. For instance, a web browser (which contains an HTTP client) sends requests to servers on port 80, which is the well-known HTTP port.

Still, SIP can be used to establish TCP connections or SCTP associations without using well-known ports. A session description describes the type of data to be transported and the protocol to be used. This way, reliable connections are established in the same way as unreliable connections, such as audio and video streams. Figure 15.1 shows an SDP session description with a TCP connection that will transport images using T.38 (T.38 is the format used to send faxes). The *passive* direction attribute indicates that the recipient of this session description will be the *active* part and, so, will initiate the TCP connection.

```
v=0
o=Alice 2790844676 2867892807 IN IP4 192.0.0.1
s=I will send you a fax
c=IN IP4 192.0.0.1
t=0 0
m=image 20000 TCP t38
a=direction:passive
```

Figure 15.1: A TCP connection in SDP

15.2 Unreliable Media Transport

UDP is the transport protocol used to send media unreliably. Senders send UDP packets and hope that the recipient receives enough of them to be able to understand the message (e.g., to understand a voice message). Since lost packets are not retransmitted, applications introduce redundancy in the data to be transferred in order to be able to tolerate a certain level of packet loss. Some audio codecs, for instance, spread the information about every talk burst over several packets.

Although UDP is widely used it has a big disadvantage when it comes to transporting large amounts of data: it does not provide any congestion control mechanism. UDP senders do not slow down sending data even when the network is severely congested, making the congestion even worse.

The increase in the number of applications using UDP to send media, especially audio and video, made the IETF realize that we need an unreliable transport protocol that includes congestion control. One of the proposals was to create an SCTP extension for unreliable delivery. Nevertheless, the TCP-like congestion control mechanisms of SCTP do not suit some types of multimedia traffic. DCCP (Datagram Congestion Control Protocol, specified in the Internet-Draft "Datagram Congestion Control Protocol" [153]) supports different types of congestion control and has been designed with multimedia traffic in mind.

15.2.1 DCCP

DCCP is an unreliable transport protocol that provides connection establishment and termination as well as negotiation of congestion control algorithms.

DCCP connections are established with a three-way handshake, during which the characteristics of the connection are negotiated. In particular, DCCP peers negotiate the congestion control algorithm to be used. Congestion control algorithms are identified by their CCIDs (Congestion Control Identifiers). At present, there are three CCIDs to choose from: sender-based congestion control, TCP-like congestion control, and TFRC (TCP-Friendly Rate Control) congestion control.

TFRC (specified in RFC 3448 [114]) is especially suitable for multimedia traffic, because it avoids abrupt changes in the data rate while sharing the available bandwidth with TCP flows fairly. Nevertheless, DCCP is still a fairly new protocol and, so, is not widely used at present.

15.2.2 RTP

RTP (Real-time Transport Protocol, specified in RFC 3550 [225]) allows transporting real-time media, such as audio and video, over unreliable transports, such as UDP and DCCP. It is always used in conjunction with RTCP (RTP Control Protocol), which provides quality-of-service statistics and information to perform inter-media synchronization.

The main purpose of RTP is to allow receivers to play out media at the proper pace, given that IP networks do not keep the timing relationship of the data being transported: that is, IP networks introduce jitter. If we send two IP packets to the same destination separated by 10 ms, nothing ensures that the second packet arrives at the destination exactly 10 ms after the first one. The second packet may arrive right after the first one, much later, or even before the first one. Consequently, receivers cannot rely on the arrival times of the packets to recover the timing relationship of the media. They use RTP timestamps for this purpose.

Receivers place incoming RTP packets in a buffer according to their timestamps and start playing them. If a packet with a particular timestamp needs to be played and still has not arrived, the receiver uses interpolation techniques to fill the gap (e.g., in the case of audio it may play the last audio packet for a longer time). If this packet is received afterwards, it is simply discarded. Figure 15.2 shows a few packets in a buffer. The packet with timestamp 40 has not arrived yet. If it does not arrive by the time it needs to be played, when it eventually arrives it will be discarded.

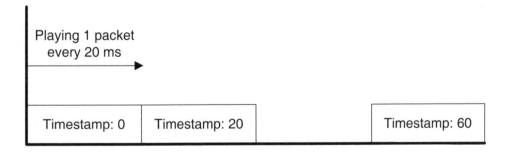

Figure 15.2: Receiver's buffer

The receiver in Figure 15.2 needs to make an important decision: when to start playing media to the user. If it starts as soon as the packet with timestamp 0 is received, it runs the risk that the packet with timestamp 20 might not arrive in time to be played (this would decrease the quality of the media being played out). On the other hand, if the receiver waits many seconds before playing the media, the delay would be so big that the users would be unable to maintain a normal conversation, and it would have to implement a much larger buffer.

Different implementations use different parameters to decide the lengths of their buffers: long buffers cause long delays but good quality while short buffers cause short delays but poor quality. Let's use an example to discover what would be a good value for the buffer of a particular receiver. The graph in Figure 15.3 shows the delay experienced by packets sent from a sender to a receiver. In this example, most of the packets experience a delay of around 50 ms, some experience smaller delays, and a few experience much larger delays (lost packets have an infinite delay). A good tradeoff for the receiver would be to start playing packets 100 ms after they are sent. This way, only the few packets that appear in the tail of the distribution (which have a too long delay) would be discarded when they arrive.

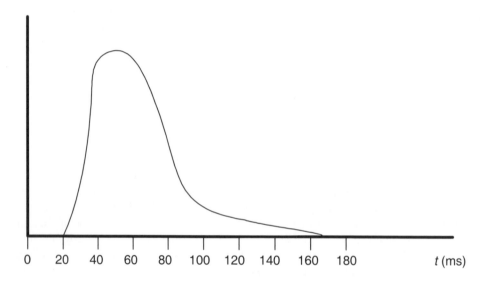

Figure 15.3: Packet arrival distribution

In addition to timestamps, RTP packets carry sequence numbers. Recipients use them to figure out how many packets get lost in the network during a transmission. If the network drops too many packets at a particular time, peers may decide to use a different codec, one that provides better quality under heavy packet loss (i.e., a codec with more redundancy).

RTP packets also carry binary sender identifiers and the payload type. Binary sender identifiers are used in conferences to identify the current speaker, and the payload type identifies the encoding and transport format of the data carried in the RTP packet. Payload types are numeric values that identify a particular codec and are typically negotiated (or simply assigned) using a session description protocol. There are two kinds of payload types: static and dynamic.

Static payload types are numbers that always correspond to the same codec. For example, payload type 0 corresponds to the G.711 μ-law audio codec (RFC 3551 [224] defines some audio and video static payload types) and payload type 31 corresponds to the video codec H.261.

Dynamic payload types are typically negotiated using the offer/answer model. They identify a particular codec within a particular session. The SDP session description in Figure 15.4 contains an audio stream that can be encoded using G.711 μ-law (static payload type 0) audio codec or using a 16-bit linear stereo codec sampled at 16 kHz (dynamic payload type 98, specified in the a=rtpmap line). The recipient of this audio stream will receive RTP packets whose payload types will be 0 or 98 and, based on this session description, will decode them.

```
v=0
o=Alice 2790844676 2867892807 IN IP4 192.0.0.1
s=Let's talk
c=IN IP4 192.0.0.1
t=0 0
m=audio 20000 RTP/AVP 0 98
a=rtpmap:98 L16/16000/2
```

Figure 15.4: SDP with static and dynamic payload types

Figure 15.5 shows all the protocol headers of an RTP packet carrying an audio sample. The transport protocol used in this example is UDP.

Link-layer Header	IP Header	UDP Header	RTP Header	Audio Sample	Optional Padding

Figure 15.5: Protocol headers of an RTP packet

15.2.3 RTCP

RTCP (RTP Control Protocol) is a protocol that is always used together with RTP. It provides quality-of-service statistics, information to perform inter-media synchronization, and mappings between RTP binary sender identifiers and human-readable names. RTCP messages are sent by both RTP senders and RTP receivers.

To develop quality-of-service statistics, RTP senders report (using RTCP) the number of RTP packets they have sent to the network and RTP receivers report the number of packets they have received. This way it is easy to deduce the packet loss rate that the session is experiencing.

RTP senders use RTCP to provide a mapping between the timestamps of their media streams and a wall clock. This way, receivers can synchronize the play-out of different media streams. References to a wall-clock are needed because clock frequencies used for the timestamps of different media streams may be different and the initial value of the timestamps is random. So, by only inspecting the timestamps of two media streams, it is impossible to determine which samples should be played at the same time. A common example of this type of inter-media synchronization is lip synch; that is, audio–video synchronization. Figure 15.6 shows how RTCP mappings help the receiver perform lip synch.

RTCP messages also provide mappings between RTP binary sender identifiers and human-readable names. This is particularly useful in conferences where the media from all

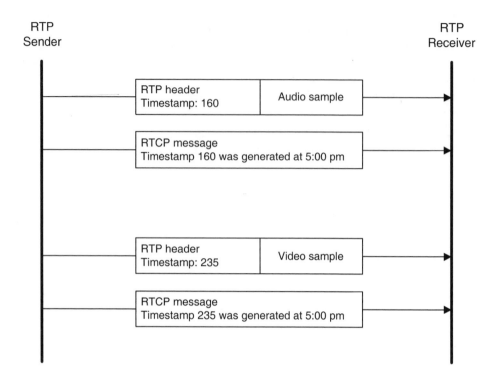

Figure 15.6: Lip synch using RTCP

the participants is received at the same transport address. This way the users can be informed about who is speaking at every moment.

When SDP is used, RTP packets are normally sent a port with an even number and RTCP messages are sent to the consecutive odd port. For instance, a user agent that generates the session description in Figure 15.4 will receive RTP packets carrying audio samples on UDP port 20000 and RTCP packets related to the audio stream on UDP port 20001.

15.2.4   SRTP

SRTP (Secure RTP, specified in RFC 3711 [53]) provides confidentiality, message authentication, and replay protection to RTP and RTCP traffic. Figure 15.7 shows which portions of an RTP packet are authenticated and which are encrypted.

Peers using SRTP to exchange media use a key management protocol (as described in Section 8.8) to come up with a master key, which is used to generate session keys. Session keys are typically refreshed periodically so that attackers do not have access to large amounts of traffic encrypted under the same key.

15.3   Media Transport in the IMS

The IMS uses RTP over UDP to transport media unreliably. DCCP may be used in the future, but at present it is not mature enough and widespread enough for the IMS.

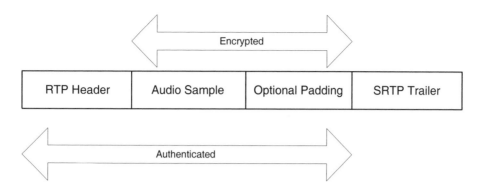

Figure 15.7: SRTP protection

Regarding security the IMS does not provide any kind of security at the media level. It is assumed that the traffic on the radio access is encrypted at lower layers and that the core IMS network is trustworthy. So, SRTP is not supported.

When it comes to reliable transport protocols the natural choice would be to use TCP. It has been a stable protocol for many years and is supported virtually everywhere. In addition to this there are TCP extensions specifically designed to make its flow control mechanisms more suitable for radio links (e.g., RFC 3522 [157] defines the Eifel detection algorithm).

Part IV

Building Services with the IMS

Now that we have introduced all the technologies used in the IMS (see Parts II and III), we are ready to describe how we can use those technologies to achieve the IMS's main goal: providing services.

In previous sections we looked at how the IMS provides some services (e.g., multimedia sessions). Other services do not need extra standardization and can be provided with all the tools we have described (e.g., a voicemail service). In addition, there are services that require extra standardization. It is these services that Part IV focuses on.

We have chosen some of the most significant services that will be initially provided by the IMS: presence, instant messaging, and Push-to-Talk. Still, the reader should keep in mind that this is not, by any means, an exhaustive list of IMS services. Many other new and innovative services will be developed in the future using the IMS infrastructure. For instance, 3GPP and the IETF are currently working on standards for conferencing. However, we have decided not to include them in this part because they are still at an early stage of development.

Chapter 16

The Presence Service on the Internet

Presence is one of those basic services that in the future is likely to become omnipresent. On the one hand, the presence service is able to provide an extensive customized amount of information to a set of users. On the other hand, third-party services are able to read and understand presence information, so that the service provided to the user is modified (actually, we should say customized) according to the user's needs and preference expressed in the presence information.

16.1   Overview of the Presence Service

Presence is the service that allows a user to be informed about the reachability, availability, and willingness of communication of another user. The presence service is able to indicate whether other users are online or not and if they are online, whether they are idle or busy (e.g., attending a meeting or engaged in a phone call). Additionally, the presence service allows users to give details of their communication means and capabilities (e.g., whether they have audio, video, instant messaging, etc. capabilities and in which terminal those capabilities are present).

The presence framework defines various roles, as shown in Figure 16.1. The person who is providing presence information to the presence service is called a *presence entity*, or for short, a *presentity*. In Figure 16.1 Alice plays the role of a presentity. The presentity is supplying *presence information* (i.e., the set of attributes that characterize the properties of a presentity, such as status, capabilities, communication address, etc.). A given presentity has several devices known as *Presence User Agents* (PUA) which provide information about her presence.

Figure 16.1 shows three PUAs: an IMS terminal, a laptop, and a desktop computer. Each has a piece of information about Alice, the presentity. The laptop knows whether Alice is logged on or not, as does the desktop computer. The IMS terminal knows Alice's registration status and whether she is engaged or not in any type of communication. They can have even richer presence information, such as what time Alice will be back from lunch, whether Alice is available for videoconferences, or whether she only wants to receive voice calls right now. All the PUAs send their pieces of information to a *presence agent* (PA). The PA gathers all the information received and obtains a complete picture of Alice's presence.

The 3G IP Multimedia Subsystem (IMS) Second Edition   Gonzalo Camarillo and Miguel A. García-Martín
© 2006 John Wiley & Sons, Ltd

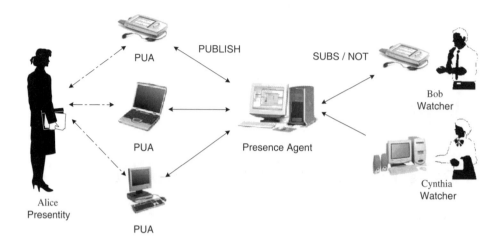

Figure 16.1: SIP Presence architecture

A Presence Agent can be an integral part of a *Presence Server* (PS). A Presence Server is a functional entity that acts as either a PA or as a proxy server for SUBSCRIBE requests.

Figure 16.1 also shows two *watchers*: Bob and Cynthia. A watcher is an entity that requests (from the PA) presence information about a presentity or watcher information about her watchers. There are several types of watchers. A fetcher is a watcher that retrieves the current presentity's presence information from the PA. A subscribed watcher asks to be notified about future changes in the presentity's presence information, so that the subscribed watcher has an accurate updated view of the presentity's presence information.

Typically, applications combine the watcher and presentity functionalities in a single piece of software, thus hiding the functional distinction of presence publication from presence information acquisition by the end-user. However, since both functions are different and governed by different procedures we treat them separately.

The presence service is a particular application built on top of the SIP event notification framework (we described the SIP event notification framework in Section 4.15). The framework allows a PUA to subscribe to or fetch (using a SIP SUBSCRIBE transaction) the presentity's presence information. The subscription state is kept in the presentity's PA, which acts as a notifier (according to the SIP event notification framework). The PA notifies (using SIP NOTIFY transactions) all the subscribed PUAs when a change has occurred in the presentity's presence information.

All SUBSCRIBE/NOTIFY transactions contain a SIP Event header field that identifies the actual event the subscription or notification is related to. RFC 3856 [200] defines the "presence" event package identified by the value presence in the Event header field of SUBSCRIBE and NOTIFY requests.

16.1.1   The pres URI

Traditionally, Internet technologies have used URIs to identify resources that can be accessed with a protocol (e.g., *sip*, *http*, and *ftp* URIs) or are associated to functionality (e.g., *tel* and *mailto* URIs). Presence defines a *pres* URI as identifying a presentity or a watcher. It must be noted that the *pres* URI is protocol-agnostic: therefore, there is no information indicated in the URI on how to access the resource. However, when SIP is used to access presence resources it is recommended to use *sip* or *sips* URIs as they are protocol-specific.

The syntax of the *pres* URI is:

```
PRES-URI        = "pres:" [ to ] [ headers ]
to              = mailbox
headers         = "?" header *( "&" header )
header          = hname "=" hvalue
hname           = *uric
hvalue          = *uric
```

An example of a *pres* URI is:

```
pres:alice@example.com
```

16.2 The Presence Life Cycle

As if it were a product, the presentity's supplied presence information has a life cycle. During its life cycle the presence information suffers a number of transformations, from its creation phase to its shipping and handling, storage, and the final delivery phase to the consumer (the watcher, in the case of presence).

Figure 16.2 shows a schematic representation of the life cycle of the presence information. A presentity (on the left side of the figure), has some presence information to publish. The presence information varies slightly depending on which presence user agent the presentity is using. For instance, in the figure, there could be several presence user agents (e.g., a computer, a mobile phone, a fixed phone), each one supplying different presence information. As an example, the presentity might be away from the keyboard of the computer, but engaged in a call on her mobile phone, so these details will be reflected in her presence information.

At some point in time each of these presence user agents will send a SIP PUBLISH request containing their view of the presentity's presence information in a presence document. This is the presence publication process, which is described in Section 16.10. The presence document, received at the presence agent, is fed into the merging process. The merging process, governed by a composition policy, allows the three presence documents to be merged into a unified raw view of the presentity's presence information.

The unified raw presence document is filtered to strip out all the details of information that the presentity does not want to offer to the watcher. Certainly the presentity has uploaded a presentity's policy document (typically using XCAP, see Section 16.14). The presentity's policy documented could indicate that certain watchers will not get, for example, geographical location information, while others will get it. The presence agent applies the presentity's privacy policy and gets a potential presence document. This document is still potential because it has to suffer further transformations.

Then the presence agent takes the potential presence document and applies any filter that was dictated by the watcher. This basically eliminates any extra information that the watcher is not interested in receiving. For example, a watcher may just be interested in receiving updates when the user changes his basic status information (e.g., online, offline), but neither when the geographical coordinates change nor when his activities are updated. Watchers can express their filters by sending then in the SUBSCRIBE requests that creates the subscription. We describe the event notification filtering in Section 16.15.2

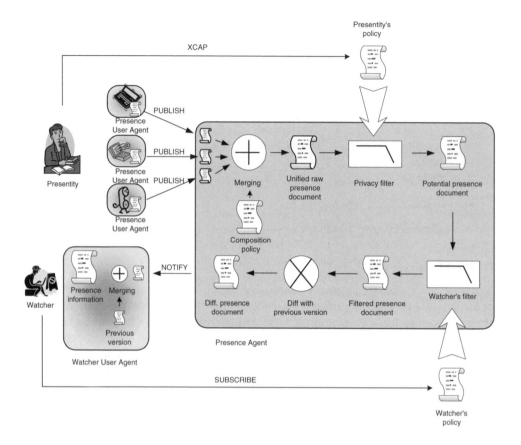

Figure 16.2: SIP presence life cycle

Once the presence agent has a filtered presence document, it may need to get a differenced presence document with respect to a previously sent copy. This is the case when partial notifications are sent, in which case, only the changes are transmitted, as opposed to full presence documents. This is so-called partial notification, which is further described in Section 16.15.1.

The presence agent then creates a NOTIFY request that contains the differenced presence document and sends it to the watcher. This is the notification process, which is described in Section 16.11. The watcher uses the previously stored version and the received version with the changes, merges them, and gets the complete presence information document that is eventually used to display to the watcher.

16.3   Presence Information Data Format

The Presence Information Data Format (PIDF) is a protocol-agnostic document that is designed to carry the semantics of presence information across two presence entities. The PIDF is specified in RFC 3863 [231]. The PIDF constitutes a common profile for presence so that various protocols, not only SIP, can use it to transport presence information.

The PIDF is designed with a minimalist approach (i.e., it includes a minimal set of features to fulfill the basic requirements). This minimal approach guarantees the reusability of the PIDF with different protocols. On the other hand, the PIDF is highly extensible, so it is possible to extend the format whenever there is a need to cross beyond the minimal model. Some extensions are being designed aimed at providing a more accurate view of the presence of a presentity.

The PIDF encodes the presence information in an XML document that can be transported, like any other MIME document, in presence publication (PUBLISH transaction) and presence subscription/notification (SUBSCRIBE/NOTIFY transactions) operations. The PIDF defines a new MIME media type `application/pidf+xml` to indicate the type of application and encoding.

16.3.1 Contents of the PIDF

A PIDF document contains the presence information of a presentity. This information consist of a number of elements, each one referred to as a `tuple`. Each tuple includes the presentity's `status` (`open` or `closed`, meaning online or offline, respectively), an optional `contact` element that provides a contact URI, an optional `note`, an optional `timestamp`, and possibly other element extensions.

It must be noted that the PIDF only defines the `open` status and the `closed` status, which for most applications is not enough. The PIDF lets extensions define other statuses such as "at home", "on the phone", "away", etc.

Figure 16.3 shows an example of the PIDF of the presentity identified as `pres:alice@example.com`. Her only tuple reveals that she is online for communications. She is providing a contact in the form of a TEL URI [220], and a note indicating that this is her cellular phone.

```
<?xml version="1.0" encoding="UTF-8"?>
<presence xmlns="urn:ietf:params:xml:ns:pidf"
          entity="pres:alice@example.com">

  <tuple id="qoica32">
    <status>
       <basic>open</basic>
    </status>
    <contact priority="0.9">tel:+1555876543</contact>
    <note>My cell phone</note>
  </tuple>
</presence>
```

Figure 16.3: Example of the PIDF

16.4 The Presence Data Model for SIP

Since the PIDF is protocol agnostic, it does not go deeply enough to identify what are the pieces of information represented in it. Certainly the PIDF represents presence information as a series of tuples, but it does not clearly indicate what a tuple is suppose to model, nor does it indicate how to map tuples to the various protocol elements available in SIP.

The Internet-Draft "A Data Model for Presence" [203] tries to cover this vacuum by providing a model that maps tuples to SIP communication systems. The model is centered around three different aspects of a presentity.

Service: A communications service, such as instant messaging or telephony, is a system for interaction between users that provides certain modalities or content.

Device: A communications device is a physical component that a user interacts with in order to make or receive communications. Examples are a phone, Personal Digital Assistance, or Personal Computer.

Person: The end-user, and for the purposes of presence, is characterized by states, such as "busy" or "sad" which impact their ability and willingness to communicate.

Figure 16.4 describes the presence data model. The model considers that a presence entity, or *presentity* is characterized by four different data components: the presentity URI, the person, the service, and the device, with each one (except for the presentity URI) containing some data associated to the person, service, or device. The presence data model stresses the importance of the presence data reported, not of the data component that reported it. As an example, a mobile phone (a device) might be reporting that the user (the person data component) is busy, independently of the fact that it is a mobile phone (a device) reporting that information.

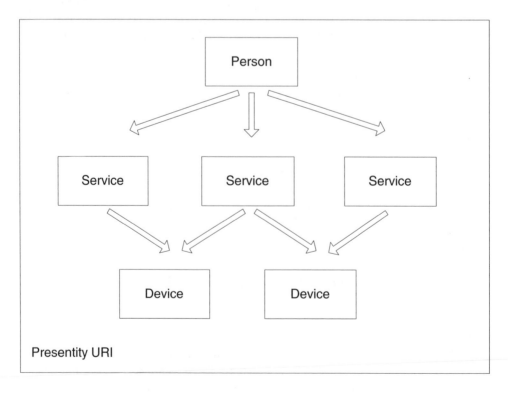

Figure 16.4: The SIP presence data model

The SIP presence data model considers that each presentity might have one or more presentity URIs. A common case is a presentity whose presence information is represented with a pres URI (specified in RFC 3859 [184]) and a SIP URI.

The person data component provides information about the user himself. Two different aspects are considered: characteristics of the user and their status. Characteristics refer to static user data that does not change often with time, such as birthday or height. Status refers to dynamic information about the user, such us the user's activities (the user is on the phone, in a meeting, etc.), or the mood (the user is sad, happy, etc.).

The SIP presence model allows only one person data component per presentity, although it allows the person component to refer to something that behaves as a person but it is not exactly a person, for example a group of assistants in a call center, or an animal.

Presentities access services, and the willingness of presentities to communicate with some services is modelled with the services data component. A service can include: videotelephony, push-to-talk, instant messaging, etc. Like the person data component, the service data component can be described in terms of characteristics (static service data) and status (dynamic service data). Characteristics of the service include, for example, the SIP methods supported by the service, or other capabilities that represent the service (e.g., audio, video). Status includes, for example, whether the user is willing to communicate with that service or not. Services can also describe a URI that can be invoked to reach the service.

The last data component is the device data component. Devices model the physical platform in which services execute. Mobile phones, personal computers, and personal digital assistants are all example of devices. Like services and persons, devices are describes in terms of characteristics and status. Characteristics include the display size, number of colors, etc. Status include the remaining battery and the geographical location of the device. Devices are identified by a device ID, which is a Uniform Resource Name (URN) [165] that temporarily uniquely identifies the device. The device ID could be an International Mobile Equipment Identity (IMEI), an Electronic Serial Number (ESN), or a Medium Access Control (MAC) address.

16.5 Mapping the SIP Presence Data Model to the PIDF

Since the PIDF was created earlier than the SIP presence data model, and since the purpose of PIDF is to become the common minimum denominator across different presence systems, there is a need to map the SIP presence data model to the existing PIDF. The idea is to reuse the PIDF in its current state where possible, and extend it when required.

Mapping of the SIP presence data model to the PIDF is achieved by reusing the existing XML elements in the PIDF, with clarified semantics according to the SIP presence data model, and by extending the PIDF with new elements to accommodate the new data components.

We describe the mapping of the SIP presence data model to the PIDF with the help of Figure 16.5. The `presence` root element contains an `entity` attribute that, in the case of the SIP presence data model, is the presentity URI.

The `tuple` element of the PIDF is used to describe the service data component of the SIP presence. A child `status` element is used to described the characteristics or dynamic information of the service. The `contact` element of the PIDF contains the service URI. The static service information are new extension elements that become children of the `tuple` element.

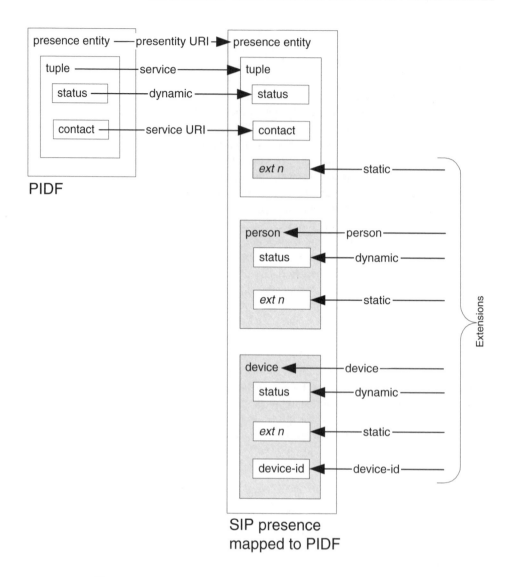

Figure 16.5: The SIP presence data model mapped to the PIDF

A new person element, a sibling of tuple, is created to contain the person data component. A new child status element carries the dynamic person information, whereas new extension elements carry the static information.

Like person, a new device element, which appears as a child of presence, is created to convey the device data component. The device element also contains a status element that contains dynamic device information, and a number of extensions that contain the static device information. A device-id element, child of device, contains the device ID.

In each of the mentioned status elements, a number of new child elements are created to contain the actual dynamic information. However, those are not represented in Figure 16.5 for the sake of clarity.

16.6 Rich Presence Information Data Format

We have just described (Section 16.3) how the PIDF document defines a minimalist model to describe the presence information of a presentity. In commercial systems this minimalist model does not give enough detailed information. For instance, Alice might be interested in informing her watchers that she is online but not willing to accept any form of communications because she is driving. Unfortunately, the PIDF alone does not provide us with the semantics to express such information.

The Rich Presence Information Data Format (RPID) is an extension to the PIDF that allows a presentity like Alice to express detailed and rich presence information to her watchers. Like the PIDF, RPID is encoded in XML. RPID is backward compatible with the PIDF. If watchers do not understand the RPID extension, they can at least get the minimal information from the PIDF document. The RPID extension is specified in the Internet-Draft "RPID: Rich Presence Extensions to the Presence Information Data Format (PIDF)" [222].

16.6.1 Contents of the RPID

A presentity like Alice can set her rich presence information by manually operating on the appropriate setting of her presence software. However, RPID allows an automaton that has access to the presentity's presence information to set such information up automatically. For instance, a calendar application can automatically set the presentity's presence information to "online – in a meeting" when the presentity's agenda indicates so. A SIP phone can automatically update the presentity's presence information to indicate that the presentity is engaged in a call when the presentity answers the phone.

Let's take a look at the type of information that the RPID is able to carry. The RPID extensions are applicable to person, services (tuple), and devices according to the presence data model.

RPID defines an `activities` element. It contains one or more `activity` elements that indicate the activity the presentity is currently doing. The specification allows the `activity` element to express that the presentity is on the phone, away, has a calendar appointment, is having breakfast, lunch, dinner, a meal, is not at work due to a national or local holiday, at work, is scheduled for a meal, in a meeting, steering a vehicle, in transit, traveling, on vacation, sleeping, just busy, in a performance, playing, presentation, watching TV, on a permanent absence, or doing some unknown activity. The list of values is expandable, so future extensions can add new values when needed.

A `class` element allows the presentity to group similar person elements, devices or services that belong to the same class. The presentity allocates the class to a tuple. The presence agent can use this information to filter tuples according to the class.

A `device-id` element contains an identifier of the device that provides a particular service. It allows us to differentiate different devices that are contributing to the same presence information. The device identifier is a Uniform Resource Name (URN) [165] that identifies the device. So, for example, it could be an International Mobile Equipment Identity (IMEI), an Electronic Serial Number (ESN), or a Medium Access Control (MAC) address.

The `mood` status element is able to indicate the mood of a person (e.g., sad, happy, afraid, confused, impressed, offended, etc.).

The RPID includes two elements that contain information related to the place where the person is located. On one side, `place-is` status elements contain properties of the place, e.g., a noisy environment, dark, quiet, too bright, uncomfortable, or inappropriate. place-type status elements allows our presentity, Alice, to indicate the type of place she is currently in. The possible initial values (usually extensible) are home, office, library, theater, hotel, restaurant, school, industrial, quiet, noisy, public, street, public area, aircraft, ship, bus, train, airport, station, mall, outdoors, bar, club, cafe, classroom, convention center, cycle, hospital, prison, underway, or unknown.

The RPID also includes a privacy element that indicates the type of media that the presentity will be able to safely receive with privacy, e.g., without third parties being able to intercept. The possible values include: audio, video, or text, indicating respectively, that the presentity can receive audio, video, or text without others intercepting that type of media.

The relationship element in the RPID indicates the type of relationship the presentity has with an alternative contact. The possible values can indicate that an alternative contact is part of her family, friend, an associate, assistant, supervisor, self, or unknown.

The service-class element indicates whether the service is an electronic service, a postal or delivery service, or describes in-person communications.

The sphere element in the RPID indicates the current state and role the presentity plays. Possible values are home, work, or unknown. This is useful information that allows the presentity to set visibility rules when she is playing a certain role. For instance, a member of the family may have access to additional information, such as a home webcam URI, when the presentity sets the sphere to home, while co-workers will not have access to this information.

The status-icon element contains a pointer (e.g., and HTTP URI) to an icon representing the person or service.

The time-offset element is able to express the offset in minutes from UTC at the user's current location.

The user-input element allows us to express human user input or the usage state of the service or device. It can contain the value either "active" or "idle", including an optional last-input attribute that indicates the origin time of such a state.

Figure 16.6 shows an example of the rich presence information that Alice provides to her watchers. The presence information is encoded according to the PIDF with the extensions defined by the presence data model and RPID. Alice is providing her presence information from a personal computer. The device is providing the idle state since a point in time. Alice indicates that she is in the away state, at home, in a quiet environment for receiving communications, and is in a happy mood.

16.7 CIPID

The Contact Information in Presence Information Data Format (CIPID) is an extension to the PIDF that provides additional information about a presentity or a tuple, such as references to her business card, home page, map, sound, display name, and icon. Typically these are extensions to the person element in the presence data model, although the tuple element can also be extended with CIPID information in some cases, such as when the information describes a service referring to another person (e.g., when the person has a given relationship different than "self"). The CIPID is specified in the Internet-Draft "CIPID: Contact Information in Presence Information Data Format" [221].

CIPID adds new card, display-name, homepage, icon, map, and sound elements to the person or tuple elements in the PIDF.

```
<?xml version="1.0" encoding="UTF-8"?>
<presence xmlns="urn:ietf:params:xml:ns:pidf"
    xmlns:xsi="http://www.w3.org/2001/XMLSchema-instance"
    xmlns:dm="urn:ietf:params:xml:ns:pidf:data-model"
    xmlns:r="urn:ietf:params:xml:ns:pidf:rpid"
    entity="pres:alice@example.com">

    <tuple id="3bfua">
      <status>
        <basic>open</basic>
      </status>
      <r:deviceID>urn:device:001349038B74</r:deviceID>
      <r:service-class><r:electronic/></r:service-class>
      <r:status-icon>
            http://www.example.com/alice/icon.jpg
      </r:status-icon>
      <contact priority="0.8">
            sip:alice@pc.example.com
      </contact>
      <timestamp>2005-06-05T07:52:14Z</timestamp>
    </tuple>

    <dm:device id="vjsa43">
      <r:user-input idle-threshold="300"
          last-input="2005-06-03T00:23:21">idle</r:user-input>
      <dm:deviceID>urn:device:001349038B74</dm:deviceID>
      <dm:note>PC</dm:note>
    </dm:device>

    <dm:person id="alice">
      <r:activities><r:away/></r:activities>
      <r:mood><r:happy/></r:mood>
      <r:place-is>
        <r:audio>
          <r:quiet/>
        </r:audio>
        <r:video>
            <r:quiet/>
        </r:video>
      </r:place-is>
      <r:place-type>
          <r:home/>
      </r:place-type>
      <r:privacy>
          <r:audio/>
          <r:video/>
          <r:text/>
      </r:privacy>
      <r:sphere>
          <r:home/>
      </r:sphere>
    </dm:person>
    <note>I have some visitors at home</note>
</presence>
```

Figure 16.6: Example of the RPID

The `card` element contains a URI that points to a business card stored in LDIF (LDAP Data Interchange Format, specified in RFC 2849 [110]) or vCard (specified in RFC 2426 [80]) format.

The `display-name` element adds a name to the tuple or presentity that the presentity suggests to the watcher's user interface to display.

The `homepage` element contains a URI that points to the web home page of the presentity. The `icon` element contains a URI that points to an image of the presentity.

The `map` element contains a URI that points to a tuple's or presentity's map. It could be a GIF or PNG file, but also a Geographical Information System (GIS) document.

The `sound` element contains a URI that points to a tuple's or presentity's sound. The format of such a file is not standardized, but it is recommended to support MP3.

Figure 16.7 shows an example of Alice's PIDF document that includes CIPID information in the `person` element. Alice is publishing pointers to her business card, home page, icon, map, and sound.

```xml
<?xml version="1.0" encoding="UTF-8"?>
<presence xmlns="urn:ietf:params:xml:ns:pidf"
        xmlnl:dm="urn:ietf:params:xml:ns:pidf:data-model"
        xmlns:ci="urn:ietf:params:xml:ns:pidf:cipid"
        entity="pres:alice@example.com">

    <tuple id="39dsaq">
      <status>
        <basic>open</basic>
      </status>
      <contact>sip:alice@ws3.example.com</contact>
      <timestamp>2005-06-12T18:53:02</timestamp>
    </tuple>

    <dm:person id="dpoia1">
      <ci:card>http://example.com/alice/card.vcd</ci:card>
      <ci:homepage>http://example.com/alice</ci:homepage>
      <ci:icon>http://example.com/alice/icon.gif</ci:icon>
      <ci:map>http://example.com/alice/map.png</ci:map>
      <ci:sound>http://example.com/alice/mysound.mp3</ci:sound>
      <dm:timestamp>2005-06-12T18:53:02</dm:timestamp>
    </dm:person>
</presence>
```

Figure 16.7: Example of the CIPID

16.8   Timed Presence Extension to the PIDF

We have seen that the PIDF together with the RPID provides the current status of a presentity. However, they cannot provide information about past or future actions that the presentity had taken or will take. For instance, a presentity may start a meeting in the next half an hour. If a presentity publishes this information, watchers may decide to postpone interaction with the presentity until that meeting is over. The Timed Presence extension is specified in the Internet-Draft "Timed Presence Extensions to the Presence Information Data Format (PIDF) to Indicate Status Information for Past and Future Time Intervals" [223] and allows a presentity to express what they are going to be doing in the immediate future or actions that took place in the near past.

A new `timed-status` element that contains information about the starting time of the event is added to the PIDF `tuple` element, or the data model `person` or `device` elements. The starting time of the event is encoded in a `from` attribute, whereas an optional `until` attribute indicates the time when the event will stop.

Figure 16.8 shows an example of the timed status extension. Alice is publishing that she will be offline from 13:00 to 15:00. Let's imagine that it is 12:45 when a watcher gets access to this information. The watcher wants to interact (by a call or instant messaging) with Alice, but the interaction may take more than 15 minutes. Since Alice will be offline in 15 minutes, perhaps due to a scheduled meeting, the watcher is able to delay the interaction with Alice until 15:00 when she will most likely be online again.

```
<?xml version="1.0" encoding="UTF-8"?>
<presence xmlns="urn:ietf:params:xml:ns:pidf"
          xmlns:ts="urn:ietf:params:xml:ns:pidf:timed-status"
          entity="pres:alice@example.com">

    <tuple id="qoica32">
      <status>
         <basic>open</basic>
      </status>
      <ts:timed-status from="2004-02-15T13:00:00.000+02:00"
          until="2004-02-15T15:00:00.000+02:00">
         <basic>closed</basic>
      </ts:timed-status>
      <contact>sip:alice@example.com</contact>
    </tuple>
</presence>
```

Figure 16.8: Example of the *timed status* extension

16.9 Presence Capabilities

We have seen in previous sections how presentities can express their presence status including online status, contact address, device capabilities, etc. When we described the basic operation of SIP we explored the Caller Preferences and User Agent Capabilities extension to SIP (see Section 4.12). This extension is concerned with the registration and session establishment processes in SIP. It seems natural to mimic that extension for presence publication and watcher notification procedures, so that presentities could indicate in their presence information the same features that they would otherwise express in a registration.

Any watcher getting a notification about the presentity's presence status could also get the information about the presentity's supported features. This has the advantage that, in case a watcher wants to initiate any form of communication with a presentity, the watcher knows in advance the capabilities supported at the remote end. For instance, if Alice knows that Bob is using an application (service) that does not support text but does support audio and video, she may want to initiate an audiovisual session, rather than sending an instant message.

This is exactly what the presence capabilities extension does. The Internet-Draft "User Agent Capability Extension to the Presence Information Data Format (PIDF)" [155] provides an extension to the PIDF that maps the caller preferences features (defined in RFC 3840 [217]) to new XML elements that are part of a PIDF document. These new elements express whether the presentity is reachable on a mobile or fixed device, whether it supports audio or video capabilities, the list of supported SIP methods and SIP event packages, etc.

In addition to all the features defined in the Caller Preferences (RFC 3840 [217]), the Presence Capabilities allow us to express a few other features, such as "type", "message", and "language", which are not defined in the Caller Preferences documentation.

Presence capabilities are subdivided into service capabilities and device capabilities. Service capabilities characterize features related to services, for example, whether the service supports audio, video, messaging, full duplex operation, etc. Device capabilities are features that characterize a physical device, for example, whether a device is mobile or fixed. As the reader can expect, the presence capabilities are built on top of the presence data model, and as such, expand the `tuple` and `device` elements with service capabilities and device capabilities, respectively.

So presence capabilities define two new elements that act as containers of service and device capabilities: these are the `servcaps` and `devcaps` elements, respectively. Each of these elements contain a collection of new elements representing a feature that characterizes the service or the device. Let us take a deeper look at the service and device capabilities.

16.9.1   *Service Capabilities*

Service capabilities are enclosed in a `servcaps` XML element. The `servcaps` element has to be a child of the `tuple` XML element defined in the Presence Information Data Format (PIDF), since `tuple` elements are meant to describe services, according to the presence data model. A `servcaps` XML element can contain a number of `audio`, `application`, `data`, `control`, `video`, `text`, `message`, `type`, `automata`, `class`, `duplex`, `description`, `event-packages`, `priority`, `methods`, `extensions`, `schemes`, `actor`, `isfocus`, and `languages` elements.

The `audio`, `application`, `data`, `control`, `video`, `text`, and `message` elements are boolean indications (true or false) of the support of the service for audio, application, data, control, video, text, or message streams, respectively. Note that these elements refer to the type of media streams supported by the service, not by the device. So if there are two services (e.g., two applications) running on the same device, and one supports only audio media streams, it will indicate this, even when the device supports other capabilities (e.g., video).

The `type` element indicates possible MIME types that the service is able to accept. These MIME types are typically indicated in a `Content-Type` header field in SIP.

The `class` element indicates whether the service is used for `business` communications or `personal` communications.

The `duplex` element indicates whether a communications service can simultaneously send and receive media (value of `full`), alternate between sending and receiving media (value of `half`), can only receive media (value of `receive-only`), or only send media (value of `send-only`).

The `description` element contains a textual description of the service.

The `event-packages` element contains a list of SIP event package supported by the service. Like in the types of supported media streams, the `event-packages` elements refer to the SIP event packages supported by the service (`tuple` XML element) under description, not all the SIP event packages supported by the union of all the applications running in the device.

The `priority` element indicates the SIP priorities that the service is able to handle. Priorities are expressed as integers, and the application can indicate ranges of supported and non-supported priorities.

The `methods` element indicates the list of supported (and if available, the non-supported too) SIP methods in the service. Like the rest of the service capabilities, this element refers to the list of supported methods in the service (`tuple` element in the PIDF) where it appears, and it does not refer to the list of supported methods by the union of all the applications or services running in the device.

The `extensions` element indicates the list of supported (and non-supported as well) SIP extensions by the service. The element contains a list of option-tags that can appear in `Supported` or `Require` SIP header fields.

The `schemes` element indicates the list of URI schemes that the service is able to handle (e.g., sip, sips, tel, etc.).

The `actor` element allows the presentity to indicate whether the type of entity residing behind the service, for example, if it is the principal associated with the service, an attendant of substitute of the principal, a message taker (such as a voice mail system), or some person or automata that can provide further information about the principal.

The `isfocus` element indicates that the service is a centralized conference server.

The `languages` element indicates the ability of the service to display human languages.

16.9.2 Device Capabilities

Device capabilities are enclosed in a `devcaps` XML element. The `devcaps` element has to be a child of the `device` XML element defined in the presence data model. A `devcaps` XML element can contain `mobility`, `priority`, and `description` elements.

The `mobility` element merely indicates whether the device is `fixed` or `mobile`.

The `priority` reflects the priorities of the call that the device is willing to handle. It can contain a range of values that are accepted by the device.

The `description` element indicates a textual description of the device.

16.9.3 An Example of the Presence Capabilities Document

Figures 16.9 and 16.10 show an example of a PIDF document extended with the presence data model and the presence capabilities (the example has been split into two parts for presentation purposes).

The presentity, Alice, is indicating her presence capabilities related to both the service (i.e., the `servcaps` element under the `tuple` element), and the device capabilities (i.e., the `devcaps` element that extends the `device` element defined in the presence data model). Alice describes the capabilities that her service supports: audio, video, text, and full duplex capabilities. She then describes the list of SIP methods, event packages, SIP extensions, and URI schemes that her service is able to handle. In the devices capabilities section she indicates she is using a mobile device.

```xml
<?xml version="1.0" encoding="UTF-8"?>
<presence xmlns="urn:ietf:params:xml:ns:pidf"
   xmlns:xsi="http://www.w3.org/2001/XMLSchema-instance"
   xmlns:cap="urn:ietf:params:xml:ns:pidf:caps"
   xmlns:dm="urn:ietf:params:xml:ns:pidf:data-model"
   entity="pres:alice@example.com">

   <tuple id="3bfua">
      <status>
         <basic>open</basic>
         <cap:servcaps>
            <cap:audio>true</cap:audio>
            <cap:video>true</cap:video>
            <cap:message>true</cap:message>
            <cap:duplex>
               <cap:supported>
                  <cap:full/>
               </cap:supported>
            </cap:duplex>
            <cap:description>General service</cap:description>
            <cap:methods>
               <cap:supported>
                  <cap:INVITE/>
                  <cap:ACK/>
                  <cap:CANCEL/>
                  <cap:BYE/>
                  <cap:MESSAGE/>
                  <cap:PRACK/>
                  <cap:UPDATE/>
                </cap:supported>
            </cap:methods>
            <cap:event-packages>
               <cap:supported>
                  <cap:reg/>
               </cap:supported>
            </cap:event-packages>
            </cap:methods>
            <cap:extensions>
               <cap:supported>
                  <cap:100rel/>
                  <cap:precondition/>
               </cap:supported>
            </cap:extensions>
```

Figure 16.9: Example of the presence capabilities extension, part I

```
                    <cap:schemes>
                        <cap:supported>
                            <cap:s>sip</cap:s>
                            <cap:s>sips</cap:s>
                        </cap:supported>
                    </cap:schemes>
                </cap:servcaps>
            </status>
            <contact>sip:alice@example.com</contact>
        </tuple>

    <dm:device>
        <cap:devcaps>
            <cap:mobility>
                <cap:supported>
                    <cap:mobile/>
                </cap:supported>
            </cap:mobility>
        </cap:devcaps>
        <dm:deviceID>urn:device:039fa209</dm:deviceID>
    </dm:device>
    </presence>
```

Figure 16.10: Example of the presence capabilities extension, part II

16.10   Presence Publication

Sections 16.3 and 16.6 described the contents of an XML document that conveys the actual presence information. However, we still have not described how a PUA makes this information available to the PA.

An obvious mechanism to use in this interface is the REGISTER method. REGISTER transactions provide the current location (IP address, not to be confused with geographical location) of the user. Therefore, when users are not registered the PA sets their presence to "offline" and when they are registered the PA sets their presence to "online". On the other hand, the semantics of the REGISTER method are very clear: REGISTER binds an Address-Of-Record (public identity) with a contact address. Therefore, it does not seem appropriate to overload these semantics for the purpose of presence publication. Consequently, we need another mechanism that allows PUAs to upload presence information (e.g., PIDF/RPID documents) to a PA.

The IETF defined the SIP PUBLISH method in RFC 3903 [167]. The purpose of a PUBLISH request is to publish the event state used within the framework for SIP-specific event notification (RFC 3265 [198]). Thus, the PUBLISH method is not only used for presence publication, it is generic enough to be used to publish any state associated with an event package. However, we focus in this section on presence publication.

Figure 16.11 shows a typical flow used to publish presence information: the PUA sends a PUBLISH request that contains a PIDF/RPID document to the PA. The PA acknowledges with a 200 (OK) response.

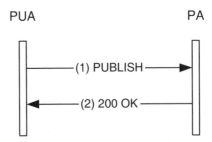

Figure 16.11: Publication of presence information

16.11   Presence Subscription and Notification

The interface defined between a watcher and a PA allows a watcher to subscribe to the presence information of a presentity. Presence subscription is implemented with a SIP SUBSCRIBE transaction. The subscription can be a simple fetch operation, whereby the watcher just wants to get the current presence information of a presentity, but does not want to be informed about future changes to such information. A subscription can last for a period of time. In this case the watcher gets updates of the presentity's presence information whenever that information changes. If a watcher wants to keep the subscription active they need to renew it prior to its expiration.

Figure 16.12 shows an example flow. A watcher sends a SUBSCRIBE request (1) to the PUA, the request including an `Event` header field set to `presence`, indicating the subscription to the presence information of a particular presentity. The PA authenticates and authorizes the watcher and answers with a 200 (OK) response (2), followed by a NOTIFY request (3), which in most cases does not contain presence information, just the status of the subscription. The watcher answers with a 200 (OK) response (4). Whenever the presentity's presence information changes the PA sends a new NOTIFY request (5) that includes a presence document (PIDF, RPID, etc.). The watcher replies with a 200 (OK) response (6).

16.12   Watcher Information

In the previous sections we addressed two main problems that the presence service solves: how presentities can publish their presence information and how watchers can be updated when changes in such presence information occur. There is a further problem that we have not yet addressed: how can a presentity like Alice (or any other authorized observer) be informed of the watchers who are subscribed to her presence information. Typically, presentities are interested in knowing who is watching their presence information. Besides the presentity, it is also possible that other services or authorized observers get notifications of the watchers of a particular presentity.

In order to solve this problem the IETF has created a *Watcher info* event template-package defined in RFC 3857 [202]. Event template-packages are event packages that can be applied to any other event package. The watcher info event template-package provides the subscriber with information about who is watching the subscribed resource. If the watcher info event template-package is applied to presence the value of the `Event` header field is set to `presence.winfo`.

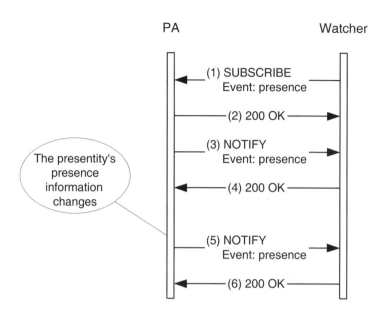

Figure 16.12: Subscription and notification of presence information

Let's see how watcher info works with the help of Figure 16.13. A subscriber, typically but not necessarily a PUA, sends a SUBSCRIBE request (1) to their PA with an `Event` header field set to `presence.winfo`. The PA authenticates and authorizes the subscription and answers with a 200 (OK) response (2). The PA also sends a NOTIFY request (3) to indicate the status of the subscription. This NOTIFY can also contain an XML document containing the list of watchers of the presence information of the presentity. The PA will keep the PUA updated, using NOTIFY requests (5), about changes in the list of watchers. That is, it will inform Alice every time a new watcher subscribes or unsubscribes to the presentity's presence information.

Being informed about watchers is the motivation behind creating watcher info functionality, but it is not the only one. Perhaps the main motivation is related to presence authorization. When a watcher subscribes to a presentity's presence information the presentity needs to authorize the subscription. In order for the presentity to be able to authorize watcher subscriptions the presentity must be aware of who is requesting to watch the presentity's presence information and what is the subscription status of each of these watchers. The watcher info event package provides the means to transport this information.

The watcher info event package is encoded in XML. The package defines a new MIME type `application/watcherinfo+xml`. When a SIP request or response contains a `Content-Type` header field set to `application/watcherinfo+xml`, it is indicating that the body is an XML document that contains watcher information.

Figure 16.14 shows an example of a watcher info XML document, where the presentity is Alice and the watcher is Bob. Bob has a subscription to Alice's presence, but it is in the `pending` state, and is just waiting for Alice to authorize the subscription. Watcher subscription authorization is done by means other than SIP (we describe watcher subscription authorization in Section 16.14).

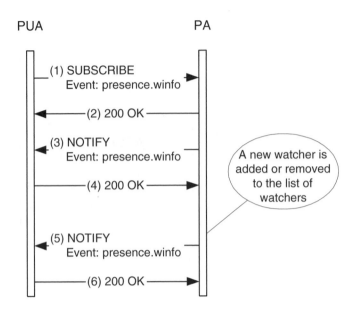

Figure 16.13: A PUA compiles the list of her watchers

```
<?xml version="1.0"?>
<watcherinfo xmlns="urn:ietf:params:xml:ns:watcherinfo"
             version="0" state="full">
   <watcher-list resource="sip:alice@example.com"
                 package="presence">
       <watcher id="342sd2" event="subscribe" status="pending">
                sip:bob@example.com</watcher>
   </watcher-list>
</watcherinfo>
```

Figure 16.14: Example of a watcher info XML document

16.13   URI-list Services and Resource Lists

We saw in Section 16.11 that every time a watcher wants to subscribe to the presence information of a presentity the watcher needs to exchange a SUBSCRIBE transaction and a NOTIFY transaction with the presentity's PUA, just to set up the subscription. If the PUA challenges the watcher the number of SIP transactions may be even bigger. At a minimum the watcher always exchanges four SIP messages with the presentity's PA for each presentity the watcher is subscribed to.

For example, Alice wants to know the presence status of her friends Bob, David, and Peter. She acts as a watcher and sends a SUBSCRIBE request to each of their presence agents (1), (3), (5), as shown in Figure 16.15. Later, she receives a NOTIFY request from each of them (7), (9), (11). If the watcher is subscribed to, say, the presence information of 100 presentities, the watcher's presence application needs to process a minimum of 400 SIP messages (perhaps more) every time the watcher initializes the presence application. And this is just to set up the subscriptions. Typically, NOTIFY requests do not contain the presentity's presence information at this stage. Obviously, this mechanism does not scale well, particularly in wireless environments: the number of messages required to set up the subscriptions should be independent of the number of presentities in the watcher's list.

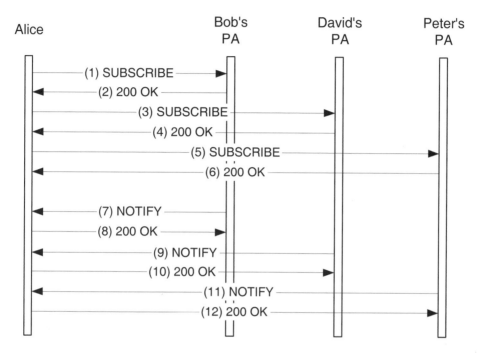

Figure 16.15: Long presence list without a URI-list service

In order to solve this problem the IETF has created the concept of *resource lists*. A resource list is a list of SIP URIs that is stored in a new functional entity called the *Resource List Server* (RLS), sometimes known as a *URI-list service* for SUBSCRIBE requests (see Section 20.2.1 for a general description of URI-list services). The list is addressed with its own SIP URI. When the concept of resource lists is applied to the presence service it is also called *presence lists*. A presence list contains a list of all the presentities a watcher is subscribed to.

A SIP URI-list service receives a request from a user agent and forwards it to multiple users. SIP URI-list services used for subscriptions (described in the Internet-Draft "A Session Initiation Protocol (SIP) Event Notification Extension for Resource Lists" [197]) also summarize the presence information received in NOTIFY requests. Figure 16.16 shows how this type of URI-list service works.

Instead of sending a SUBSCRIBE request to every user in her presence list, Alice sends a single SUBSCRIBE request (1) addressed to her presence list. The request is received by the SIP URI-list service, or Resource List Server. Alice has previously provided the URI-list service, using an out-of-band configuration mechanism of her choice, with her presence list (which is a list of the presentity's URIs). The URI-list service sends a SUBSCRIBE request to every URI in the list (3), (5), (7). Later, when the URI-list service receives the NOTIFY

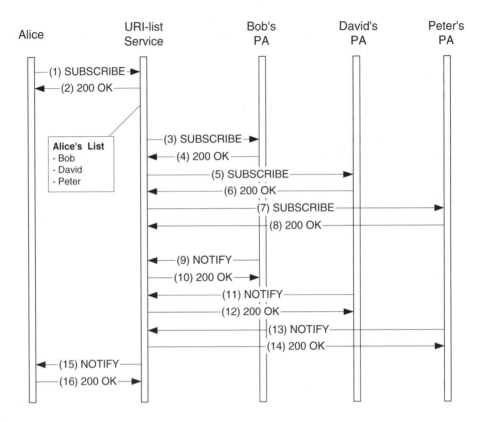

Figure 16.16: Resource list through a URI-list service

requests from them (9), (11), (13), it aggregates the presence information and sends a single NOTIFY request (15) to Alice. This mechanism saves a considerable amount of bandwidth on Alice's access network; moreover, it requires much less processing power in the PUA.

The RLS mechanism defines a new MIME type `application/rlmi+xml`. This document contains a list of subscribed resources (presentities) that points to each of the PIDF documents of the presentities. As a consequence, NOTIFY requests contain a `multipart/related` document that comprises a collection of other embedded XML documents.

Usage of presence lists rather than individual subscriptions has the side effect of not storing the presence list locally in the presence application (e.g., the computer or the IMS terminal), but in a network node. When Alice uses her presence application from a different terminal than usual, she can get access to her presence list, since it is stored in the network. In her new terminal she just needs to configure the URI of her own presence list to get access to it, rather than configuring one by one all the URIs of the presentities. This allows seamless terminal roaming. For instance, it allows Alice to become a watcher when she is logged on to a public computer, such as an Internet kiosk.

16.14   XML Configuration Access Protocol

So far we have seen mechanisms to subscribe, notify, and publish presence information and Section 16.13 introduced the concept of the Resource List Server, a server that hosts resource lists.

Whenever there is a list there is a need to manage the list. The owner of the list may need to add new URIs at some point in time. On other occasions they will need to remove or just modify existing URIs in the list. All of these operations are known as list management. In Section 16.13 we indicated that resource lists are managed by some out-of-band mechanism. Certainly, a web page where the list owner logs on is a good candidate to provide a configuration interface. However, a simple web page does not offer a standard interface for configuration, because, although HTTP is standardized, the actual contents of the HTML body included in the HTTP responses is not. This is not the desired behavior, as it complicates interoperability between different terminal vendors and Resource List Servers. Especially in mobile environments, where the size of the screen of a mobile device, the input devices, and the number of colors in the display may vary from vendor to vendor, it might get difficult to render the configuration web page in an appropriate way.

The IETF is working on a solution to standardize a management interface between terminals and servers. The idea is to standardize the mechanisms to manipulate the contents of a list, rather than the presentation of a web page as HTML does. The protocol on that interface is referred to as XCAP (XML Configuration Access Protocol), defined in the Internet-Draft "The Extensible Markup Language (XML) Configuration Access Protocol (XCAP)" [210]. XCAP provides a client with the means to add, modify, and delete XML configuration data of any kind stored in a server, such as users in a presence list, authorization policies (e.g., a list of authorized watchers), or a list of participants in a conference. XCAP does not control the user interface (e.g., the graphical representation of the list).

As the name indicates, XCAP is encoded in XML. XCAP is just a set of conventions that tell us how to work with a remotely stored XML document. XCAP uses Hypertext Transfer Protocol 1.1 (HTTP, specified in RFC 2616 [101]) as the transport protocol. Figure 16.17 shows a schematic representation of the protocol stack used by XCAP.

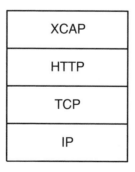

Figure 16.17: The XCAP protocol stack

XCAP defines conventions that map XML documents and their components to HTTP URIs. It also defines the rules that govern how modification of one resource affects another. Additionally, XCAP also defines the authorization policies associated with access to resources.

XCAP implements a set of operations that are mapped to regular HTTP 1.1 methods. The operations also set some HTTP headers to a particular value. XCAP provides the client with the following operations: create a new document, replace an existing document, delete an existing document, fetch a document, create a new element in an existing document, replace an existing element in a document, delete an element in a document, fetch an element in a document, create an attribute in an existing element of a document, replace an attribute in the document, delete an attribute from the document, and fetch an attribute of a document.

Figure 16.18 shows an example of XCAP in which Alice is sending an HTTP PUT request to create a new presence list. The list contains the members of her family. Two URIs are initially added to the list: Bob's and Cynthia's.

```
PUT http://www.example.com/pr-lists/alice/fr.xml HTTP/1.1
Content-Type: application/resource-lists+xml
Content-Length: 460


<?xml version="1.0" encoding="UTF-8"?>
<resource-lists
     xmlns:xsi="http://www.w3.org/2001/XMLSchema-instance">
 <list name="family" uri="sip:family@example.com"
                   subscribeable="true">
    <entry name="Bob" uri="sip:bob@home1.com">
      <display-name>Bob</display-name>
    </entry>
    <entry name="Cynthia" uri="sip:cynthia@example.com">
      <display-name>Cynthia</display-name>
    </entry>
 </list>
</resource-lists>
```

Figure 16.18: Example of XCAP

16.14.1   XCAP Application Usage

As we have just seen, XCAP is a generic protocol that can be used for a number of purposes related to the configuration of XML documents stored in a server, but there are other applications of XCAP. For example, presentities can use XCAP to configure their presence lists stored in a RLS. Presentities can also use XCAP to provide authorization to their watchers, so that watchers can see all or part of the presentity's presence information. In an application that is not related to presence but to centralized conferences, the creator of a dial-out conference can use XCAP to configure the list of participants of the conference. In general, whenever there is a need to configure a list remotely stored in XML format, XCAP is a good protocol to fulfill these needs.

Because of this versatility the XCAP specification introduces the concept of *application usages*. An application usage defines how a particular application uses XCAP to achieve the desired functionality. For instance, each of the previously mentioned applications of XCAP – presence list management, authorization policies, and conference list management – constitute an application usage on its own. Each application usage is identified by an

AUID (Application Usage ID) that uniquely identifies the application usage. There are two types of AUIDs: standard (i.e., application usages standardized in the IETF) and vendor-proprietary (i.e., are private application usages).

The IETF has defined a number of XCAP application usages related to the presence service.

XCAP application usage for resource list: defined in the Internet-Draft "Extensible Markup Language (XML) Formats for Representing Resource Lists" [207], this provides the means to manipulate resource lists that are typically used as presence lists.

XCAP application usage for presence authorization: defined in the Internet-Draft "Presence Authorization Rules" [209], this allows a client to specify presence authorization rules, i.e., the set of rules that provides some watchers with the permissions to access certain subsets of the presentity's presence information.

16.15   Presence Optimizations

We have discussed the presence service in this chapter, and we have seen several examples of presence publication, subscription, and notification. Presence publication and presence notification operations contain a PIDF/RPID XML document. PIDF/RPID documents are naturally large because they are rich in information. A watcher who is subscribed to a number of presentities may get one of these XML documents every time the presentity's presence information changes. When presence information reaches a small device that has constraints in memory, processing capabilities, battery lifetime, and available bandwidth, the device may be overwhelmed by the large amount of information and might not be able to acquire or process it in real time.

3GPP engineers, aware of the constraints of cellular devices and networks, have worked with IETF engineers, the experts in protocol design, in order to find solutions to optimize the amount of presence information transmitted to presence clients. Obviously, there has to be a compromise between the amount of information sent, the frequency of the notifications, and the bandwidth used to send that information. Sending less information in presence documents may lead to users not getting a good experience with presence systems used from wireless terminals. Sending presence information less periodically will lead to an inaccurate presence view of the presentities. It is important that the user gets accurate and rich presence information while the bandwidth used is reduced.

We have described the usage of Resource List Servers, a mechanism that solves not only bandwidth problems but also terminal-roaming problems. In Section 4.15.1.1 we described a mechanism to throttle and control the rate of notifications. This mechanism is not only applicable to presence, but to any event package controlled by the event notification framework. Apart from these mechanisms, there are other-presence-specific mechanisms that are intended to reduce the amount of presence information transmitted to watchers: partial notification and event filtering. These are still under design in the IETF at the time of writing. The requirements are stable, but the solution is still under development.

16.15.1   Partial Notification of Presence Information

A partial notification is a notification that may carry a subset of the presentity's presence information. The partial notification mechanism defines a new XML body that is able to transport partial or full state. The XML body is quite similar in structure to the PIDF, with the addition of a few new elements that indicate the version number, whether the document contains a full or a partial state, and whether a tuple has been completely removed from the presentity's presence information. The new XML body is identified with a MIME type `application/pidf-partial+xml`.

The mechanism is illustrated in Figure 16.19. A watcher subscribes to a presentity's presence information. The SUBSCRIBE request (1) contains an `Accept` header field indicating the support for both the PIDF document and the partial PIDF document. A weight or preference is also indicated through the q parameter. The PA installs the subscription and sends a first NOTIFY request (3) that contains an XML document containing the full presence state, although encoded according to the rules of the partial notification. Therefore, the `Content-Type` header field of this NOTIFY is set to `application/pidf-partial+xml`. Later, when there is a need to update the watcher with new information the PA sends a new NOTIFY request (5), but in this case it contains only the changes (additions, modifications, or deletions) with respect to the full-state document.

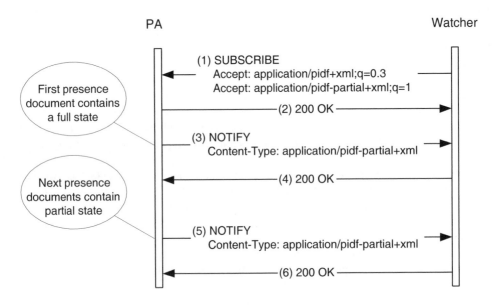

Figure 16.19: Partial notification

16.15.2   Event Notification Filtering

In daily usage a PA may deliver a lot of presence-related information about a presentity to one or more watchers. It may happen that the watchers are not interested in all of that information, but just in a subset of it. For instance, Alice may be interested to know when Bob is online or offline, but she may not be interested in knowing his detailed status (i.e., whether he is in a meeting, out to lunch or speaking on the phone). In another scenario Alice may just be interested in knowing details of Bob's instant messaging capabilities, but she may not be interested in the communication address of phones, email, web pages, etc.

If a watcher is only interested in a subset of the presence information available for a particular presentity and, especially, if the watcher has a narrow bandwidth connectivity channel, then it seems a waste of bandwidth, processing power, and even battery for the PA to send the whole presentity's presence information to the watcher. Therefore, it seems natural for a watcher to be able to specify a set of rules that filters the presence information to just those pieces that are of interest to the watcher.

Event notification filtering allows watchers to specify a set of filters that are applicable to the notification of events they are subscribed to. Filters can be applicable to elements of the XML presence information, attributes, extensions, transitions between two states, etc.

On sending a SUBSCRIBE request for the presence event of a particular presentity the watcher may include a new XML body that specifies the filters to be applied to that subscription. The local policy of the presentity's PA overrides the filter, so the filter just gives information to the PA on these parts of the presence information that the watcher is interested in.

Chapter 17

The Presence Service in the IMS

Chapter 16 gave an overview of the presence service on the Internet, as defined by the IETF. This chapter focuses on the usage of the presence service in the IMS. We explore the IMS architecture that supports the presence service and the applicability of presence to the IMS.

17.1 The Foundation of Services

When we described the presence service on the Internet we unveiled a few of the powerful and rich possibilities that the presence service can offer to both end-users and other services.

On the one hand, end-users benefit from the presence service since they decide what information related to presence they want to provide to a list of authorized watchers. Presentities can decide the information they want to publish, such as communication address, capabilities of the terminals, or availability to establish a communication. Watchers get that information in real time and decide how and when to interact with the presentity. The possibilities enrich both the communication and the end-user experience of always being in touch with their relatives, friends, and co-workers.

On the other hand, presence information is not only available to end-users but also to other services. These other services can benefit from the presence information supplied. For instance, an answering machine server is interested in knowing when the user is online to send them an instant message announcing that they have pending voicemails stored in the server. A video server can benefit by adapting the bandwidth of the streaming video to the characteristics of the network where the presentity's device is connected. For all of these reasons we refer to the presence service as the foundation for service provision, as depicted in Figure 17.1.

17.2 Presence Architecture in the IMS

3GPP defined in 3GPP TS 23.141 [33] an architecture to support the presence service in the IMS. The presence service is included in 3GPP Release 6.

Figure 17.2 depicts the presence IMS architecture that maps the already defined roles in presence to existing functional entities in the IMS. The IMS terminal plays the role of both a watcher and a PUA (Presence User Agent). The Presence Agent (PA) is an Application Server (AS) located in the home network. In the IMS the PA is typically referred to as a Presence Server (PS), although effectively it is a PA, according to the IETF definition.

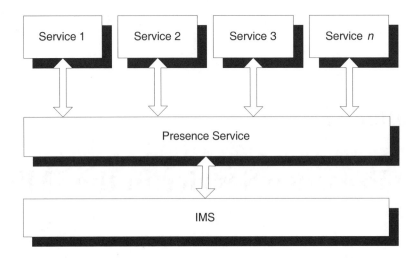

Figure 17.1: Presence service: the foundation of all the services

A Resource List Server (RLS) is also implemented as an Application Server. Any AS providing any other type of service can be acting as a watcher of the presence information of the presentity.

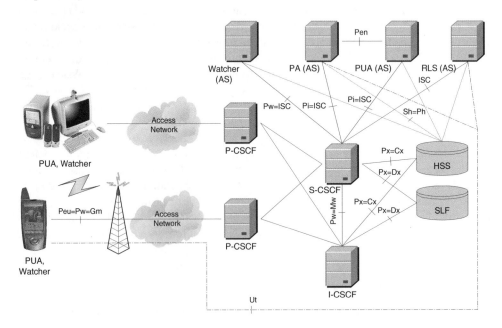

Figure 17.2: SIP-based presence architecture in the IMS

Most of the interfaces have a name starting with a "P" (e.g., *Pw*, *Pi*, *Px*), but most of them are existing IMS SIP or Diameter interfaces that map to a presence-oriented function. The *Pen* interface deserves special attention, since it allows an Application Server that is acting as a PUA to publish presence information to the presentity's PA. The PUA acquires the presence information from any possible source of information, such as the HLR (Home Location Register), the MSC/VLR (Mobile Switching Center/Visited Location Register) in circuit-switched networks, the SGSN (Serving GPRS Support Node), the GGSN (Gateway GPRS Support Node) in GPRS networks, or the S-CSCF through IMS registration.

The other new interface is the so-called *Ut* interface. This interface is defined between the IMS terminal and any application server, such as a PA or an RLS. The *Ut* interface allows the user to get involved in configuration and data manipulation, such as configuration of presence lists, authorization of watchers, etc. The protocol on this interface is XCAP (XML Configuration Access Protocol) with one or more specific application usages that depend on the particular application. We described XCAP in Section 16.14 and XCAP application usage in Section 16.14.1.

17.3 Watcher Subscription

A user like Alice can act as a watcher from her IMS terminal. When she starts the presence application she will subscribe to the presence information of her presentities.

Although the IMS allows the watcher, Alice, to subscribe individually to each of the presentities, as mentioned in Chapter 16, on most occasions the watcher will subscribe to her own presence list hosted in an RLS in her home network.

The flow is illustrated in Figure 17.3. The watcher application residing in the IMS terminal sends a SUBSCRIBE request (1) addressed to her list (e.g., `sip:alice-list@ home1.net`). The SUBSCRIBE request contains an `Event` header field set to `eventlist` to indicate that the subscription is addressed to a list rather than a single presentity. The request (2) is received at the S-CSCF, which evaluates the initial filter criteria. One of those criteria indicates that the request (3) ought to be forwarded to an Application Server that happens to be an RLS. The RLS, after verifying the identity of the subscriber and authorizing the subscription, sends a 200 (OK) response (4). The RLS also sends a NOTIFY request (7), although it does not contain any presence information at this stage. The RLS subscribes one by one to all the presentities listed in the resource list and, when enough information has been received, generates another NOTIFY request (13) that includes a presence document with the aggregated presence information received from the presentities' PUAs.

Figure 17.4 shows the RLS subscribing to one of the presentities contained in the resource list. The RLS sends a SUBSCRIBE request (1) addressed to a presentity in the list. The request contains an `Event` header field set to `presence`. The request is forwarded via the S-CSCF in the RLS home network (2) to the I-CSCF in the presentity's network. The I-CSCF queries the HSS using the Diameter protocol, (3), (4) in Figure 17.4, to locate the S-CSCF allocated to the presentity and forwards the SUBSCRIBE request (5) to the allocated S-CSCF. The S-CSCF evaluates the initial filter criteria where there is a criterion indicating that the request ought to be forwarded to the presentity's PA (6). After sending the 200 (OK) response (7) the PA sends a NOTIFY request (11) that contains the presentity's presence information. In the example in Figure 17.3 neither the presentity's S-CSCF nor the presentity's I-CSCF record the route. Therefore, the PA sends the NOTIFY request (11) directly to the next node that recorded the route: the S-CSCF in the RLS network. The lack of this S-CSCF record route avoids the S-CSCF becoming locked when the presentity is not registered. So, if the presentity registers later there will be an S-CSCF allocation that may lead to a different S-CSCF.

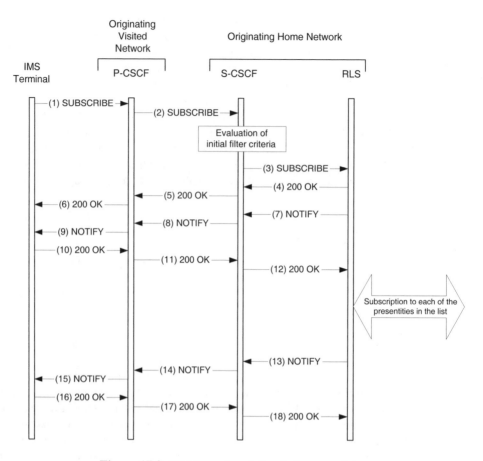

Figure 17.3: Watcher subscription to her own list

17.4 Subscription to Watcher Information

When the presence application is started in the IMS terminal, PUAs subscribe to their own watcher information state, so they can be updated on who is watching their presence information and with the state of the subscriptions to their presence states. We described the watcher information subscription in Section 16.12.

The high-level flow is similar to the one we presented in Figure 17.3. The Event header field in the SUBSCRIBE request (1) is set to the value presence.winfo to indicate to the PA the state the presentity is interested in.

17.5 Presence Publication

When the IMS presence application starts, it publishes the current presentity's presence information. Figure 17.5 shows the flow and, as we can see, there is not much difference from the mechanism we explained in Section 16.10. The IMS terminal sends a PUBLISH request (1) that includes an Event header set to presence. The S-CSCF receives the request (2) and evaluates the initial filter criteria for the presentity. One of the initial filter criteria indicates

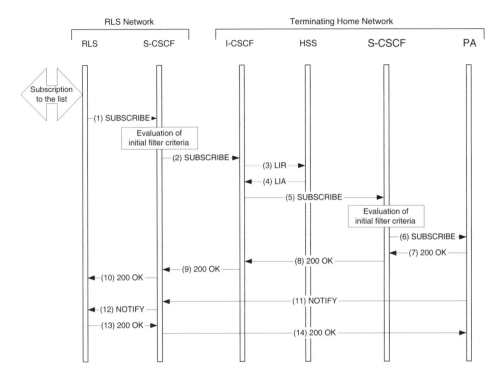

Figure 17.4: The RLS subscribes to a presentity

that PUBLISH requests containing an Event header set to presence ought to be forwarded to the PA where the presentity's presence information is stored. So, the S-CSCF forwards the PUBLISH request (3) to that Application Server. The PA authorizes the publication and sends a 200 (OK) response (4).

Apart from the IMS terminal, other PUAs can publish presence information related to the presentity. For instance, an S-CSCF can send a third-party REGISTER request to the presentity's PA to indicate that the presentity has registered with the IMS. Any other entity that contains presence information for that presentity can publish it to the PA or to a Presence Network Agent, that in turn publishes it to the presentity's PA.

17.6 Presence Optimizations

We discussed in Section 4.15.1 a few optimizations that are applicable to the event notification framework in SIP. These are generic optimizations that apply to any event package. In Section 16.15 we discussed some other optimizations that are applicable to the presence event package. These optimizations were standardized in the IETF in co-operation with engineers from a wireless background. The result is a standard that is not only applicable to any Internet host, but also to wireless devices.

Consequently, the IMS networks and terminals implement all the extensions that provide presence optimization.

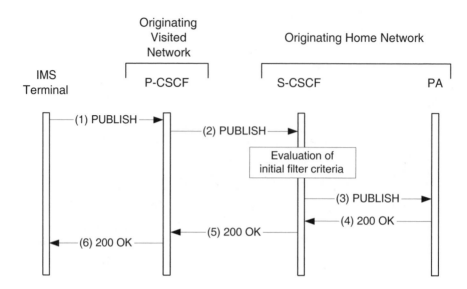

Figure 17.5: The IMS terminal publishing presence information

17.7   The *Ut* Interface

As mentioned before, 3GPP defined the *Ut* interface that runs between an IMS terminal and an Application Server. When we refer to the presence service, the Application Server is either a PUA or an RLS.

The *Ut* interface is not used for live traffic. Instead, it provides the user with the ability to configure resource lists (such as presence lists), authorize watchers in the PUA, and do any other type of data manipulation.

The *Ut* interface is implemented with HTTP 1.1 (Hypertext Transfer Protocol, specified in RFC 2616 [101]). On top of it there is the XML Configuration Access Protocol (XCAP), specified in the Internet-Draft "The Extensible Markup Language (XML) Configuration Access Protocol (XCAP)" [210], and on top of XCAP there is an XCAP usage that depends on the particular aim.

Chapter 18

Instant Messaging on the Internet

Instant messaging is one of today's most popular services. Many youngsters (and not-so-young people) use the service to keep in touch with their relatives, friends, co-workers, etc. Millions of instant messages are sent every day. So, it will come as no surprise that such a popular service is already supported in the IMS.

Instant messaging is the service that allows a user to send some content to another user in near-real time. Due to the real-time characteristics of instant messages the content is typically not stored in network nodes, as often happens with other services such as email.

The content in an instant message is typically a text message, but can be an HTML page, a picture, a file containing a song, a video clip, or any generic file.

The instant messaging service combines perfectly with the presence service, since presence allows a user to be informed when other users become available (e.g., connect to the network). Then, users can send instant messages to their friends and start some sort of messaging conversation.

18.1 The *im* URI

Like presence, mail, or AAA functions an instant messaging service can be identified by an *im* URI. Like the *pres* URI the *im* URI does not define the protocol used to access an instant message resource. So, whenever SIP is the protocol used to send the instant message it is recommended to use *sip* or *sips* URIs.

The syntax of the *im* URI is:

```
IM-URI      = "im:" [ to ] [ headers ]
to          = mailbox
headers     = "?" header *( "&" header )
header      = hname "=" hvalue
hname       = *urlc
hvalue      = *urlc
```

An example of an *im* URI is:

```
im:alice@example.com
```

18.2   Modes of Instant Messages

There are two modes of operation of the instant message service, depending on whether they are stand-alone instant messages or part of a session of instant messages.

We refer to a *pager-mode* instant message as one that is sent as a stand-alone message, not having any relation with previous or future instant messages. This mode of instant messaging is referred to as "pager mode" because the model resembles the way a two-way pager works. The model is also similar to the SMS (Short Message Service) in cellular networks.

We refer to a *session-based* instant message as one that is sent as part of an existing session, typically established with a SIP INVITE request.

Both models have different requirements and constraints; hence their implementation is different. The following sections describe the implementation of both models.

18.3   Pager-mode Instant Messaging

The IETF has created an extension to SIP that allows a SIP UA to send an instant message to another UA. The extension consists of a new SIP method named *MESSAGE*. The SIP MESSAGE method, which is specified in RFC 3428 [76], is able to transport any kind of payload in the body of the message, formatted with an appropriate MIME type.

Figure 18.1 illustrates the mode of operation. A UAC sends a MESSAGE request (1) to a proxy. The detailed contents of the request are shown in Figure 18.2. The proxy forwards the MESSAGE request (2) like any other SIP request, even when the proxy does not support or understand the SIP MESSAGE method. Eventually, the UAS will receive it and answer with a 200 (OK) response (3) that is forwarded (4) to the UAC.

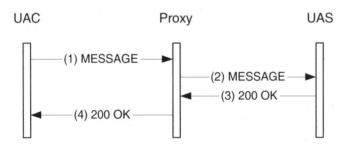

Figure 18.1: Pager-mode instant message with a MESSAGE request

Like OPTIONS or REGISTER, MESSAGE requests do not create a SIP dialog. They are stand-alone requests. However, they can be sent as part of an existing dialog (e.g., created by a SIP INVITE request).

18.3.1   Congestion Control with MESSAGE

One of the problems with SIP derives from the fact that any proxy can change the transport protocol from TCP to UDP, SCTP, or other transport protocols and vice versa. UDP is notorious for not offering any congestion control, whereas TCP and SCTP do offer congestion control. If a UA is sending a large instant message over a transport protocol that does not offer congestion control, the network proxies can become congested and stop processing other SIP requests like INVITE, SUBSCRIBE, etc. Even if a UA sends a large SIP MESSAGE over a

```
MESSAGE sip:bob@example.org SIP/2.0
Via: SIP/2.0/TCP alicepc.example.com;branch=z9hG4bK776sgd43d
Max-Forwards: 70
From: Alice <sip:alice@example.com>;tag=48912
To: Bob <sip:bob@example.org>
Call-ID: a3d3hdj5ws223ns6lk8djds
Cseq: 1 MESSAGE
Content-Type: text/plain
Content-Length: 31

Hi, what is going on today?
```

Figure 18.2: (1) MESSAGE

transport protocol that implements end-to-end congestion control (e.g., TCP, SCTP), the next proxy can switch to UDP and congestion can occur.

At the time of writing, SIP does not offer a mechanism for a UA to indicate that all proxies in the path must use a transport protocol that implements end-to-end congestion control.[14] Consequently, a limit has been placed on the SIP MESSAGE method such that MESSAGE requests cannot exceed the MTU (Maximum Transmission Unit) minus 200 bytes. If the MTU is not known to the UAC this limit is 1300 bytes.

A solution to sending SIP MESSAGE requests with large bodies is to use the content indirection mechanism. The UAC uses HTTP – or any other protocol that runs over a congestion-controlled transport protocol – to store the body of the SIP request in a server. Then, the UAC inserts a link to the URI where the payload is stored, instead of sending the whole body embedded in the SIP MESSAGE request. When the UAS receives the SIP request, it uses HTTP or the appropriate protocol to download the body. We described the content indirection mechanism in more detail in Section 4.17.

Another solution to getting around the size limit problem with MESSAGE is to use the session-based instant message mode rather than pager mode. Let's take a look at session-based instant messaging.

18.4 Session-based Instant Messaging

The session-based instant message mode uses the SIP INVITE method to establish a session where the media plane is not audio or video, but an exchange of instant messages.

When a UAC establishes a session to send and receive audio or video, the media is sent via the Real-Time Transport Protocol (RTP, specified in RFC 3550 [225]). However, when the UAC establishes a session to send and receive instant messages the actual media (the collection of instant messages) are sent over the Message Session Relay Protocol (MSRP, specified in the Internet-Draft "The Message Session Relay Protocol" [75]).

MSRP is a simple text-based protocol whose main characteristic is that it runs over transport protocols that offer congestion control, such as TCP (RFC 793 [190]), SCTP (RFC 2960 [230]), and TLS over TCP (RFC 2246 [82]). Explicitly, MSRP does not run over UDP (RFC 768 [188]) or any other transport protocol that does not offer end-to-end

---

[14]The IETF is working on an extension to SIP that will allow a UA to request proxies to use a transport protocol that supports end-to-end congestion control.

congestion control. Due to this, the main characteristic of MSRP is not imposing a restriction on the size of an instant message.

Another characteristic of MSRP is that it runs on the media plane. Therefore, MSRP messages do not traverse SIP proxies. This is an advantage, since SIP proxies are not bothered with proxying large instant messages.

MSRP supports instant messages to traverse zero, one, or two MRSP relays. MSRP relays play an important role when one of the endpoints is located behind a NAT (Network Address Translator). They are also helpful for network administrators when configuring firewall traversal. Additionally, MSRP relays can provide logging and statistical usage. For historical reasons, the behavior of MSRP relays is specified in the Internet-Draft "Relay Extensions for the Message Sessions Relay Protocol (MSRP)" [146]).

18.4.1   The MSRP and MSRPS URLs

MSRP defines two new URLs to address MSRP resources: *msrp* and *msrps*. Both URLs are similar in concept, but the *msrps* URL indicates a requirement for a secure TLS connection over TCP. They have the following structure:

```
"msrp://" [userinfo "@"] hostport ["/" session-id] ";" transport
"msrps://" [userinfo "@"] hostport ["/" session-id]  ";" transport
```

where `userinfo` and `hostport` are specified in RFC 2396 [54], and `session-id` and `transport` are strings of characters defined by MSRP.

An example of an MSRP URL is:

```
msrp://alice@pc.example.com/dslkj2nd;tcp
```

18.4.2   MSRP Overview

As is the case in SIP, MSRP is a text-based protocol whose messages are either requests or responses. But unlike SIP, not every MSRP request is answered by an MSRP response. Like SIP, MSRP defines methods that indicate the semantics of a request. There are currently three methods defined in MSRP.

- SEND: sends an instant message of any arbitrary length from one endpoint to another.

- REPORT: an endpoint or a relay provides message delivery notifications.

- AUTH: used by endpoints to authenticate to relays.

Like SIP, MSRP requests and responses contain a number of headers that describe, for example, the message identification, the path to the receiver or to the sender, or the content type.

MSRP SEND requests typically carry a body encoded according to MIME-encoding rules. At a minimum, all the MSRP implementations support the `message/cpim` MIME-type format (specified in RFC 3862 [152]). Typically, MSRP implementations support other MIME-type formats, such as `text/plain` or `text/html`.

The MSRP standards define new extensions to SDP (Session Description Protocol, specified in RFC 2327 [115]) to be able to express the media type `message` and associated attributes.

Figure 18.3 shows a high-level flow of the interleaving that takes place between SIP and MSRP when a session is established. The endpoint that originates the session, Alice, sends a SIP INVITE request (1) that contains an SDP offer indicating the message media type and the support for MSRP. An example of this INVITE request is shown in Figure 18.4. We do not show any potential SIP proxy server that is in the path for the sake of clarity.

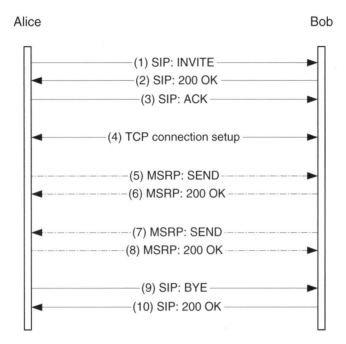

Figure 18.3: Establishment of a session of instant messages

The MSRP address of the originator of the session is indicated in the MSRP URL contained in the a=path line in the SDP of the INVITE request (1). Bob answers with a 200 OK response (2) that also contains his MSRP URL in the a=path line of the SDP. Figure 18.5 shows the complete SIP response. The session is acknowledged with the SIP ACK request (3).

Once the session is established, the SDP offerer (Alice, in our example), establishes a TCP connection (4) to the answerer's MSRP URL (Bob's MSRP URL) and sends an immediate SEND request (5) that may or may not have a payload. Figure 18.6 shows an MSRP SEND request (5) that includes an actual instant message.

Then Bob replies with an MSRP 200 response (6) to confirm the reception of the message. Figure 18.7 shows the 200 OK response.

Any of the endpoints is able to close the session at any time. When an endpoint wants to close the session it sends a SIP BYE request (9), as it would do with any other session it wanted to close.

18.4.3 Extensions to SDP due to MSRP

MSRP extends SDP by adding new forms of encoding existing lines (for example, the m= line), and by adding new attributes (a= lines) to describe MSRP sessions.

```
INVITE sip:Bob.Brown@example.org SIP/2.0
Via: SIP/2.0/UDP ws1.example.com:5060;branch=z9hG4bK74gh5
Max-Forwards: 70
From: Alice <sip:Alice.Smith@example.com>;tag=hx34576sl
To: Bob <sip:Bob.Brown@example.org>
Call-ID: 2098adkj20
Cseq: 22 INVITE
Contact: <sip:alice@192.0.100.2>
Content-Type: application/sdp
Content-Length: 220

v=0
o=alice 2890844526 2890844526 IN IP4 ws1.example.com
s=-
c=IN IP4 ws1.example.com
t=0 0
m=message 8231 msrp/tcp *
a=accept-types:message/cpim text/plain text/html
a=path:msrp://ws1.example.com:8231/9s9cpl;tcp
```

Figure 18.4: INVITE request (1)

```
SIP/2.0 200 OK
Via: SIP/2.0/UDP ws1.example.com:5060;branch=z9hG4bK74gh5
From: Alice <sip:Alice.Smith@example.com>;tag=hx34576sl
To: Bob <sip:Bob.Brown@example.org>;tag=ba03s
Call-ID: 6328776298220188511@192.0.100.2
Cseq: 1 INVITE
Contact: <sip:bob@bob.example.org>
Content-Type: application/sdp
Content-Length: 215

v=0
o=bob 2890844528 2890844528 IN IP4 bob.example.org
s=-
c=IN IP4 bob.example.org
t=0 0
m=message 7283 msrp/tcp *
a=accept-types:message/cpim text/plain text/html
a=path:msrp://bob.example.org:7283/d9s9a;tcp
```

Figure 18.5: 200 (OK) response (2)

Figure 18.8 shows an example of the SDP generated by an MSRP endpoint. The first five lines (v=, o=, s=, c=, and t=) are created as in regular SDP. In particular, in spite of the presence of an MSRP URL later, the c= lines contain the same hostname or IP address that later will be encoded in the MSRP URL.

```
MSRP 230cmqj SEND
To-Path: msrp://bob.example.org:7283/d9s9a;tcp
From-Path: msrp://ws1.example.com:8231/9s9cpl;tcp
Message-ID: 309203
Byte-Range: 1-20/20
Content-Type: text/plain

This is Alice typing
-------230cmqj$
```

Figure 18.6: An instant message carried in an MSRP SEND request (5)

```
MSRP 230cmqj 200 OK
To-Path: msrp://ws1.example.com:8231/9s9cpl;tcp
From-Path: msrp://bob.example.org:7283/d9s9a;tcp
Message-ID: 309203
Byte-Range: 1-20/20
-------230cmqj$
```

Figure 18.7: MSRP 200 response (6)

```
v=0
o=bob 2890844528 2890844528 IN IP4 bob.example.org
s=-
c=IN IP4 bob.example.org
t=0 0
m=message 7283 msrp/tcp *
a=accept-types:message/cpim
a=accept-wrapped-types:text/plain text/html *
a=path:msrp://bob.example.org:7283/d9s9a;tcp
a=max-size=8000
```

Figure 18.8: SDP describing an MSRP media

The m= line contains a message media type to indicate the type of media. The port number is set to the real port number that, otherwise, also appears encoded later in the MSRP URL. The transport protocol is set to msrp/tcp, or msrp/tls/tcp, as required. The last item in the m= line would indicate the format list of the media (e.g., the codec in audio or video media types). MSRP sets this item to an asterisk "*".

The list of supported MIME body types in MSRP SEND messages is encoded in a new a=accepted-types line. MSRP mandates to support, at least, message/cpim (for messages encoded according to RFC 3862 [152]) and text/plain (for pure text messages). Although it is common to support other types, such us text/html, image/jpeg, etc. An asterisk "*" at the end of the list indicates that the list of supported MIME body types is not comprehensive (perhaps due to the large number of supported types), and that the peer endpoint may attempt to send instant messages that include other types.

The a=accept-wrapped-types describes a list of MIME body types that are only accepted when they are wrapped in another MIME body type. The encoding is the same as the `a=accept-types` line. This allows, for example, a gateway, to force messages to be wrapped in the type identified in the `a=accept-types`, and still describe general types that are accepted inside the wrapper (in the `a=accept-wrapped-types` line). For example, the SDP in Figure 18.8 indicates that the MSRP endpoint accepts MSRP messages that include a Message/CPIM body which, in turn, includes a plain text or HTML body (or any other non-declared type, due to the presence of an asterisk in the `a=accept-wrapped-types` line).

The new `a=path` line indicates the MSRP URL of the endpoint that is creating the SDP. If the endpoint requires a relay to operate, then the line contains two entries, one describing the endpoint and the other describing the relay. Note that there is some redundancy in the information, since the `a=path` contains the MSRP URL, and the `c=` and `m=` lines contain the hostname (or IP address) and the port number, respectively. MSRP prefers to use URLs to identify endpoints, since they are a very rich means of describing the connection. The `c=` and `m=` lines are populated in a traditional way to allow backward compatibility with some nodes that they require to access the IP address and port number of the endpoint (the P-CSCF is an example of it).

The new `a=max-size` line indicates the maximum size message the endpoint wishes to receive. The value is indicated in octets and provides a hint to the receiver of the SDP.

18.4.4   MSRP Core Functionality

MSRP is able to transport instant messages between two endpoints, perhaps through some relays in the path. Instant messages are, in this context, any MIME-encoded body of any arbitrary length, for instance, a pure text message, a large video file, or an image. Essentially, anything that can be MIME-encoded can be transported in MSRP.

Due to the fact that some instant messages can be large in nature (think of a large video file), and due to the users' requirement to be able to share a single TCP connection to simultaneously send a large instant message without disturbing other existing text conversations, MSRP provides a chunking and rechunking mechanism. In the case where a message is chunked, the endpoint splits the payload into several chunks and sends each chunk in a SEND request. Both endpoints and relays are able to split the contents of an instant message into a number of chunks, of no longer than 2048 octets. The receiver endpoint can glue all the received chunks to compose the original instant message.

Additionally, MSRP assumes that messages might be interruptible, meaning that a user may decide to abort the transmission of a message prior to the completion of its transmission.

In a different scenario, a user might want to send a message whose length is not known at the time the message transmission starts. This might be the case, for example, for a video file that is being created at the time the instant message transmission starts.

In order to support all of these requirements, MSRP uses a boundary-based framing mechanism. SEND requests include, right after the body, a boundary string that identifies the end of the SEND request. Additionally, the boundary string also contains a flag indicating whether the request contains a complete instant message, a chunk, or an aborted message.

Figure 18.9 shows an example of a simple MSRP message. The first line contains the protocol name `MSRP` followed by unique boundary string "bnsk1s" and the method `SEND`. The unique boundary string identifies the end of the message, as we will see soon.

The SEND request in Figure 18.9 also includes `From-Path` and `To-Path` headers that contain the full MSRP URLs of the sender and the receiver, respectively, for this particular session. Potential relays (serving either the sender or the recipient) would also be listed in the `From-Path` or `To-Path` headers.

```
MSRP bnsk1s SEND
To-Path: msrp://alice.example.com:4423/xodj2;tcp
From-Path: msrp://bob.example.org:15000/vnskq;tcp
Message-ID: 003293
Byte-Range: 1-22/22
Content-Type: text/plain

Hi, how are you today?
-------bnsk1s$
```

Figure 18.9: MSRP SEND request

Next in the SEND request we find the `Message-ID` header, which contains a unique message identifier within the session. If a message is split into several chunks, each of the SEND requests has the same `Message-ID` value. Finally, responses also have the same `Message-ID` value as the corresponding request.

The example in Figure 18.9 also shows a `Byte-Range` header that indicates the range of bytes of the body transmitted, and the total number of bytes of the complete message (if known). In our example the `Byte-Range` header indicates that messages ranging from 1 to 22 of a total of 22 bytes are being sent (hence, it is a complete message). `Byte-Range` headers are also present in 200 OK responses to the SEND request for the purpose of acknowledging the range of bytes received at the receiving endpoint.

A `Content-Type` header indicates the MIME type that is encoded in the body data. In our example plain text is sent.

Then an empty line separates the MSRP header from the actual body data. After the body data, seven dashes "-" plus the boundary string declared in the request line follow. In the example shown in Figure 18.9 the boundary string is "bnsk1s". The receiver of a SEND request just needs to search for seven dashes "-" and the boundary string to find the end of the body. The last character after the boundary string is a flag that can have any of the following values.

- $: the body data contains a complete message or the last chunk of a message.

- +: the body data contains a chunk that does not complete the message.

- #: the body contains an aborted message; the endpoint will not send the remaining bytes or chunks of the message.

Figure 18.10 shows two SEND requests, each one containing a chunk that makes a complete message. The boundary strings are different in each message, but the `Message-ID` value is the same. The `Byte-Range` header indicates the number of bytes sent in each chunk and its position with respect the complete message. The flag at the end of the first message indicates that more chunks will be sent in other SEND requests. In the last message the flag indicates that it is the last chunk of a message.

18.4.5 Status and Reports

MSRP introduces the concept of status and reports in instant messaging. On one hand, an MSRP endpoint might be interested in the *transaction status*, which refers to the status of

```
MSRP ea1dof SEND
To-Path: msrp://alice.example.com:4423/xodj2;tcp
From-Path: msrp://bob.example.org:15000/vnskq;tcp
Message-ID: 459874
Byte-Range: 1-11/22
Content-Type: text/plain

Hi, how are
-------ea1dof+

MSRP ea1eeo SEND
To-Path: msrp://alice.example.com:4423/xodj2;tcp
From-Path: msrp://bob.example.org:15000/vnskq;tcp
Message-ID: 459874
Byte-Range: 12-22/22
Content-Type: text/plain

 you today?
-------ea1eeo$
```

Figure 18.10: MSRP SEND request

the delivery of an instant message to a next hop (a relay or an endpoint). On the other hand, an MSRP endpoint might be interested in receiving the status of the delivery of the instant message at the other end, in what is called the *request status*.

The transaction status is typically sent in MRSP responses, although on some occasions it can be carried in REPORT requests. In contrast, the request status is always carried in REPORT requests.

The sender of the SEND request governs the type of status the user is interested in receiving. MSRP provides the sender with two headers that control the request of status indications: the Success-Report and Failure-Report headers.

Success-Report can take the values "yes" or "no". A value of "yes" means that the receiver will generate a REPORT request when the last chunk of the message or a complete message is received. A value of "no" means that the recipient will not generate a report of the successful reception of the message. The default value, in the absent of a Success-Report header, is "no".

Failure-Report can take the values "yes", "no", or "partial". A value of "yes" means that the receiver will generate an error response if the transaction fails. In some cases (e.g., a gateway) it can generate a 200 OK response, and then, later when a response from the other system is received, it can generate a REPORT request. A value of "no" means that the recipient will not generate any response, not even a successful response. A value of "partial" indicates that the recipient will not generate 200 OK responses, but it will generate other failure responses. In the absence of a Failure-Report header, the default value of "yes" is assumed.

With all of these values in place, all the different scenarios can be accommodated. For example, a system administrator who wants to send an instant message to all users indicating that the system is shutting down probably is not interested in getting failure reports, so he would set the `Failure-Report` header to "no". In another example, an online securities trading system will most likely set the `Success-Report` header to "yes". But in a public Internet chat system, where performance is important, the `Success-Report` might be set to "no".

The request of success reports is illustrated in Figure 18.11. Alice sends an MSRP SEND request (1) whose `Success-Report` header is set to "yes". We assume that the request contains a complete message. The 200 OK response (2) constitutes a transaction report and it is generated by Alice's relay, which relays the request (3) to Bob's relay. Eventually Bob receives the SEND request (5). Then Bob honors the `Success-Report` header and generates a REPORT request (7).

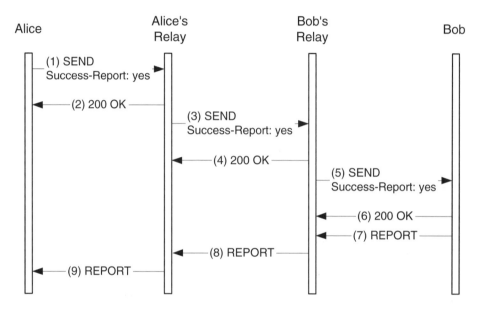

Figure 18.11: Request for a success report

Let us take a look at the MSRP REPORT request (7). Figure 18.12 shows an example of it. REPORT requests keep the value of the `Message-ID` header of the SEND request the report belongs to. The `Byte-Range` header indicates the range of bytes to which the report applies.

```
MSRP 439dscd REPORT
To-Path: msrp://ws1.example.com:8231/9s9cpl;tcp
From-Path: msrp://bob.example.org:7283/d9s9a;tcp
Message-ID: 309203
Byte-Range: 1-22/22
Status: 000 200 OK
-------439dscd$
```

Figure 18.12: REPORT request (7)

A new `Status` header contains the status of the request. The first three digits in the value indicate the namespace of the status. The namespace indicates the context of the rest of the information present in the value of the Status header. Only namespace "000" is standardized at the time of writing and it indicates that the rest of the value in the header correspond to the status code of a transaction response code. In our example the rest of the Status header value is populated with a "200 OK", indicating that the REPORT request has the same meaning as a "200 OK" response, i.e., the SEND request has been successfully received.

Figure 18.13 shows an example where Bob receives a SEND request (4) with the Failure-Report header set to "yes". If the transaction does not exist at Bob's endpoint, he generates a 481 "Session does not exist" response (6) that is received at Bob's relay. Since Bob's relay had already acknowledged the SEND request (3) with a 200 OK response (4), and now it has received further information that the request was not successfully received by Bob, the relay generates a REPORT request (7) that contains a Status request set to the value "000 481 Session does not exist" (the 000 indicates the namespace of the MSRP responses, and the rest indicates the actual response received by the relay). When Alice receives the REPORT request (8) it can determine that the request failed due to the value of the Status header.

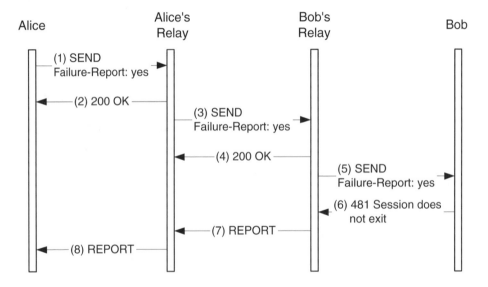

Figure 18.13: Request for a failure report

18.4.6 MSRP Relays

Although Figures 18.11 and 18.13 have intuitively indicated the usage of relays in MSRP, we have not given a formal description. MSRP relays are specified in the Internet-Draft "Relay Extensions for the Message Sessions Relay Protocol (MSRP)" [146].

An MSRP relay is a specialized node in transiting MSRP messages between two other MSRP nodes (endpoints or other relays). MSRP relays, which are located in the media plane, must not be confused with SIP proxies, which are located in the signaling plane. MSRP does not offer a mechanism for an endpoint to discover its relay, so it is assumed that each endpoint is provisioned with its MSRP relay, in a similar way as HTTP proxies are configured in HTML browsers.

When an endpoint wants to make usage of an MSRP relay, it first opens a TLS connection towards its relay, authenticates (by sending an AUTH request), and if authentication is successful the relay provides the endpoint with an MSRPS URL that the endpoint can use for its MSRP sessions.

Figure 18.14 shows the detailed flow of signals. First, the endpoint opens the TLS connection (1) towards the relay. After that, the endpoint sends an AUTH request (2), which is answered by the relay with a 401 response (3) that contains the name of a realm where a username and password combination should be valid. The endpoint then builds a new AUTH request (4) that contains a valid username and password combination in that realm. If they are correct, the MSRP relay answers with a 200 OK response (5). At any time the endpoint can initiate an MSRP session by first sending a SIP INVITE request (6) that contains an SDP offer that declares both the MSRP relay URL and the MSRP endpoint URL.

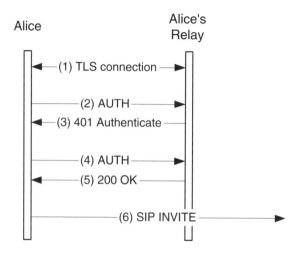

Figure 18.14: Authentication to an MSRP relay

Let us take a look at the authentication and authorization details in MSRP. Figure 18.15 shows an example of an AUTH request that Alice sends to her MSRP relay. The To-Path and From-Path contain the relay and Alice's MSRPS URL, respectively. It is assumed that the relay URL has either been provisioned to Alice's endpoint or has been learnt somehow (MSRP does not specify how to learn the relay MSRPS URL).

```
MSRP w39sn AUTH
To-Path: msrps://alice@relay.example.com:3233;tcp
From-Path: msrps://alice.example.com:9892/98cjs;tcp
Message-ID: 020391
-------w39sn$
```

Figure 18.15: AUTH request (2)

The MSRP relay answers with a 401 (Authenticate) response (3), as shown in Figure 18.16. The response includes a WWW-Authenticate header containing the realm where the username and password should be valid. The WWW-Authenticate header is imported from HTTP authentication, and it is specified in RFC 2617 [102]. Unlike SIP, where the only authentication mechanism imported from HTTP is Digest, MSRP only imports the Basic authentication mechanism from HTTP. Basic authentication is simpler than Digest, but usernames and passwords are sent in the clear. While that is a major drawback for SIP, it is not for MSRP, since the endpoint-to-relay connection is protected and encrypted with TLS. Thus, even Basic authentication sends usernames and password openly within a channel, the channel is encrypted, and eavesdroppers will not be able to guess them.

```
MSRP w39sn 401 Authenticate
From-Path: msrps://alice@relay.example.com:3233;tcp
To-Path: msrps://alice.example.com:9892/98cjs;tcp
Message-ID: 020391
WWW-Authenticate: Basic realm="relay.example.com"
-------w39sn$
```

Figure 18.16: 401 (Authenticate) response (3)

Then the endpoint creates the response and sends a new AUTH request (4) to the relay. The `Authorization` header, also imported from RFC 2617 [102] contains a Base64 encoded string (specified in RFC 3548 [148]) that contains the username and the password. Figure 18.17 shows an example of this request.

```
MSRP p2pe3 AUTH
To-Path: msrps://alice@relay.example.com:3233;tcp
From-Path: msrps://alice.example.com:9892/98cjs;tcp
Message-ID: 929195
Authorization: Basic bWlndWVsLmdhcmNpYTp3cm90ZSB0aGlzIGNoYXB0ZXI=
-------p2pe3$
```

Figure 18.17: AUTH request (4)

When the MSRP relay receives this new AUTH request (3), it decodes the `Authorization` header value and verifies that the username and password combination is valid in the administrative realm. If everything is correct, the MSRP relay creates a 200 (OK) response (5). This response contains a `Use-Path` header that contains one or more MSRPS URLs of the relay or relays that the endpoint has to use for this session. The MSRPS URLs returned are "session" URLs (i.e., they contain a `session-id` path, and they are valid only for one unique session). Figure 18.18 shows an example of the 200 (OK) response (5).

```
MSRP p2pe3 200 OK
From-Path: msrps://alice@relay.example.com:3233;tcp
To-Path: msrps://alice.example.com:9892/98cjs;tcp
Message-ID: 929195
Use-Path: msrps://relay.example.com:3233/uwduqd3s;tcp
-------p2pe3$
```

Figure 18.18: 200 (OK) response (5)

Then, at any time, the endpoint can create an INVITE request (6) to establish the MSRP session. This INVITE request is very similar to the one in Figure 18.4. The differences are subtle: the m= line indicates the usage of TLS; and the a=path line in the SDP now contains two or more MSRPS URLs, the first one pertaining to the relay (learnt from the Use-Path header) and the other belonging to the endpoint. Figure 18.19 shows an example of this INVITE request (6).[15]

```
INVITE sip:Bob.Brown@example.org SIP/2.0
Via: SIP/2.0/UDP ws1.example.com:5060;branch=z9hG4bK74g3d
Max-Forwards: 70
From: Alice <sip:Alice.Smith@example.com>;tag=329s8a
To: Bob <sip:Bob.Brown@example.org>
Call-ID: 438fw34kjaljs
Cseq: 56 INVITE
Contact: <sip:alice@192.0.100.2>
Content-Type: application/sdp
Content-Length: 274

v=0
o=alice 2890844526 2890844526 IN IP4 ws1.example.com
s=-
c=IN IP4 ws1.example.com
t=0 0
m=message 8231 msrp/tls/tcp *
a=accept-types:message/cpim text/plain text/html
a=path:msrps://relay.example.com:3233/uwduqd3s;tcp
        msrps://alice.example.com:9892/98cjs;tcp
```

Figure 18.19: INVITE request (6)

Let us take a look at an end-to-end example. Figure 18.20 depicts the flow of information in an end-to-end MSRP session. We have omitted potential SIP proxies for the sake of clarity, and we have assumed that both Alice and Bob are using MSRP relays, and both have authenticated to their respective relay.

The flow begins when Alice sends a SIP INVITE request (1) that is similar to the one described in Figure 18.19. The SDP in the INVITE request (1) contains the MSRPS URLs of both Alice and her relay. Bob receives the INVITE request (1) and replies with a 200 OK response (2) that also contains Bob's MSRPS URL and that for his relay. At this point, both Alice and Bob are able to send an MSRP SEND request that contains some content.

An example of such an MSRP SEND request is shown in Figure 18.21. The To-Path header includes the two MSRP relays and Bob's endpoint.

When Alice's relay receives the SEND request (3), it verifies that the session (in Alice's MSRP relay URL) is correct and bound to the TLS connection from where the request was received. Alice's MSRP relay answers (typically with a 200 OK response) and then tries to forward the request. Hence, the 200 OK merely indicates the successful reception of the SEND request at the next hop rather than the reception of the request at the remote endpoint.

Here we can see the usefulness of REPORT requests, either in success or failure circumstances. In successful cases, Bob's endpoint will send a REPORT request, after

---

[15]Note that the URLs contained in the a=path line should be listed in a single line, separated by just a single blank space. However, for presentation purposes, we show them separated by a carriage return and a few spaces.

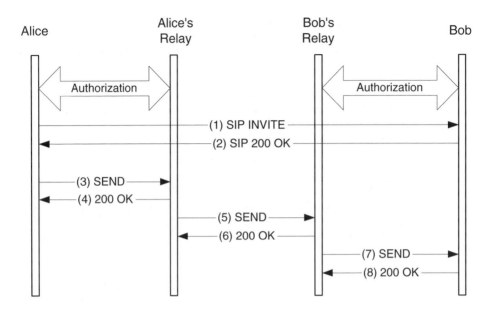

Figure 18.20: End-to-end session establishment with MSRP relays

```
MSRP 230cmqj SEND
To-Path: msrps://relay.example.com:3233/uwduqd3s;tcp
         msrps://otherrelay.example.org:23153/b8s8d;tcp
         msrp://bob.example.org:7283/d9s9a;tcp
From-Path: msrp://ws1.example.com:8231/9s9cpl;tcp
Message-ID: 193254
Byte-Range: 1-20/20
Content-Type: text/plain

This is Alice typing
-------230cmqj$
```

Figure 18.21: MSRP SEND request (3)

receiving the SEND request (7), that will be forwarded to Alice. In failure circumstances a relay that detects an error generates a REPORT request to the sender.

Chapter 19

The Instant Messaging Service in the IMS

Chapter 18 described the basic components of the instant messaging service. We learnt that there are two modes of operation: pager mode and session-based mode. This chapter analyzes how these two modes are applied to the IMS. We explore the basic call flows and present examples of services that can be enriched with instant message capabilities.

19.1   Pager-mode Instant Messaging in the IMS

The pager-mode instant messaging service was introduced with the first phase of IMS that came as part of Release 5 of the 3GPP specifications. 3GPP TS 23.228 [23] already contains requirements for Application Servers and S-CSCFs to be able to send textual information to an IMS terminal. 3GPP TS 24.229 [16] introduces support for the MESSAGE method extension. The specification mandates IMS terminals to implement the MESSAGE method (specified in RFC 3428 [76]) and to allow implementation to be an optional feature in S-CSCFs and ASs (e.g., if required by the service). Obviously, pager-mode instant messages are subject to the constraints (e.g., message size, etc.) that we described in Section 18.3.

The main purpose of a pager-mode instant message is to allow the S-CSCF or Application Servers to send short instant messages to the IMS terminal. Since the MESSAGE method is already implemented in IMS terminals, users are able to send pager-mode instant messages to other IMS users. The flow is simple, as depicted in Figure 19.1.

An example of a service provided with the SIP MESSAGE method is shown in Figure 19.2. An Application Server is the controller of a voicemail system. The Application Server is interested to know when the user successfully logs on to the IMS, so that the AS can inform the user that there are pending voicemails to be retrieved. The service is implemented as follows: the user registers with the IMS as usual, (1)–(20) in Figure 19.2. When the registration is complete the S-CSCF evaluates the initial filter criteria. One of the initial filter criteria indicates that the S-CSCF should do a third-party registration with a particular Application Server. The S-CSCF then sends a third-party REGISTER request (21) to the indicated AS. The purpose of the third-party REGISTER request is not to register the user with the AS, but instead to indicate to the AS that the user has just registered with the S-CSCF. Upon receipt of the REGISTER request (21) the AS generates a MESSAGE request (23) that

The 3G IP Multimedia Subsystem (IMS) Second Edition   Gonzalo Camarillo and Miguel A. García-Martín
© 2006 John Wiley & Sons, Ltd

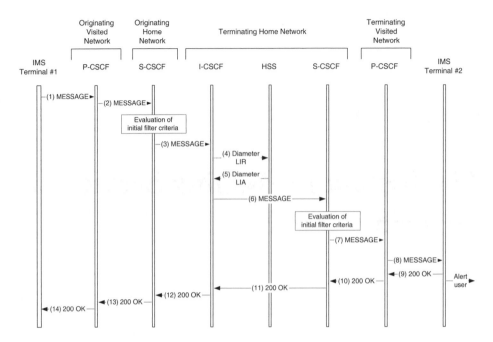

Figure 19.1: Pager-mode instant messaging in the IMS

contains some informative text, maybe a link to a website that holds the transcription of the pending messages to retrieve or any other information that the service designer considers appropriate. The MESSAGE request is forwarded via the S-CSCF and P-CSCF as any other SIP message.

19.2   Session-based Instant Messaging in the IMS

The session-based instant messaging service was introduced in Release 6 of the 3GPP specifications. The detailed protocol specification is described in 3GPP TS 24.247 [25]. Session-based instant messaging was not included in Release 5 because at the time 3GPP closed Release 5 the IETF had just started work on session-based instant messaging. Therefore, the functionality of session-based instant messaging was postponed until Release 6.

We described in Section 18.4 how to establish a session of instant messages with an INVITE request that contains provisions in SDP for the instant message media. The Message Session Relay Protocol (MSRP) is the actual protocol used to transport the messages.

In the IMS, MSRP is implemented in the IMS terminals. Additionally, the MRFP may also implement MSRP. The reason behind this is that there are two different scenarios for establishing a session of instant messages. In the first scenario, which we show in Figure 19.3, an IMS terminal establishes a session toward another endpoint. SIP messages traverse regular IMS nodes (P-CSCFs, S-CSCFs, perhaps ASs, etc.). MSRP is then sent end to end. The only difference from a basic session setup consists in the absence of the precondition extension requirement in the INVITE request, if session-based messaging is the only media stream

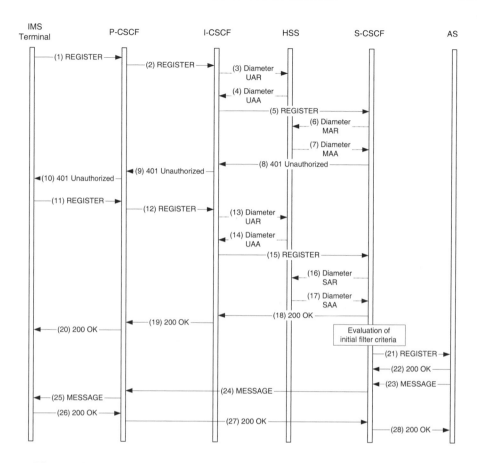

Figure 19.2: Example of a service provided with pager-mode instant messages

declared in SDP. This is the reason for not having 183 (Session Progress) responses, PRACK, or UPDATE requests in the flow.

In the second scenario the MRFC and MRFP are intermediaries in the network. This might be due to operator constraints, such as the ability to generate charging events depending on the size of the message or on any additional contents in MSRP SEND messages. Another reason might be that the MRF is acting as a multiparty conference unit for instant messages (also known as chat rooms). Figure 19.4 shows the flow for a multiparty conference. For the sake of simplicity we assume that both users who join the conference belong to the same network operator, although they may have allocated different S-CSCFs and P-CSCFs. In the figure a first user sends an INVITE request (1) that traverses their allocated P-CSCF and S-CSCF. Their S-CSCF forwards the INVITE request (5) to the MRFC. The MRFC, which controls the MRFP by means of the H.248 protocol, creates a new termination for the user. The MRFP initiates a TCP connection to the IMS terminal and sends an MSRP VISIT request (8). The IMS terminal acknowledges it with an MSRP 200 (OK) response (9). H.248 acts (10) to inform the MRFC that the new MSRP connection has been established, at which point the MRFC sends the 200 OK response (11) to complete the session setup.

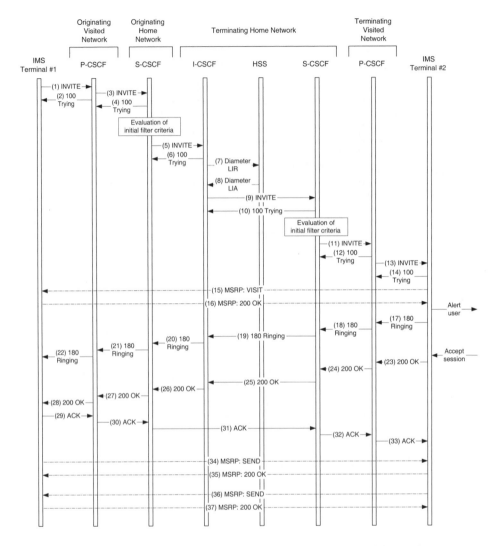

Figure 19.3: Session-based instant messages: end-to-end MSRP session

Later, a second user joins the conference and establishes another session with the MRFC. At any time any of the users can send an instant message that is transported over an MSRP SEND request: for instance, when the second user sends an MSRP SEND request (33) the MRFP sends a copy of it (35) to the remaining participants of the conference.

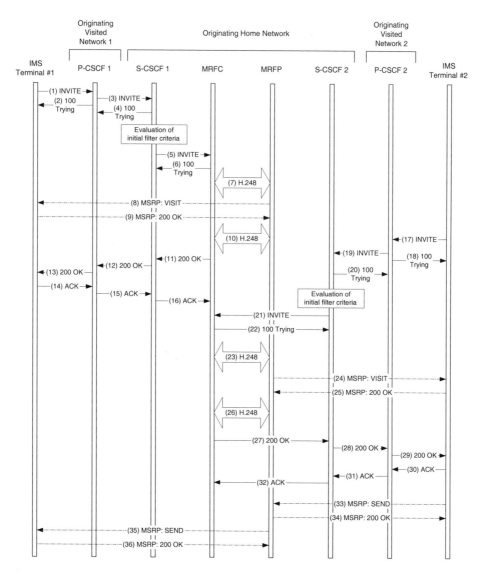

Figure 19.4: A multi-party session-based conference (chat server)

Chapter 20

Push-to-Talk over Cellular

We expect Push-to-Talk over Cellular (PoC) to be the first IMS-based service to be provided by many mobile operators because it does not require the deployment of new radio technologies. PoC can run on top of low-bandwidth and high-delay links. These links would be inappropriate for running other types of services, such as voice calls.

PoC is a walkie-talkie type of service. Users press (and hold) a button when they want to say something, but they do not start speaking until their terminal tells them to do so (usually by beeping). At this point, users say whatever they want to say and signal the end of their speech by releasing the button.

Unlike regular voice calls, which are full-duplex, PoC is a half-duplex service; that is, only one user can speak at a time.

PoC sessions can have more than two participants. At a given time, one user speaks and the rest listen (as in the two-party case). A simple way of understanding a multiparty PoC is a group of friends going to the movies. One at a time, they take turns to tell the rest which movie they want to watch, at which point they can make the final choice (usually after some extra rounds of discussions).

20.1 PoC Standardization

There are several incompatible PoC specifications at present. Many are not based on the IMS, but consist of proprietary solutions implemented by a single vendor. As a result these PoC solutions generally cannot interoperate with equipment from other vendors.

Many operators willing to provide PoC services felt uncomfortable with the situation just described and asked a few vendors for a standard solution based on the IMS. As a consequence, a group of vendors teamed up to develop an open PoC industry standard. These vendors were Ericsson, Motorola, Nokia, and Siemens. The result of this collaboration was a set of publicly available PoC specifications.

It was clear that a widely-accepted PoC standard which took into account the requirements of most of the industry was needed. The industry standard was a good starting point, but there was still a long way to go before having a fully-featured PoC service. This situation prompted OMA (Open Mobile Alliance) to create the PoC working group to start working on the OMA PoC service. (For a description of OMA, its structure, and the different types of recommendations it produces, see Section 2.6.)

The 3G IP Multimedia Subsystem (IMS) Second Edition   Gonzalo Camarillo and Miguel A. García-Martín
© 2006 John Wiley & Sons, Ltd

OMA decided to base its PoC service on the IMS. So, the consortium that developed the PoC industry standard provided OMA with their PoC specifications, which were based on the IMS as well. These specifications were taken as the starting point for the OMA PoC standard.

At the same time the IETF started working on some building blocks that were missing in SIP and in the conferencing architecture to be able to provide a fully-featured PoC service. These building blocks were needed by OMA for its PoC service.

Section 20.2 covers the building blocks developed by the IETF that are relevant to PoC. Section 20.3 describes the PoC service as standardized by OMA.

As you have probably noticed, we have not introduced a chapter called "PoC on the Internet" as we have done with other services covered in this book. Instead, we describe the relevant IETF specifications in Section 20.2. We have chosen to do so because the IETF has not defined a PoC framework. The IETF has a conferencing framework and, from the IETF perspective, PoC is just a conference. A conference that uses a set of extensions (e.g., conference establishment using request-contained lists) and a particular floor control policy, but a conference nevertheless.

20.2   IETF Work Relevant to PoC

Given that a PoC session is, at the end of the day, a conference, all the work developed in the IETF on conferencing is very relevant to PoC. There are two main IETF Working Groups (WGs) involved in this conferencing work: SIPPING and XCON.

The SIPPING WG has developed a set of extensions to establish conferences using SIP. However, conferencing-related issues that do not have to do with SIP (e.g., floor control) are outside the scope of the SIPPING WG. These issues are typically handled by the XCON WG, which focuses on centralized conferences.

In this section we focus on a set of extensions to SIP, which are referred to as URI-list services (see Section 20.2.1). The IETF developed these extensions after noticing that some of the OMA PoC requirements related to multiparty sessions could not be met by existing IETF technology.

In addition the IETF has developed two SIP extensions that only apply to the OMA PoC service: an event package to discover the settings of a PoC terminal (specified in the Internet-Draft "A SIP Event Package and Data Format for Various Settings in Support of the PoC Service" [105]) and a set of SIP header fields (specified in the Internet-Drafts "The P-Answer-State Header Extension to SIP for OMA PoC" [46] and "Requesting Answering and Alerting Modes for SIP" [234]).

20.2.1   *URI-list Services*

Some services involve multiple very similar transactions. For example, a user may need to send a page-mode instant message to a number of friends telling them at what time they should meet in the movie theater. The user would send one message to each friend; however, all the messages would have the same contents (e.g., "Let's meet at seven."). If the user of our example sits on a low-bandwidth access, sending all of those messages can take a while. URI-list services were designed for this type of situation (their framework is specified in the Internet-Draft "Requirements and Framework for Session Initiation Protocol (SIP) Uniform Resource Identifier (URI)-List Services" [69]).

Servers providing a URI-list service perform a similar transaction towards all the members (identified by URIs) of a list provided by the user agent invoking the service. In our previous example the user agent would send the server a MESSAGE request with two bodies: one body with the contents of the instant message (e.g., "Let's meet at seven.") and another body with an XML-encoded list of URIs. On receiving this MESSAGE request the server would send a MESSAGE request to each of the URIs on the list, as shown in Figure 20.1.

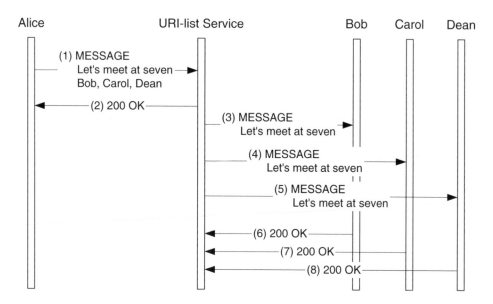

Figure 20.1: URI-list service for MESSAGE

Figure 20.2 shows how the message flow would have looked like if Alice had not used a URI-list service to send her messages to her friends. Alice's access network sees six messages (Figure 20.2) instead of two (Figure 20.1). Moreover, if the contents of the instant message or the number of recipients had been larger, the savings would have been enormous.

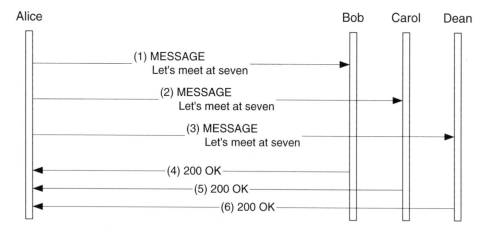

Figure 20.2: Multiple MESSAGEs without using a URI-list service

In addition to the URI-list service for MESSAGE (specified in the Internet-Draft "Multiple-Recipient MESSAGE Requests in SIP" [108]), there are URI-list services defined for methods such as INVITE (specified in the Internet-Draft "Conference Establishment Using Request-Contained Lists in SIP" [66]) and SUBSCRIBE (specified in the Internet-Draft "Subscriptions to Request-Contained Resource Lists in SIP" [70]). The INVITE URI-list service can be used to establish a conference with multiple participants and the SUBSCRIBE URI-list service can be used to subscribe to the presence information of several users.

20.2.1.1 Multiple REFER

An extension that may seem like a URI-service but is not is the multiple-REFER extension (specified in the Internet-Draft "Refering to Multiple Resources in SIP" [63]). The difference between this extension and the URI-list services is that, while a URI-list service replicates the same transaction towards a set of users, the recipient of a multiple REFER executes the transaction identified by the `Refer-To` header field (which does not need to be a REFER transaction). For example, a user may send a REFER to a conference focus so that the focus INVITEs a number of new users into the conference, as shown in Figure 20.3.

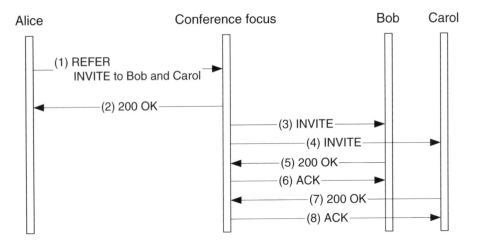

Figure 20.3: Multiple REFER

Since multiple REFERs are typically used within the context of an application, the user generating it has generally an application-specific means to discover the result of the transactions initiated by the server (in our example, the INVITE transactions initiated by the conference focus). In the conferencing example, the user may use the conference event package to discover which users were brought into the conference successfully. Consequently, multiple REFERs are normally combined with an extension (specified in the Internet-Draft "Suppression of SIP REFER Method Implicit Subscription" [154]) that eliminates the implicit subscription that is usually linked to REFER. Once again, the subscription to the results of the transactions initiated by the server is eliminated by that extension. That is why Figure 20.3 does not show any NOTIFY requests from the conference focus to Alice (see Section 4.18 for a discussion of the REFER method and its implicit subscription).

Both URI-list services and multiple-REFER are used by PoC. URI-list services are used to establish multiparty PoC sessions (INVITE URI-list service) and to send multiple-recipient page-mode instant messages (MESSAGE URI-list service). Multiple-REFER is used to invite multiple users to a PoC session.

20.2.1.2   URI-list Format

A user agent using a URI-list service or generating a multiple REFER needs to include a list of URIs in its request. The default format for these lists is supposed to be service-specific, but effectively, all the URI-list services defined so far and multiple-REFER use the same URI-list format. This format is based on XML and is a simplified version (e.g., hierarchical lists are not allowed) of the general format for representing resource lists (specified in the Internet-Draft "XML Formats for Representing Resource Lists" [207]).

Figure 20.4 shows an example of an INVITE request that carries two body parts: an SDP session description and a URI list in the XML-based format just described.

20.2.1.3   Consent-based Communications

As we have already stated, URI-list services allow user agents using low-bandwidth accesses to request the generation of a potentially large number of transactions towards a set of URIs. While this type of service is a great tool for implementing services such as PoC, servers providing URI-list services could be used as traffic amplifiers to launch DoS attacks.

An attacker would just need to generate a single request with a URI list containing the URIs of the victims. The attacker would send this request to a URI-list service. The URI-list service would then flood the members of the list with undesired traffic.

This type of attack is similar to a form of email SPAM, where the attacker places the email address of the victim on a distribution list. The victim keeps receiving the messages sent to the distribution list but has no means to unsubscribe from the list.

In order to avoid this type of attack in SIP, URI-list services need to obtain permission from the recipients before sending them any traffic. Effectively, they implement a form of consent-based communication (as specified in the Internet-Draft "A Framework for Consent-Based Communications in SIP" [204]) where entities need to agree to communicate before the actual communication takes place.

At the time of writing the extensions needed to implement a framework for consent-based communications in SIP are still being developed. Therefore, at this point, OMA does not use this framework in PoC.

20.2.2   Event Package for PoC Settings

In the PoC service, situations in which the network needs to be informed about the settings of a PoC terminal occur. For example, if a PoC terminal is in *don't disturb* mode (i.e., the terminal will reject all incoming session invitations) the network can reject directly any session invitation for the terminal. Sending the invitation to the terminal, only to have it rejected, would consume radio resources unnecessarily.

A terminal keeps its home PoC server updated on the terminal's settings by sending PUBLISH requests (whose bodies follow the format specified in the Internet-Draft "A SIP Event Package and Data Format for Various Settings in Support for the PoC service" [105]).

```
INVITE sip:conf-fact@example.com SIP/2.0
Via: SIP/2.0/TCP client.chicago.example.com
    ;branch=z9hG4bKhjhs8ass83
Max-Forwards: 70
To: Conf Factory <sip:conf-fact@example.com>
From: Carol <sip:carol@chicago.example.com>;tag=32331
Call-ID: d432fa84b4c76e66710
Cseq: 1 INVITE
Contact: <sip:carol@client.chicago.example.com>
Allow: INVITE, ACK, CANCEL, OPTIONS, BYE, REFER,
    SUBSCRIBE, NOTIFY
Allow-Events: dialog
Accept: application/sdp, message/sipfrag
Require: recipient-list-invite
Content-Type: multipart/mixed;boundary="boundary1"
Content-Length: 690

--boundary1
Content-Type: application/sdp
v=0
o=carol 2890844526 2890842807 IN IP4 chicago.example.com
s=-
c=IN IP4 192.0.2.1
t=0 0
m=audio 20000 RTP/AVP 0
a=rtpmap:0 PCMU/8000
m=video 20002 RTP/AVP 31
a=rtpmap:31 H261/90000

--boundary1
Content-Type: application/resource-lists+xml
Content-Disposition: recipient-list

<?xml version="1.0" encoding="UTF-8"?>
<resource-lists xmlns="urn:ietf:params:xml:ns:resource-lists"
                xmlns:xsi="http://www.w3.org/2001/XMLSchema-instance">
  <list>
    <entry uri="sip:bill@example.com" />
    <entry uri="sip:joe@example.org" />
    <entry uri="sip:ted@example.net" />
  </list>
</resource-lists>
--boundary1--
```

Figure 20.4: INVITE with two body parts

20.2.3 SIP Header Fields

The PoC server defines the concept of answer mode, which can be set to automatic or manual. A terminal in automatic answer mode accepts session invitations automatically, without any user intervention. The `Answer-Mode` header field (specified in the Internet-Draft "Requesting Answering and Alerting Modes for SIP" [234]) carries information related to the answer mode. The `Answer-Mode` header field can be inserted in an INVITE request to request a particular answer mode from the callee. Additionally, the callee can insert this header field in a response to indicate which answer mode was actually applied.

The `Alert-Mode` header field specified in the Internet-Draft "Requesting Answering and Alerting Modes for SIP" [234]) can be inserted in an INVITE request to request the user agent server to alert or not to alert the callee. The `P-Answer-State` header field can be included in a response to an INVITE to indicate which entity (the user agent server or an intermediary) generated the response. The usage of both header fields is further described in the following sections.

20.3 Architecture

In this section we look at the architecture of the PoC service (as specified in the Candidate Enabler Release Package for PoC Version 1.0 [175]). Figure 20.5 indicates the nodes involved in PoC and the interfaces between them.

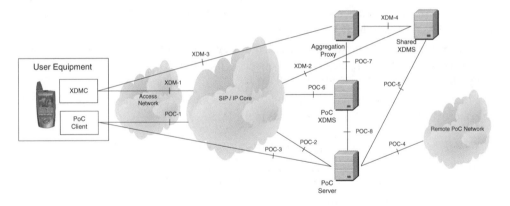

Figure 20.5: PoC architecture

The User Equipment contains two logical elements: the PoC client and the XDMC (XML Document Management Client). The PoC client uses SIP to communicate with the SIP/IP Core over the POC-1 interface, and RTP and TBCP (Talk Burst Control Protocol) to communicate with the PoC server over the POC-3 interface. TBCP is a floor control protocol based on RTCP that is used to signal which user is allowed to speak at a given time.

The XDMC uses SIP to communicate with the SIP/IP Core over the XDM-1 interface and XCAP to communicate with the Aggregation Proxy over the XDM-3 interface. The XDM-3 interface is used to perform document management (e.g., set up a URI list with the user's golf buddies) and the XDM-2 interface is used to subscribe to changes in documents that are stored in the network.

The Aggregation Proxy acts as a single point for the XDMC to contact the network. The Aggregation Proxy performs user authentication and routes the XCAP messages from the XDMC (received over the XDM-3 interface) to the appropriate server (the PoC XDMS or the Shared XDMS). The Aggregation proxy uses XCAP to communicate with the PoC XDMS over the POC-7 interface and with the Shared XDMS over the XDM-4 interface.

The PoC XDMS manages documents that are specific to PoC (e.g., the members of a PoC group). On the other hand, the Shared XDMS manages documents that are needed by PoC but that may be shared with other services (e.g., presence-related documents).

The PoC XDMS uses XCAP to communicate with the Aggregation Proxy over the POC-7 interface and with the PoC Server over the POC-8 interface. The PoC XDMS uses SIP to communicate with the SIP/IP Core over the POC-6 interface.

The Shared XDMS uses XCAP to communicate with the Aggregation Proxy over the XDM-4 interface and with the PoC Server over the POC-5 interface. The PoC XDMS uses SIP to communicate with the SIP/IP Core over the XDM-2 interface. Since the documents managed by the Shared XDMS may be shared with other services, the Shared XDMS has additional interfaces towards those services. For example, the interface between the Shared XDMS and the Presence Server, which does not appear in Figure 20.5, is referred to as PRS-5.

The PoC Server uses SIP to communicate with the SIP/IP Core over the POC-2 interface. In addition, it uses RTP and TBCP to communicate with the PoC Client over the POC-3 interface and with other PoC networks over the POC-4 interface. Furthermore, the PoC server uses XCAP to communicate with the PoC XDMS over the POC-8 interface and with the Shared XDMS over the POC-5 interface.

The SIP/IP Core can be realized in different ways. Nevertheless, we expect that it will usually be realized by using the IMS. Consequently, the SIP/IP Core cloud would correspond to the IMS (as described in 3GPP TR 23.979 [4]).

Table 20.1 shows the protocols used in the different interfaces and the corresponding IMS interface when the SIP/IP Core corresponds to the IMS.

Table 20.1: PoC interfaces

Interface	Protocol	Corresponding IMS interface
POC-1	SIP	Gm
POC-2	SIP	ISC
POC-3	RTP / TBCP	Mb
POC-4	RTP / TBCP	Mb
POC-5	XCAP	Not Applicable
POC-6	SIP	ISC
POC-7	XCAP	Not Applicable
POC-8	XCAP	Not Applicable
XDM-1	SIP	ISC
XDM-2	SIP	ISC
XDM-3	XCAP	Ut
XDM-4	XCAP	Not Applicable

20.4   Registration

In order to use the PoC service, a terminal needs to register to the PoC service. When the terminal performs IMS registration, it adds the +g.poc.talkburst and +g.poc.groupad feature tags to the Contact header field of the REGISTER request. On receiving these feature tags, the S-CSCF performs a third-party registration towards the PoC server of the domain (i.e., the user's home PoC server).

The +g.poc.talkburst feature tag indicates that the terminal can handle PoC sessions. The +g.poc.groupad feature tag indicates that the terminal can handle group advertisement (see Section 20.8).

20.5   PoC Server Roles

A PoC server within a session can perform two roles: Controlling PoC Function or Participating PoC Function. A given PoC server in a given PoC session will be performing one or both roles. However, only one PoC server in a session performs the Controlling PoC Function. This server may or may not perform the Participating PoC Function as well. The rest of the PoC servers, assuming that there are several PoC servers involved in the session, will only perform the Participating PoC Function.

The PoC server performing the Controlling PoC Function is usually referred to as the controlling PoC server. Similarly, the PoC server performing the Participating PoC Function is usually referred to as the participating PoC server.

Figure 20.6 shows a PoC session with one controlling and four participating PoC servers. Each of the PoC servers is in a different domain. In order to simplify this and other figures in this chapter, we do not show all the elements involved in the session. That is, we do not show the IMS nodes between the different PoC entities.

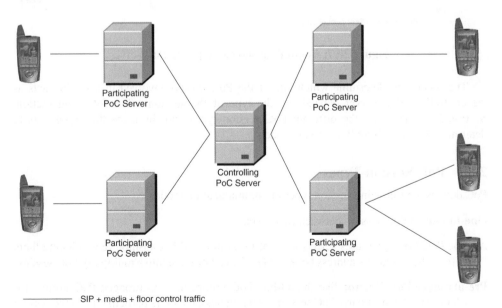

SIP + media + floor control traffic

Figure 20.6: PoC session with a central controlling PoC server

The controlling PoC server provides centralized PoC session handling. This includes media mixing, centralized floor control, and policy enforcement for participation in group sessions. The participating PoC server exchanges SIP signaling with the client and with the controlling PoC server and, optionally, relays media and floor control messages between them. When a participating PoC server chooses not to be on the media path, clients exchange media and floor control traffic directly with the controlling PoC server.

Figure 20.7 shows another PoC session where the controlling PoC server is co-located with a participating PoC server. That is, the same PoC server performs both roles at the same time.

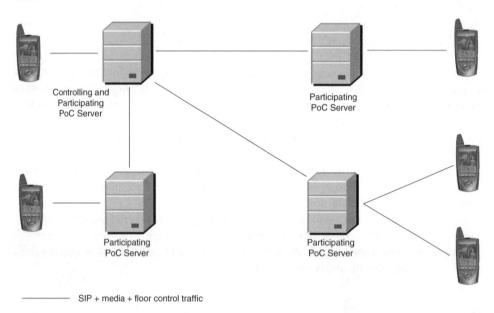

SIP + media + floor control traffic

Figure 20.7: Controlling and participating PoC server

The process to determine which one of the PoC servers involved in a session acts as the controlling PoC server depends on the type of the session. The following section, Section 20.6, describes the different PoC session types and discusses the procedures to determine the controlling PoC server in each.

20.6   PoC Session Types

PoC defines the following session types or communication modes:

One-to-one: a PoC session between two users.

Ad-hoc PoC Group: a user selects a set of users in an ad-hoc fashion (e.g., picking them from the terminal's address book) and invites all of them into a multiparty PoC session.

Pre-arranged PoC Group: like the ad-hoc PoC group, the pre-arranged PoC group also consists of a multiparty PoC session. Nevertheless, the users participating in the session are selected beforehand, not in an ad-hoc manner when it is established. That is, a pre-arranged PoC group includes a predefined set of users (e.g., the user's golf buddies).

Chat PoC Group: chat PoC groups are also multiparty PoC sessions. However, when a user joins a chat PoC group, no invitations are sent to other users. Conversely, when a user joins a pre-arranged PoC group, all the users that belong to that PoC group are invited to the PoC session.

These PoC session types are classified into two forms of PoC sessions: one-to-one and one-to-many. One-to-many PoC sessions include ad-hoc, pre-arranged, and chat PoC groups. Additionally, some pre-arranged PoC groups may use a special media mixing policy whereby a user (called the distinguished participant) can talk to the whole group and listen to the answers from each individual user (called ordinary participants). However, the rest of the users (the ordinary participants) cannot talk or listen to each other. They only talk and listen to the distinguished participant. When this form of mixing is used in a pre-arranged PoC group, the session is referred to as a one-to-many-to-one PoC session.

There are scenarios where one-to-many-to-one PoC sessions are useful. For example, a taxi dispatcher needs to inform all the drivers about customers waiting for a taxi, but the individual drivers answer only to the dispatcher. Any given driver does not hear the answers from any other driver to the dispatcher.

Now, let's look at the signaling involved in the establishment of the different session types. Additionally, we will discuss which PoC server is selected as the controlling PoC server in each session type. However, before looking at these issues, we would like to provide an important clarification.

PoC defines two session establishment types: using *on-demand* signaling and using a *pre-established session*. Both session establishment types are discussed in Section 20.9. In the following message flows, we discuss session establishment procedures using only *on-demand* signaling. The use of a pre-established session is an optimization, which is described in Section 20.9.

20.6.1 One-to-one PoC Sessions

In one-to-one PoC sessions, the controlling PoC server is the inviting user's PoC server. That is, this PoC server is, at the same time, the participating PoC server of the inviting user and the controlling PoC server for the session.

Figure 20.8 shows the message flow for one-to-one PoC session establishment. In this message flow, we have included the SIP/IP Core in order to illustrate how filter criteria are used to route requests to the PoC servers. In subsequent message flows, we do not show the SIP/IP Core for the sake of clarity.

The PoC terminal generates an INVITE (1) which is addressed to the callee and has an SDP session description in its body. The INVITE (1) also carries an `Accept-Contact` header field with the `+g.poc.talkburst` feature tag in it.

On receiving the INVITE, the originating user's S-CSCF evaluates the initial filter criteria for the user. According to the filter criteria, an outgoing INVITE request with the `+g.poc.talkburst` feature tag should be routed to the PoC server of the domain.

When the PoC server receives the INVITE (3), it forwards it (5) towards the home domain of the callee.

The S-CSCF of the terminating user receives the INVITE (5) and evaluates the initial filter criteria for the terminating user. According to the filter criteria, an incoming INVITE request with the `+g.poc.talkburst` feature tag should be routed to the PoC server of the domain.

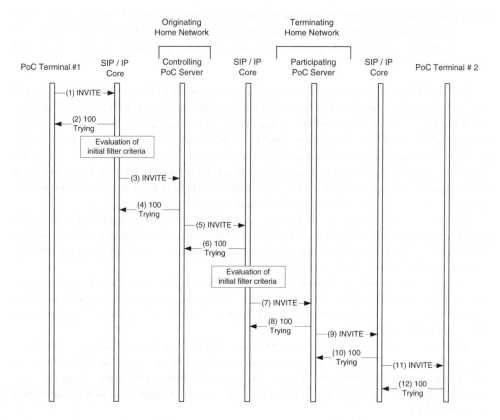

Figure 20.8: One-to-one PoC session establishment

When the PoC server receives the INVITE (7), it generates a new INVITE (9) that will be routed by the SIP/IP Core towards the terminating user.

20.6.2 Ad-hoc PoC Group

In ad-hoc PoC group sessions, the controlling PoC server is the PoC server of the inviting user. That is, this PoC server is, at the same time, the inviting user's participating PoC server and the controlling PoC server for the session.

Figure 20.9 shows the message flow for an ad-hoc PoC session establishment. The message flow is very similar to that for one-to-one PoC sessions. The more important differences are that the INVITE (1) generated by the originating terminal is addressed to its home PoC server and that the INVITE contains two body parts: an SDP session description and a URI list. The URI list contains the URIs of all the callees.

On receiving INVITE (1), the controlling PoC server generates an INVITE towards each of the URIs in the URI list. This results in two INVITEs: (3) and (4). These INVITEs

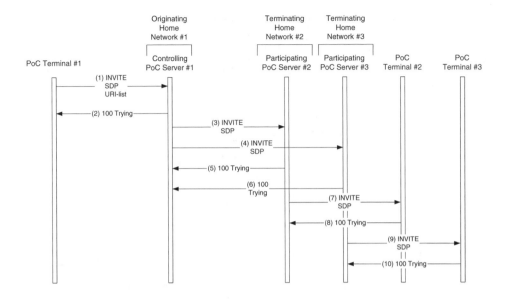

Figure 20.9: Ad-hoc PoC group session establishment

contain a single body part: an SDP session description. The participating PoC servers route these INVITEs towards the terminating terminals.

20.6.3   Pre-arranged PoC Group

In pre-arranged PoC group sessions, the controlling PoC server is the PoC server hosting the pre-arranged PoC group. That is, the controlling PoC server is the PoC server of the domain that owns the URI that identifies the pre-arranged PoC group.

Figure 20.10 shows the message flow for pre-arranged PoC group session establishment. In this example, the inviting user's participating PoC server is not the controlling PoC server because the pre-arranged PoC group is hosted in another domain.

The INVITE (1) generated by the inviting terminal does not carry a URI list because the members of the pre-arranged PoC group have been previously set up in the network. The Request-URI of this INVITE (1) identifies the pre-arranged PoC group at the controlling PoC server.

The inviting user's PoC server behaves as a participating PoC server and, thus, relays the INVITE (3) to the controlling PoC server. On receiving this INVITE (3), the controlling PoC server invites all the members of the pre-arranged PoC group.

20.6.4   Chat PoC Group

In Chat PoC group sessions, the controlling PoC server is the PoC server hosting the chat PoC group. That is, the controlling PoC server is the PoC server of the domain that owns the URI that identifies the chat PoC group.

Figure 20.11 shows the message flow for pre-arranged PoC session establishment. In this example, the participating PoC server of the inviting user is not the controlling PoC server because the chat PoC group is hosted in another domain.

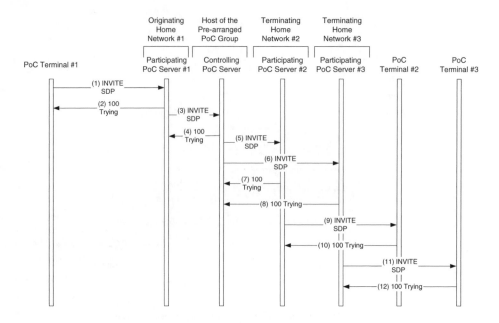

Figure 20.10: Pre-arranged PoC group session establishment

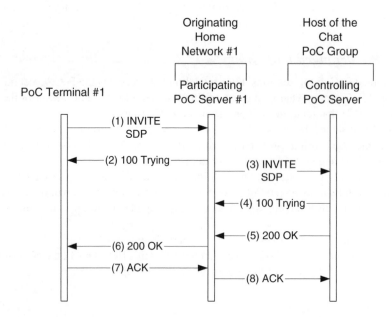

Figure 20.11: Chat PoC group session establishment

The INVITE (1) generated by the inviting terminal does not carry a URI list because joining a chat room does not trigger any invitations to other users. The Request-URI of this INVITE (1) identifies the chat PoC group at the controlling PoC server.

The PoC server of the inviting user behaves as a participating PoC server and, thus, relays the INVITE (3) to the controlling PoC server. On receiving this INVITE (3), the controlling PoC server returns a 200 (OK) response (5) accepting the user into the chat PoC group session. Note that in this case, as opposed to what happens in ad-hoc and pre-arranged PoC group sessions, the controlling PoC server does not invite any users on receiving an INVITE request.

20.7 Adding Users to a PoC Session

New users can be added to an ongoing PoC session in two ways: the new user sends an INVITE request to the URI of the session or the controlling PoC server sends an INVITE request to the new user.

Participants in a PoC session can have the controlling PoC server send an INVITE request to the new user by sending a REFER request to the controlling PoC server. Figure 20.12 shows how a controlling PoC server receives a multiple REFER and, as a consequence, invites two new users to an ongoing PoC session.

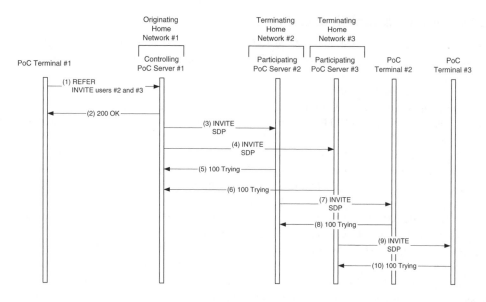

Figure 20.12: Bringing new users into a PoC session

Different session types have different authorization policies regarding which users can be added to an ongoing PoC session. In one-to-one and ad-hoc PoC sessions, new users can be added to an ongoing session only if they are invited by a participant. That is, a random user cannot send an INVITE request to the URI of an ongoing PoC session and get it accepted. The only users where the controlling PoC server of an ongoing one-to-one or ad-hoc session accepts an incoming INVITE request from are users that leave the session for some reason and later rejoin.

Participation in pre-arranged PoC group sessions is limited to members of the pre-arranged PoC group. That is, the controlling PoC server will not allow any other users to join the PoC session.

Participation in chat PoC group sessions may be limited to a set of users or open to everyone. Users can join a chat PoC group session by sending an INVITE to the URI of the chat PoC group, as described in Figure 20.11, or by accepting an invitation from the controlling PoC server.

20.8   Group Advertisements

In previous sections, we have seen that in order to join a pre-arranged and a chat PoC group session, a user needs to know its URI. Users can obtain URIs identifying PoC group sessions in many ways, such as in an email, in a phone conversation, from a piece of paper, or in a face-to-face meeting. Additionally, PoC defines a means for PoC users to exchange these URIs: users can exchange URIs of PoC group sessions using *group advertisements*.

A group advertisement is a MESSAGE request that carries an XML body that contains the URI of the PoC group session and, optionally, information about the session (e.g., the topics its members usually talk about). Additionally, the MESSAGE request carries a +g.poc.groupad feature tag in an `Accept-Contact` header field.

20.9   Session Establishment Types

All the message flows we have discussed so far use so-called *on-demand* signaling. Nevertheless, PoC defines two session establishment types: using *on-demand* signaling and using a *pre-established session*.

When on-demand signaling is used, terminals generate INVITE requests to create new PoC sessions or join chat PoC groups. On the terminating side, the terminating user's PoC server relays incoming INVITE requests to the user. Figures 20.8–20.11 are examples of on-demand signaling.

The use of pre-established sessions results in an optimization that allows a more rapid session establishment at the terminating side. The terminal using the pre-established session establishes a session (using an INVITE request) with its home PoC server, typically right after registration. As in a regular session establishment, the terminal and its home PoC server negotiate all the parameters needed to exchange media and floor control protocol messages. Nevertheless, they do not exchange media immediately.

At a later point, when the home PoC server receives an INVITE for the user, there is no need to use any SIP signaling towards the terminal. The session that was established previously is used to deliver media and floor control messages to the terminal. The floor control protocol carries all the information the terminal needs about the incoming INVITE received by the PoC server.

It is also possible to use pre-established session at the originating side. However, the pre-established session does not eliminate the need for SIP signaling at the originating side. It only replaces the typical INVITE transaction used to establish a new session with a REFER transaction that instructs the PoC server to generate an INVITE.

Figure 20.13 shows a message flow where both the originating and the terminating sides use pre-established sessions. The terminals set up their pre-established sessions using INVITE transactions (1) and (4), respectively. At a later point, the originating terminal sends a REFER request (7) to its home PoC server. This REFER request (7) prompts the home PoC server to invite the terminating user to a one-to-one PoC session. The INVITE request (11) generated by the originating home PoC server arrives at the terminating PoC server, which generates a final response (12) without contacting the terminating terminal.

The originating terminal is informed about the result of the invitation in a NOTIFY request (14).

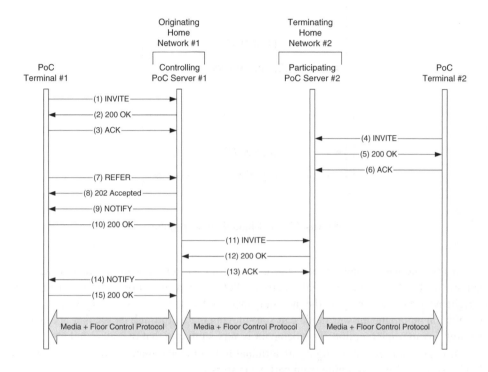

Figure 20.13: Pre-established session

Note that when REFER is used, the user agent server needs to generate a NOTIFY request immediately after accepting the REFER request (see Section 4.18). That is why the PoC server sends the first NOTIFY request (9).

It is possible to eliminate all of these NOTIFY requests by using an extension to REFER (specified in the Internet-Draft "Suppression of SIP REFER Method Implicit Subscription" [154]) that eliminates the implicit subscription that is usually linked to REFER. When terminals apply this extension to a REFER request, they use the floor control protocol to discover when the terminating user answers.

20.10 Answer Modes

PoC defines two answer modes: manual and automatic. The manual answer mode is the answer mode used in traditional telephones. When the terminal receives a PoC session invitation, the terminal alerts the user (usually by ringing). At that point, the user decides to accept or to reject the invitation.

Figure 20.14 shows the message flow for the manual answer mode. Note that PoC terminals do not use reliable provisional responses (the use of reliable provisional responses between PoC servers is optional). That is why there is no PRACK transaction after the 180 (Ringing) (2) response from the terminal.

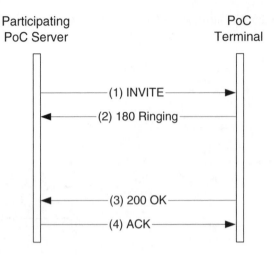

Figure 20.14: Manual answer mode

In the automatic answer mode, the user sets up the terminal to accept PoC sessions automatically. When the terminal receives a PoC session invitation, the terminal accepts it right away and starts playing the incoming media related to that session. This means that a terminal in automatic answer mode can start playing media through its speaker at any point without any user intervention. This behavior is very similar to that of walkie-talkies.

Of course, a user can configure its terminal to use the automatic answer mode only for PoC session invitations coming from particular users.

Figure 20.15 shows the message flow for the automatic answer mode. The terminal responds directly with a 200 (OK) (2) final response.

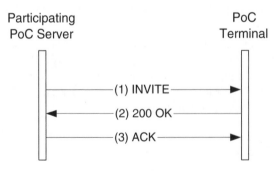

Figure 20.15: Automatic answer mode

PoC callers can request terminals at the callee side to apply a particular answer mode by using the `Alert-Mode` SIP header field (specified in the Internet-Draft "Requesting Answering and Alerting Modes for SIP" [234]). In addition to the manual and automatic answer modes, an authorized PoC caller can request *Manual Answer Override* (MAO). When an authorized PoC caller requests MAO, the callee's terminal will answer automatically even if it is configured in manual answer mode. A situation where MAO may be useful is an emergency when the user has to be alerted about something urgently.

20.11 Right-to-speak Indication Types

A controlling PoC server sending an INVITE request to a participating PoC server expects, following standard SIP procedures, to receive a response. If this response contains a session description, the controlling PoC server can start sending media using the media parameters just received.

In a regular SIP session, such a response is generated by the user agent server, and usually consists of a 183 (Session Progress) or a 200 (OK) response. Nevertheless, in PoC, the entity generating such a response is not always the terminal (i.e., the user agent server). The participating PoC server can act as a B2BUA and answer on behalf of the terminal.

The situation where a controlling PoC server receives a response that was originally generated by the terminating terminal is referred to as a *confirmed* answer state. The situation where a controlling PoC server receives a response that was generated by the participating PoC server without contacting the terminating terminal is referred to as an *unconfirmed* answer state.

Figure 20.16 shows a message flow with a confirmed answer state. When the participating PoC server receives a 200 (OK) response (5) from the terminal, it relays it to the controlling PoC server.

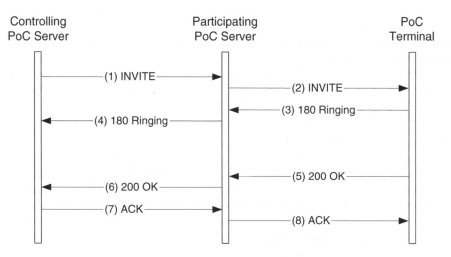

Figure 20.16: Confirmed answer state

Figure 20.17 shows a message flow with an unconfirmed answer state. The participating PoC server generates a 183 (Session Progress) response (3) before even contacting the terminal. This response carries a P-Answer-State header field (which is specified in the Internet-Draft "The P-Answer-State Header Extension to SIP for OMA PoC" [46]) with the value unconfirmed.

The reason why a participating PoC server may want to provide use of the unconfirmed answer state is that the participating PoC server may be pretty sure that the terminal will accept the session anyway in a short period of time.

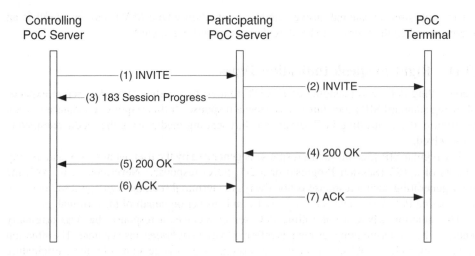

Figure 20.17: Unconfirmed answer state

There are two situations where a participating PoC server can be pretty sure that the terminal will accept a session: when MAO is used or when the terminal is configured in automatic answer mode. A terminal informs its participating PoC server of the terminal's current answer mode by sending PUBLISH requests (whose bodies follow the format specified in the Internet-Draft "A SIP Event Package and Data Format for Various Settings in Support for the PoC Service" [105]).

On receiving a response indicating an unconfirmed answer state, the controlling PoC server can provide the inviting PoC terminal with an unconfirmed right-to-speak indication in a 200 (OK) response with a `P-Answer-State` header field with the value `unconfirmed`. Unconfirmed right-to-speak indications are an optimization that allows callers to start speaking a little earlier than using traditional confirmed indications.

Note that when a PoC server generates an unconfirmed right-to-speak indication, it needs to be ready to buffer incoming media in case the terminating terminal takes a little longer than expected to accept the session. If a PoC server does not support media buffering, it cannot generate unconfirmed right-to-speak indications.

20.12   Participant Information

A user taking part in a PoC group session may want to be informed about who is participating in the session. The user may want to know who accepted the invitation and who rejected it.

Users get this type of information by having their terminals subscribe to the conference state event package (specified in the Internet-Draft "A SIP Event Package for Conference State" [205]). The controlling PoC server provides information about the participants in a PoC session by sending NOTIFY requests with XML-encoded bodies.

20.13   Barring and Instant Personal Alerts

A user who does not want to receive PoC session invitations can use *PoC session barring*. The terminal informs its home PoC server about this setting using a PUBLISH request (see Section 20.2.2). A PoC server does not relay any PoC session invitation to a terminal using PoC session barring.

When a user invites another user who is using PoC session barring to a PoC session, the inviting user will get an error response. In this case, the inviting user can send an *Instant Personal Alert* to the user using PoC session barring. An Instant Personal Alert requests its receiver to call back (at a better time) the sender of the Instant Personal Alert. Instant Personal Alerts are implemented using MESSAGE requests.

Still, a user may not want to even receive Instant Personal Alerts. In this case, the user would use the *Instant Personal Alert Barring* feature. The terminal informs its home PoC server about this setting using a PUBLISH request (see Section 20.2.2). A PoC server does not relay any Instant Personal Alert to a terminal using Instant Personal Alert barring.

20.14   The User Plane

The previous sections described what is known as the PoC control plane, which deals with the establishment of PoC sessions. In this section, we look at the PoC user plane. The PoC user plane includes the POC-3 and POC-4 interfaces in Figure 20.5. These interfaces are based on RTP (see Section 15.2.2) and RTCP (see Section 15.2.3).

As usual, RTP transports encoded-voice and RTCP provides quality feedback (RTCP also provides inter-media synchronization, but that does not apply to single-media sessions such as PoC sessions).

In addition to voice transport and quality feedback, the PoC user plane also includes floor control. In PoC terminology, floor control is referred to as Talk Burst Control, and, consequently, a floor control protocol to be used in PoC is referred to as TBCP (Talk Burst Control Protocol).

At this point, the only floor control protocol or TBCP supported by PoC is an RTCP-based protocol. However, PoC supports the negotiation of the TBCP to be used for a particular session. So, in the future it may be possible to use other TBCPs such as BFCP (which is specified in the Internet-Draft "The Binary Floor Control Protocol" [64]).

20.14.1   Media Encoding

The default audio codecs used in PoC are the same as those defined for the IMS (see Section 14.4). Consequently, 3GPP mandates that PoC clients support, at a minimum, AMR narrowband and that PoC servers support, also at a minimum, both AMR narrowband and wideband. 3GPP2 mandates both PoC clients and servers support, at a minimum, EVRC (Enhanced Variable Rate Codec).

20.14.2   Talk Burst Control Protocol

As we mentioned earlier, the only TBCP for PoC defined so far is based on RTCP. In this section, we describe this RTCP-based floor control protocol.

20.14.2.1   Message Encoding

This TBCP is an extension to RTCP that uses RTCP APP messages. RTCP APP messages are application-specific RTCP messages. That is, RTCP APP messages defined by a particular application carry an application identifier and a message subtype, which identifies the application-specific message type.

For example, the RTCP APP message used to request a floor would carry the application identifier for OMA PoC version 1 and the message subtype corresponding to the "TBCP Talk Burst Request". In our descriptions, we will refer to such a message simply as a TBCP Talk Burst Request. However, the reader should keep in mind that the message is encoded as a RTCP APP message with a particular message subtype.

Of course, different message subtypes have different structures. For example, a TBCP Talk Burst Request message will not have the same application-specific fields as a TBCP Talk Burst Granted message.

Figure 20.18 shows the structure of an RTCP APP message. The message contains the following fields.

Ver: the current version is 2.

P: padding.

Message Subtype: padding.

Packet Type (APP): it identifies this message as an RTCP APP message.

Length: length of the message.

SSRC/CSRC: it identifies the originator of this message.

Application Name: the 4-byte identifier used for OMA PoC version 1 is "PoC1".

Application-dependent Data: the structure of this field depends on the message subtype.

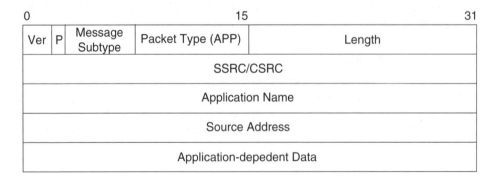

Figure 20.18: RTCP APP message

20.14.2.2 Message Reliability

TBCP does not use the standard RTCP rules to send messages. Instead, it defines a set of timers that drive message retransmissions. When a PoC entity sends a TBCP message, it starts a timer. If this timer fires before a response for the message is received, the message is retransmitted.

20.14.2.3   Message Types

TBCP performs floor control in a PoC session. The PoC client that holds the floor has permission to send media. The following TBCP messages (i.e., RTCP APP message subtypes within the "PoC1" application) are defined.

TBCP Talk Burst Request: a PoC client requests the floor from a PoC server.

TBCP Talk Burst Granted: a PoC server grants the floor to a PoC client.

TBCP Talk Burst Deny: a PoC server denies the floor to a PoC client.

TBCP Talk Burst Release: a PoC client is oversending media and releases the floor.

TBCP Talk Burst Idle: a PoC server informs the PoC clients in a session that no client holds the floor at that point.

TBCP Talk Burst Taken: a PoC server informs the PoC clients in a session that one of the clients holds the floor at that point.

TBCP Talk Burst Revoke: a PoC server revokes the floor from a PoC client.

TBCP Talk Burst Acknowledgement: a PoC client acknowledges the reception of a TBCP message from a PoC server. At this point, only the TBCP Connect and TBCP Disconnect messages require the client to acknowledge their reception.

Additionally, the following TBCP messages are used between PoC clients and servers using pre-established sessions.

TBCP Disconnect: a PoC server terminates a PoC session that was established using a pre-established session.

TBCP Connect: a PoC server sets up a PoC session that is being established using a pre-established session.

Some PoC servers support queuing. That is, besides granting or denying a talk burst request right away, the server can place the request in a queue. PoC clients and servers which support queuing use the following TBCP messages.

TBCP Talk Burst Request Queue Status Request: a PoC client requests the current queue position of a floor request.

TBCP Talk Burst Request Queue Status Response: a PoC server informs a PoC client about the current queue position of a floor request.

20.14.2.4   Message Flow

Figure 20.19 shows the message flow of a PoC session between a PoC client using on-demand signaling and a PoC client using a pre-established session. The terminating PoC client is in automatic answer mode and its participating PoC server uses a confirmed right-to-speak indication. In addition to providing TBCP message flows, this example illustrates the interactions between SIP and TBCP.

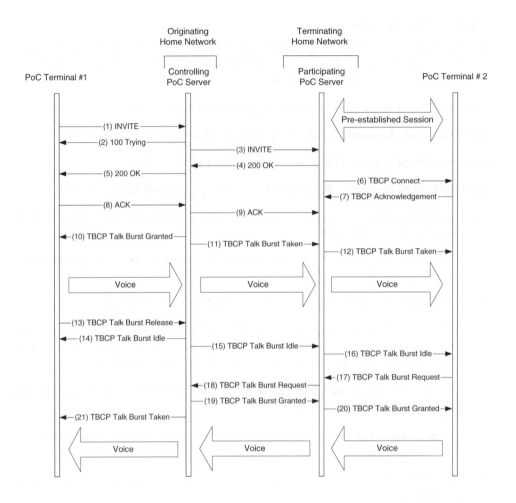

Figure 20.19: TBCP message flow

The originating PoC client generates an INVITE request (1) to establish a one-to-one PoC session. On receiving this INVITE request (1), the participating PoC server of this PoC client becomes the controlling PoC server of the session. Additionally, the PoC server considers the INVITE request as an implicit floor request. That is, the PoC server acts as if the PoC client had sent a TBCP Talk Burst Request message.

This type of INVITE request is considered as an implicit floor request to reduce the PoC session establishment time. Otherwise, the PoC client would need to issue a TBCP Talk Burst Request right after receiving the 200 (OK) response for the INVITE request.

On receiving an INVITE request (3), the participating PoC server of the terminating PoC client responds with a 200 (OK) response (4). Since this PoC server has a pre-established session with the PoC client, the PoC server informs the client using TBCP. The PoC server sends a TBCP Connect message (6) that informs the PoC client about the new PoC session. The PoC client acknowledges the reception of the TBCP Connect message (6) with a TBCP Acknowledgement message (7).

Once the controlling PoC server receives the 200 (OK) response (4), it grants the floor to the originating PoC client by sending a TBCP Talk Burst Granted message (10). The controlling PoC server also informs the terminating PoC client about the fact that the originating PoC client now holds the floor by sending a TBCP Talk Burst Taken message (11).

When the originating PoC client is done sending media, it sends a TBCP Talk Burst Release message (13) in order to release the floor. On receiving this message, the controlling PoC server informs the PoC clients in the session that the floor is idle by sending TBCP Talk Burst Idle messages (14) and (15).

At a later point, the terminating PoC client requests the floor by sending a TBCP Talk Burst Request message (17). The process followed by the controlling PoC server to grant the floor is the same as it followed before. That is, the controlling PoC server sends a TBCP Talk Burst Granted to the requester of the floor (19) and a TBCP Talk Burst Taken (21) to the other PoC client. The only different is that this time this process was triggered by a TBCP Talk Burst Request message, and previously was triggered by an implicit floor request in an INVITE request.

20.15 Simultaneous PoC Sessions

A PoC client can be involved in several PoC sessions at the same time. However, a PoC client only receives talk bursts for one of the sessions at a time. Otherwise, it could be confusing for the user. Figure 20.20 shows a PoC client, which is involved in two simultaneous PoC sessions, and its participating PoC server.

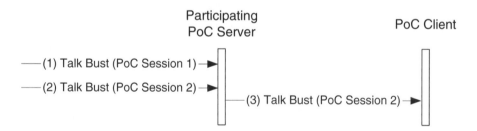

Figure 20.20: Simultaneous arrival of talk bursts

The participating PoC server receives two talk bursts, each of them coming from a different PoC session. The participating PoC server chooses to drop the talk burst that belongs to PoC Session 1 and to relay the talk burst that belongs to PoC Session 2 to the PoC client.

Of course, the PoC client needs to be able to influence the decision by the PoC server of which talk bursts should to be relayed and which ones should be dropped. To influence this type of decision, PoC clients can mark a particular PoC session as the primary PoC session or lock themselves into a particular PoC session (session marking and locking are performed by using SDP parameters in an INVITE, re-INVITE, or UPDATE request). Let us have a look at how these actions by a PoC client determine the media forwarding policy at its participating PoC server.

A participating PoC server serving a PoC client has the concept of an active PoC session. At any given point, the participating PoC server only relays to the PoC client media received from the active session. If the PoC client is involved in only one PoC session, that session is the active one. Consequently, the PoC server relays media received from that PoC session to the PoC client.

When a PoC client is involved in simultaneous PoC sessions, any of the sessions may become the active one. The participating PoC server follows a set of rules to decide which session is active. The following is a summary of those rules.

- If at any point the PoC client is granted the floor in a PoC session (because the PoC client requested it previously), that session becomes the active one.

- If the PoC client locks itself in a PoC session, that session is the active one and media from other sessions is not relayed to the client.

- If the active session is the primary session, media from other sessions is not relayed to the client.

- If the active session is the secondary session, media from other secondary sessions is not relayed to the client. However, if media from the primary session is received, the primary session becomes the active one (and so, media from this session starts being relayed to the client).

Chapter 21

Next Generation Networks

So far this book has been analyzing the IMS components and services that build the platform for delivery packet-based services to mobile users. We have also discussed that IMS is access network-independent, although 3GPP and 3GPP2 have focused on making sure that their radio access networks were ready to accept IMS services. This chapter focuses on Next Generation Networks (NGN). NGN offer access to IMS services from fixed broadband accesses such as Asymmetric Digital Subscriber Lines (ADSL). Since NGN is a broad topic, we describe the main concepts of NGN. We then focus on the IMS aspects of NGN, especially to the particulars of access to IMS from fixed broadband accesses.

This chapter focuses on the applicability of the IMS to Next Generation Networks defined by the European Telecommunication Standards Institute (ETSI). A detailed list of the ETSI specifications related to NGN can be found in Appendix B.4.

21.1 NGN Overview

We describe the general architecture of Next Generation Networks with the help of Figure 21.1. The figure schematically describes the existence of terminals that connect to an NGN. The network is divided into two main layers, namely the *service layer* and the *transport layer*. Each of the layers is composed of a number of subsystems that can be modularly plugged as required and a number of common functions. Some of the subsystems may also contain common functional elements that provide functions to more than a subsystem.

The NGN architecture allows for any distribution of the elements and subsystems in different networks. As such, it provides existence for an access network, a visited network, and a home network, each one providing a different type of service.

The transport layer is responsible for providing the layer 2 connectivity, IP connectivity, and transport control. The transport layer is further divided into the Network Attachment Subsystem (NASS), the Resource and Admission Control Subsystem (RACS), and a number of common transfer functions.

NASS is responsible for supplying the terminal with an IP address, together with configuration parameters, providing authentication at the IP layer, authorization of network access and access network configuration based on users' profiles, and the location manager at the IP layer.

The 3G IP Multimedia Subsystem (IMS) Second Edition   Gonzalo Camarillo and Miguel A. García-Martín
© 2006 John Wiley & Sons, Ltd

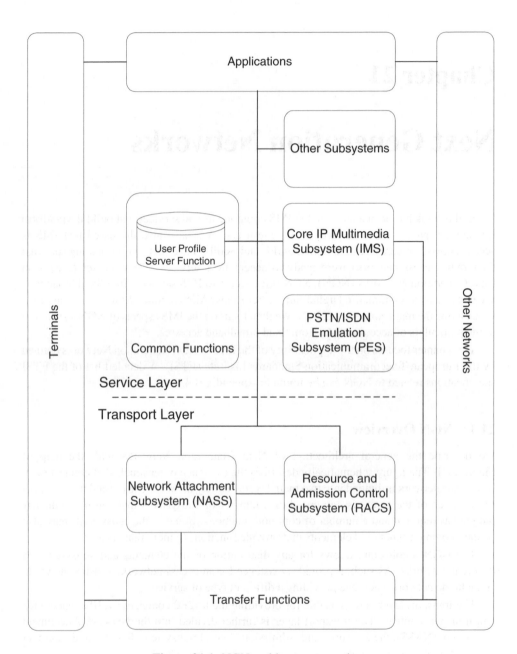

Figure 21.1: NGN architecture overview

RACS is responsible for providing resource management and admission control. Among other functions, RACS provides gate control functionality, policy enforcement, and admission control based on user profiles.

The transfer functions contain a number of functional elements that are visible and sometimes controlled by functional elements of the NASS or RACS. For instance, media gateways, border gateways, etc., are example of transfer functions.

The service layer contains a number of subsystems that provide the platform for enabling services to the user. Prior to describing each of the subsystems, we need to define the concepts of PSTN/ISDN *emulation* and PSTN/ISDN *simulation*.

The term *PSTN/ISDN emulation* is used to refer to an NGN that implements the same services that are today provided in the PSTN and ISDN. Therefore, PSTN/ISDN emulation implementations aim to replace the PSTN/ISDN core network without replacing the terminals. Users connected to an NGN providing PSTN/ISDN emulation will have exactly the same services as they have in the regular PSTN/ISDN without noticing that an NGN is actually delivering the service.

The term *PSTN/ISDN simulation* is used to refer to an NGN that is providing telecommunication services compatible with the PSTN/ISDN, but not necessarily being exactly the same. The concept also indicates not only a replacement of the PSTN/ISDN, but also a replacement of the terminals, that support further capabilities compared with a regular PSTN/ISDN phone.

So the service layer comprises a number of subsystems, of which two are defined in Release 1 of NGN; others will be defined in future releases.

The PSTN/ISDN Emulation Subsystem (PES) implements the PSTN/ISDN emulation concept. PES allows users to receive the same services that they are currently receiving in PSTN/ISDN networks with the existing PSTN/ISDN terminals. PES can be implemented either as a monolithic softswitch or as a distributed IMS. In the case of the distributed IMS, since services are kept intact, without changes, SIP requests and responses carry ISUP bodies. Network elements that provide services (e.g., Application Servers), read the ISUP body to provide a service to the user.

The core IMS implements the PSTN/ISDN simulation concept. The core IMS enables SIP-based multimedia services to NGN terminals. The core IMS enables services, some of which may be new multimedia services (such as presence, instant messaging, etc.), while others might be more traditional telephony services. The core IMS is largely based on the 3GPP IMS specifications. We describe in more detail the core IMS and its services in Section 21.2.

The service layer in NGN also provides for the existence of applications, typically implemented in Application Server Functions (ASF).

A number of common functions provide functional services to several subsystems. This is the case of the User Profile Server Function (USPF), which is a database that contains user-specific information, similar to what the HSS is to the IMS.

Other subsystems beyond the PES and the core IMS can be standardized in the future. For example, NGN could support a streaming subsystem or a content broadcasting subsystem.

21.2 The Core IMS in NGN

As we mentioned before, the core IMS is largely based on the 3GPP IMS specifications, however the core IMS in NGN considers only SIP network elements such as CSCFs, BGCF, MGCF, and MRFC. In particular, Application Servers, MRFP, MGW, user databases, etc., are considered to be outside the core IMS, although they are all present in NGN either as part of the common functions or as part any of the subsystems of the transport layer.

In this section we focus on the new functionality that has been added to the IMS due to fixed broadband access. Typically 3GPP has accepted changes to its IMS specifications to adopt new functionality due to fixed broadband access, so on most occasions the latest versions of the 3GPP specifications will describe the case of accesses to IMS over broadband fixed access.

Figure 21.2 shows the core IMS architecture and the interfaces with adjacent nodes. As expected, the core IMS architecture is very similar to the 3GPP IMS architecture, due to the fact that most of the nodes are present in the core IMS.

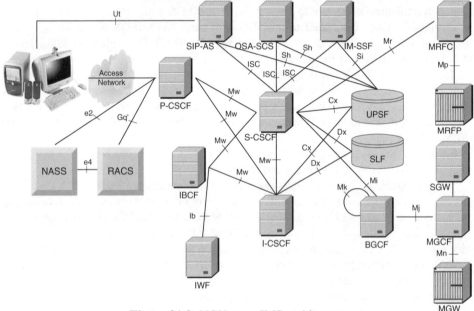

Figure 21.2: NGN: core IMS architecture

The core IMS adds new interfaces to the P-CSCF to communicate with the functional elements of the low transport layer subsystems, the RACS and NASS. The P-CSCF includes a new *Gq'* interface towards the RACS for the purpose of requesting authorization of QoS resources, reserving resources, and providing control of gates in the transport layer. The *Gq'* interface is based on the 3GPP *Gq* interface specified in 3GPP TS 29.209 [32]. The P-CSCF also implements a new *e2* interface towards the NASS for the purpose of retrieving user's location information.

The P-CSCF contains an IMS Application Level Gateway (IMS-ALG) that provides control for the network address and port translator functions located in the Transition Gateway (TrGW). We described the IMS-ALG and TrGW in Section 3.4.6, at that time as stand-alone functional entities that provide IPv4–IPv6 translation. In the case of NGN the P-CSCF embeds an IMS-ALG that controls the Transition Gateway. The goal is to support NAT traversal when the local network in the customer premises implements, for example, a private IPv4 addressing scheme.

The combination of the P-CSCF, the IMS-ALG, and the TrGW allows NAT traversal when there is a NAT located in the customer premises (e.g., when an ADSL router integrates a NAT), and when the terminal does not provide support for NAT traversal of its own. Figure 21.3 shows the IMS-ALG embedded in the P-CSCF controlling the TrGW.

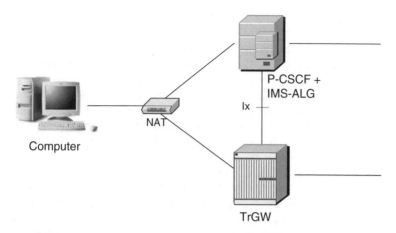

Figure 21.3: IMS-ALG embedded in a P-CSCF supports NAT traversal

The I-CSCF, S-CSCF, BGCF, and MRFC are the same as the IMS corresponding node. The MGCF in NGN keeps the same functionality as in the IMS, but it is complemented with additional functionality that provides an appropriate interworking with circuit-switched networks beyond the basic call (e.g., interworking takes place even at the service level).

Outside the core IMS, but still present in the NGN architecture are the User Profile Server Function (UPSF), the Subscriptor Locator Function (SLF), the Interconnection Border Control Function (IBCF), the Interworking Function (IWF), and Application Server Functions (ASFs).

The UPSF in NGN is similar to the HSS in IMS. The main difference lies in the fact that the HSS, since it is an evolution of the GSM HLR, includes an HLR/AUC that provides mobility management. This is not required in NGN, therefore, the UPSF is limited to the IMS-specific parts of the HSS. Besides that, the external interfaces are common to the HSS and the UPSF, and the database function for IMS users is the same in both cases.

The SLF is exactly the same function available in the IMS.

The IBCF is a new functional entity introduced by NGN. The IBCF acts as a separation between two different domains. In doing so, it may provide functionality similar to the IMS-ALG, in case there is a need for IP version interworking. The IBCF is also responsible for inserting an IWF in the path when it is needed. Additionally, the IBCF may obfuscate some SIP headers that the operator may consider dangerous to expose externally.

The IWF provides interworking between SIP and other protocols (e.g., H.323). In doing so, it could potentially provide an adaptation of SIP profiles, if a need for such case is demonstrated.

Application Server Functions (ASFs) execute services. NGN identifies two types of ASFs, namely ASF Type 1 and Type 2. The ASF Type 1 may interact with the RACS when providing a service to the user. ASF Type 2 merely relies on the call control protocol to provide the service. ASF Type 2 is functionally equivalent to the AS in the IMS.

In addition to the mentioned nodes, NGN provide for the existence of Charging and Data Collection Functions. These include data collection and mediation functions to a billing system. It is believed that NGN will reuse most of the charging functionality available in the IMS.

21.2.1   New SIP Functionality

The new functionality required by IMS to operate in an NGN comprises the usage of a few new SIP header fields, namely the `History-Info` header field and the `Reason` header field in responses.

The `History-Info` header field is specified in the Internet-Draft "An Extension to the Session Initiation Protocol for Request History Information" [52]. The extension allows proxies to record retarget or redirect information in SIP requests and responses. This allows a SIP UA to find out if the original SIP requests issued by the caller has been redirected or retargeted by network elements.

IMS already provided a small similar piece of information in the format of the `P-Called-Party-ID` header field. We have already described the `P-Called-Party-ID` header field in Section 5.7. In short, this header field allows us to retain the public user identity included in the `Request-URI` when the S-CSCF retargets and replaces the `Request-URI` with the contact information of the user. The limitation of the `P-Called-Party-ID` header field lies in the fact that it is only applicable to the last retarget, the one that takes place at the terminating S-CSCF that has got contact information from the registration of the user.

The `History-Info` header field not only records this last retarget, but also any other previous retarget that could have taken place, for instance, due to forwarding from one user to another, or due to redirections. In this way, both the caller and callee can gather information of the history of transformations that the `Request-URI` of a SIP request suffered en route to its destination.

The other extension that NGN requires for SIP is documented in the Internet-Draft "SIP Reason header extension for indicating redirection reasons" [87]. This extends the usage of the `Reason` header field, originally specified in RFC 3326 [226], to include the reason for a redirection. For instance, this extension allows us to express the fact that a session has been redirected because the callee is unavailable, busy, did not reply, as a request of the callee action, etc.

21.2.2   Unneeded Functionality

The IMS in general, and the 3GPP SIP profiled specified in 3GPP TS 24.229 [16] provide optimizations for access to the IMS from low-bandwidth environments or terminals with limited memory capabilities. These limitations are not usually present when accessing the IMS from a fixed broadband access. Therefore, the IMS contains some functionality that is disabled when the IMS operates in an NGN environment.

One of these unneeded functions is compression of SIP messages, Sigcomp. We described Sigcomp in Section 4.16. Since the goal of Sigcomp is to transmit a SIP message faster over a low-bandwidth channel, and since NGN assumes broadband access, there is no need to compress SIP.

Another function that is not really required is the extended timers that the IMS has set for the IMS terminal and the P-CSCF. The P-CSCF and the IMS terminal implement longer timers for SIP than those recommended by RFC 3261 [215]. This would allow an IMS terminal behind a low-bandwidth channel to avoid retransmissions of SIP requests and responses. Certainly, these extended timers are not required if the access is a high-bandwidth channel.

21.3 PSTN/ISDN Simulation Services

We described earlier that the core IMS enables multimedia services and traditional telephony services. In this section we focus on the traditional telephony services that, in the context of the PSTN and ISDN, are globally known as *PSTN/ISDN supplementary services*. Traditional supplementary services are: Call Forwarding, Call Hold/Resume, Connected Line Identification Presentation/Restriction, etc.

In NGN Applications Servers using the core IMS are also able to provide *PSTN/ISDN simulation services*. These are telephony services similar in nature to the *PSTN/ISDN supplementary services*, but different in the realization since SIP is the call control protocol and is quite different to the PSTN/ISDN protocols. PSTN/ISDN simulation services are a simulated version of the PSTN/ISDN supplementary services. However, the aim of the service remains.

Each PSTN/ISDN simulation service is based on the corresponding supplementary service in the PSTN/ISDN. But in many cases the name of the service is changed in NGN, since in SIP the concept of a call can be broadened with messages, subscriptions, notifications, or multimedia sessions. Therefore, most of the PSTN/ISDN simulation services refer to *communications* rather than *calls*.

In Table 21.1 we take a brief look at each of these PSTN/ISDN simulation services.

21.3.1 *Communication Diversion (CDIV) and Communication Forwarding*

The NGN Communication Diversion allows a user to divert their communications to another user. The service is specified in ETSI TS 183 004 [91] and comprises a number of diversion services: Communication Forwarding Unconditional (CFU), Communication Forwarding on Busy user (CFB), Communication Forwarding on No Reply (CFNR), Communication Deflection (CD), and Communication Forwarding on Not Logged-in (CFNL).

The service is implemented in an application server that, depending on the actual service, automatically diverts all session attempts to the given user, or after the SIP request has been sent to the user, but the user sent a busy response, did not answer, etc.

The service also implements the `History-Info` header field to record a forwarding operation.

21.3.2 *Conference (CONF)*

The Conference service provides a centralized conference server that is able to accept either multimedia participants or PSTN participants. The service is specified in ETSI TS 183 005 [95], which is largely based on 3GPP TS 24.147 [9].

The service makes extensive usage of the conference event package defined in the Internet-Draft "A SIP Event Package for Conference State" [205]. This allows participants in a conference to subscribe to the conference state and to receive notifications with the list and status of participants, types of media, topic of discussion, and other conference-related information.

The service also uses the SIP `Replaces` header field (specified in RFC 3891 [159]) in order to allow an ad-hoc conference to become a centralized one.

Table 21.1: PSTN/ISDN simulation services in NGN

Abbreviation	PSTN/ISDN simulation service	PSTN/ISDN supplementary service
CDIV	Communication Diversion	Call Diversion
CFU	Communication Forwarding Unconditional	Call Forwarding Unconditional
CFB	Communication Forwarding on Busy user	Call Forwarding Busy
CFNR	Communication Forwarding on No Reply	Call Forwarding No Reply
CD	Communication Deflection	Call Deflection
CFNL	Communication Forwarding on Not Logged-in	—
CONF	Conference	Conference Calling
MWI	Message Waiting Indication	Message Waiting Indication
OIP	Originating Identification Presentation	Calling Line Identification Presentation
OIR	Originating Identification Restriction	Calling Line Identification Restriction
TIP	Terminating Identification Presentation	Connected Line Identification Presentation
TIR	Terminating Identification Restriction	Connected Line Identification Restriction
CW	Communication Waiting	Call Waiting
HOLD	Communication Hold	Call Hold
ACR	Anonymous Communication Rejection	Anonymous Call Rejection
CB	Communication Barring	Call Barring
AoC	Advice of Charge	Advice of Charge
CCBS	Completion of Communications to Busy Subscriber	Completion of Calls to Busy Subscriber
CCNR	Completion of Communications on No Reply	Completion of Calls on No Reply
MCID	Malicious Communication Identification	Malicious Call Identification
ECT	Explicit Communication Transfer	Explicit Call Transfer

21.3.3 Message Waiting Indication (MWI)

The MWI service for NGN allows a user to be informed when he or she receives a multimedia voice mail in their multimedia voice mail server. The service is specified in ETSI TS 183 006 [97] and it is a straight implementation of RFC 3842 [158], that also describes the MWI service in the Internet.

The specification allows a user to subscribe to the `message-summary` event package of his account and to receive notifications of existing messages.

21.3.4 Originating Identification Presentation/Restriction (OIP, OIR)

The NGN OIP and OIR services allow a caller to present his identity at the callee's terminal. The service is specified in ETSI TS 183 007 [98], in spite of the fact that the service is inherent to SIP and reinforced by the definition of the `Privacy` and the `P-Asserted-Identity`/`P-Preferred-Identity` header fields, specified in RFC 3323 [183] and RFC 3325 [147], respectively. The usage of these headers is further defined in 3GPP TS 24.229 [16].

The OIP service works by the caller adding a `P-Preferred-Identity` header field that contains a registered identity for the caller. The P-CSCF verifies that the value of the header is correct and, based on the authenticated information of the IMS terminal, replaces the `P-Preferred-Identity` with a `P-Asserted-Identity` header field. If the user invokes the OIR service, then the UE inserts a `Privacy` header field that contains, for example, the "id" value. In that case the last trusted hop removes the `P-Asserted-Identity` header field, so that the callee does not receive it in the SIP request.

The service requires transitive trust between the different networks involved in a session. A more sophisticated mechanism based on cryptographic assurance of the caller's identity is not yet supported.

21.3.5 Terminating Identification Presentation/Restriction (TIP, TIR)

The NGN TIP and TIR services are counterparts of the OIP/OIR. They deliver to the caller the asserted identity of the callee. One would think in principle that the service is not so useful, because the caller is calling the callee, so the caller should know the identity of the callee in advance. Although that thought is correct, there might be scenarios of redirections, call diversions, etc., where the callee is not the one that the caller originally called. Hence, the TIP/TIR services become a bit more interesting.

The TIP/TIR services are specified in ETSI TS 183 007 [98] based on a straight implementation of the `Privacy` and `P-Asserted-Identity` header fields, specified in RFC 3323 [183] and RFC 3325 [147], respectively.

This is a service provided to the callee. If the callee, when he receives a SIP request, wants the network to provide his asserted identity, by default, for a user who is provided with the service, the P-CSCF will insert in SIP responses a `P-Asserted-Identity` header field that contains an identity of the user, in particular, the identity of the callee to which the corresponding SIP request was issued. The caller will then get the asserted identity of the callee.

If the callee invokes the TIR service, then the callee inserts a `Privacy` header field with the value set to "id" in the SIP response. The last trusted node closest to the caller will remove the `P-Asserted-Identity` header field, and therefore, the caller will not know who answered the session.

21.3.6   Communication Waiting (CW)

The Communication Waiting service, which is standardized in ETSI TS 183 009 [93], is a bit different in NGN from the traditional CW service the PSTN/ISDN. In classic networks the problem to solve is how to indicate to the user that there is a new incoming call when there are a limited number of channels (one in PSTN; two in ISDN basic rate access).

In NGN the service comprises two aspects: user-determined user busy and network-determined user busy. The former simply consists of the case where the user rejects an incoming session attempt, typically because the user is busy, engaged in another session, etc.

Network-determined user busy comprises the procedures that the network has to evaluate in order to guess that the user is busy and will not be able to accept the session attempt. One of these procedures just controls the number of simultaneous sessions that the callee has. If a threshold is passed and a new incoming session is received, the network application server will determine that the user is busy and will apply the communication waiting procedures. This aspect is covered in Release 1 of the ETSI NGN specifications.

The other procedure consists of the AS guessing that the user does not have enough bandwidth to establish the session that is proposed in the SIP request. This requires that the AS is able to know the available bandwidth of the user at a given time. This procedure is not supported in ETSI NGN Release 1.

21.3.7   Communication Hold (HOLD)

The HOLD service, similarly to the HOLD service in the PSTN, allows a user to suspend an ongoing multimedia service and to resume it at a later stage. The service is specified in ETSI TS 183 010 [92], and it is a straight implementation of the SDP offer/answer model specified in RFC 3264 [212] and already supported by 3GPP TS 24.229 [16].

Holding an ongoing session is achieved by sending a new SDP offer where each of the media streams to be held are marked as sendonly if they were previously bidirectional media streams. To resume the session, a new SDP offer is issued where each of the held media streams is marked with the default sendrecv.

The HOLD service in NGN also allows an application server to play some announcement or music to the held party. This is achieved by an AS that acts as a third-party call controller and replaces the existing session of one of the users with an AS-originated session that plays the announcement or music until the session is resumed.

21.3.8   Anonymous Communication Rejection (ACR) and Communication Barring (CB)

The ACR service allows a user to instruct the network to reject any anonymous communication. ACR is a particular case of the Communication Barring service, which allows a user to bar an incoming or outgoing communication based on some parameters. Both services are specified in ETSI TS 183 011 [90].

These services do not require any protocol behavior, other than the user indicating to the application server the rules that make some communications barred.

21.3.9   Advice of Charge (AoC)

The AoC service is able to deliver to the caller information about the costs of the session. The service is specified in ETSI TS 183 012 [89].

The service is planned to be part of ETSI NGN Release 2.

21.3.10   Completion of Communications to Busy Subscriber (CCBS) and Completion of Communications on No Reply (CCNR)

The CCBS service in NGN, which is specified in ETSI TS 183 015 [94], works when a caller calls a busy callee. The callee indicates its support for the CCBS service, and then, if the caller invokes the service, the caller gets informed when the callee is free and when the caller has a reserved time-slot to re-call.

The service also provides a queue management system, so that if several callers are waiting for a busy callee to becomes free, the service is able to indicate to the callers one by one when they can re-call.

The IETF has developed the dialog event package (specified in the Internet-Draft "An INVITE Initiated Dialog Event Package SIP" [206]) that solves the problem by providing a subscription/notification service to a dialog on a terminal. However, the dialog event package is not able to manage queues, which seems to be a requirement for NGN. The approach in NGN is to insert an AS serving the caller and another AS serving the callee. These ASs subscribe to the dialog event package of the caller and callee to provide queue support.

The CCNR service is similar to CCBS, but the service is invoked when the callee does not reply to an INVITE request. From the point of view of the implementation, it is exactly the same as for the CCBS service.

21.3.11   Malicious Communication Identification (MCID)

The NGN Malicious Communication Identification service allows a user to indicate their suspicions of a malicious communication. The network then records in a log the identities of the caller and callee and the date and time of invocation. The log is made available to the appropriate authorities (not to the invoking user). The service is specified in ETSI TS 183 016 [96].

In some cases the identity of the caller is known (due to the presence of a P-Asserted-Identity header field in the SIP request) for a request originated in a trusted network.

In some other cases the P-Asserted-Identity is not present in the request, perhaps because the request was originated in a non-trusted network, or perhaps because the request was originated in the PSTN and the calling number was not inserted in the signaling. In the latter, the application server will subscribe to an event package (yet to be designed) at the MGCF. The MGCF will issue the appropriate ISUP signaling to request the identity of the caller from the local exchange, and will notify upon arrival of the information.

Since the application server is not able to distinguish whether a SIP request has been originated in an MGCF, an IMS terminal, or SIP UA, the AS has to try to subscribe to the event package. Only if the session originated in the PSTN will the subscription succeed.

The mechanism whereby the user indicates to the AS his suspicions about a malicious communication is still to be determined, at the time of writing.

21.3.12   Explicit Communication Transfer (ECT)

The ECT service in NGN allows a party who has an ongoing communication with a second party to transfer the communication to a third party, so that when the service is completed the first party has no ongoing communications and the second and third parties have an established session (or are in the process of establishing it).

At the time of writing the mechanism to implement the service has not yet been decided, but most likely it will be based on the usage of the SIP REFER method (specified in RFC 3515 [229]).

21.3.13   *User Settings in PSTN/ISDN Simulation Services*

We have just described in Section 21.3 how a user can invoke PSTN/ISDN simulation services that are similar but not the same as the PSTN/ISDN supplementary services. Some of these services require a previous configuration before operation can take place. For example, the Communication Diversion services typically require the user to set up the URI to forward the incoming sessions. Or Communication Barring, where the barring of incoming communications requires the user to define the identity of the caller for whose incoming sessions will be barred.

NGN provides users with the ability of activate, deactivate and configure or personalize most of their PSTN/ISDN simulation services. To do this type of configuration, the *Ut* interface is used, and XCAP is the protocol running over that interface. We have already described XCAP in Section 16.14 in the context of the presence service. Now we have another application of XCAP for activating, deactivating, and configuring services.

ETSI has defined in ETSI TS 183 023 [99] a `simservs` modular XML document. This document contains common parts, applicable to all the services, and a collection of XML subdocument trees, each one representing a PSTN/ISDN simulation service. This approach allows us to store and then later manipulate the `simservs` XML document in a centralized server (e.g., an AS that provides all the PSTN/ISDN simulation services), or in a distributed fashion (e.g., when the AS hosts a few PSTN/ISDN simulation services). The XML schema that defines the layout of the `simservs` XML document and its constraints is defined in ETSI TS 183 023 [99]. The subdocuments that are service-specific are defined in the specification that defines the service.

A new application usage is defined, under the private vendor tree, where ETSI is considered as a vendor. The new application usage defines the rules and conventions used in manipulating `simservs` XML documents.

Users can then invoke XCAP operations (e.g., HTTP PUT, GET, DELETE) to activate and deactivate the service, and to configure the operational parameters of that service.

Appendix A

The 3GPP2 IMS

So far, we have focused on the 3GPP IMS in all of our descriptions. However, we have pointed out the areas where the 3GPP2 IMS differs from the 3GPP IMS in the chapters corresponding to each particular area. In this appendix we provide a list of their differences for ease of reference.

A.1 An Introduction to 3GPP2

So far, we have focused on the IMS as specified by 3GPP. 3GPP is committed to the evolution of the GSM standard toward a third-generation mobile system. 3GPP created the IMS, which is part of Release 5 of the 3GPP specifications. Although the IMS is independent of the access network, 3GPP has focused on access provided by WCDMA networks and the GPRS packet core network.

At almost the same time, 3GPP2 was chartered to evolve the ANSI/TIA/EIA-41 standards into a third-generation system based on CDMA2000® access technology and the 3GPP2 wireless IP network. 3GPP2 created a similar multimedia subsystem, which is specified in Release A of the 3GPP2 specifications. This initial release of the IMS 3GPP2 specifications is based on the IMS specified in Release 5 of the 3GPP specifications. 3GPP2 IMS specifications follow as much as possible their 3GPP IMS counterparts. Nevertheless, due to the nature of the dissimilar access technologies and substantial differences in the core networks, the two IMS networks are not exactly the same. This chapter provides a summary of the main differences between 3GPP2 IMS specifications and their 3GPP IMS counterparts.

A.2 The Multimedia Domain (MMD)

3GPP2 provides for the existence of a multimedia domain that gives IP packet data support and multimedia services to users over a wireless IP network. The MMD is further divided into the PDS (Packet Data Subsystem) and the IMS (IP Multimedia Subsystem). The PDS provides terminals with general IP packet connectivity (similar to what GPRS provides in GSM/3GPP networks). The IMS provides the terminals with IP multimedia capabilities similar to those offered by the 3GPP IMS.

The 3G IP Multimedia Subsystem (IMS) Second Edition   Gonzalo Camarillo and Miguel A. García-Martín
© 2006 John Wiley & Sons, Ltd

A.3 Architecture of the 3GPP2 IMS

Figure A.1 gives a 3GPP2 architecture overview of the IMS. 3GPP2 took the approach of assigning a number to each interface; but, since the 3GPP2 IMS is based on the IMS designed by 3GPP, most of the 3GPP2 interfaces have a 3GPP counterpart name. The figure shows both names when applicable. It does not show the complete list of interfaces and architectural elements, just the most relevant ones. A complete architectural figure can be found in 3GPP2 TS X.S0013-000 [44].

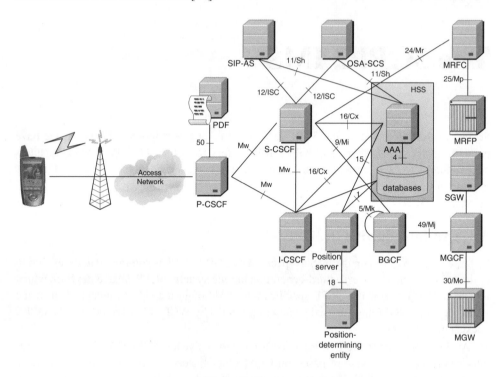

Figure A.1: 3GPP2 IMS architecture overview

The reader familiar with the 3GPP architecture that was introduced in Section 3.4 has probably noticed the similarities between both IMSs. Nevertheless, there are some differences between the two IMS architectures, the most relevant being the following.

- Mobility management: 3GPP2 IMS is built on top of mobile IP, whereas 3GPP manages the mobility through GPRS, which provides a layer 2 tunneling mechanism.

- Home Subscriber Server: the 3GPP HSS is the aggregation of the 3GPP2 Home AAA and its databases.

- Speech codec: AMR is the default narrow-bandwidth speech coded in the 3GPP IMS. 3GPP2 defines the EVRC and SMV codecs.

- P-CSCF discovery: 3GPP supports P-CSCF discovery through PDP context activation and through DHCP. 3GPP2 does not support PDP context activation, since that is a GPRS-specific procedure.

- *Si* interface: 3GPP defines the *Si* interface between the HSS and the IM-SSF (part of CAMEL). This interface is not needed in 3GPP2, since 3GPP2 does not have CAMEL-inherited systems.

- Smart card: 3GPP defines a UICC that contains a USIM or ISIM application that contains configuration and security data. 3GPP2 IMS terminals store configuration and security data in the IMS terminal itself or in an R-UIM (Removable User Identity Module).

- P-CSCF location: in the 3GPP IMS the P-CSCF and the GGSN are always located in the same network. In the 3GPP2 IMS the GGSN's counterpart is named the PDSN (Packet Data Service Node). In the 3GPP2 IMS the P-CSCF and the PDSN may be located in different networks.

- Anchored GGSN/PDSN: in 3GPP networks the GGSN is allocated prior to registration and the GGSN does not change during this allocation. In 3GPP2 networks the PDSN may change, even when an ongoing session is taking place. The allocated PDF may also change when the terminal moves from one PDSN to another.

- *Go* interface: the *Go* interface is not supported in the first release of the 3GPP2 IMS specifications.

- *Sh* interface: in the 3GPP IMS the *Sh* interface may provide 3GPP-specific location information that is not supported in 3GPP2.

- Position location: the 3GPP2 IMS contains a Position Server and a PDE (Position Determining Entity) that provide positioning information. These entities do not have counterparts in the 3GPP IMS.

Table B.1: 3GPP IMS-related specifications (part 1)

Spec#	Title	R#	Observations
21.905	Vocabulary for 3GPP Specifications	R5	
22.066	Support of Mobile Number Portability (MNP); Stage 1	R6	
22.101	Service aspects; Service principles	R5	
22.141	Presence service; Stage 1	R6	
22.228	Service requirements for the Internet Protocol (IP) multimedia core network subsystem; Stage 1	R5	
22.250	IP Multimedia Subsystem (IMS) Group Management; Stage 1	R6	
22.340	IP Multimedia Subsystem (IMS) messaging; Stage 1	R6	
22.800	IMS Subscription and access scenarios	R6	
23.002	Network Architecture	R5	
23.003	Numbering, Addressing and Identification	R5	
23.125	Overall high-level functionality and architecture impacts of flow-based charging; Stage 2	R6	
23.141	Presence service; Architecture and functional description; Stage 2	R6	
23.207	End-to-end Quality of Service (QoS) concept and architecture	R5	
23.218	IP Multimedia (IM) session handling; IM call model; Stage 2	R5	
23.221	Architectural requirements	R5	
23.271	Location Services (LCS); Functional description; Stage 2	R6	
23.278	Customised Applications for Mobile network Enhanced Logic (CAMEL) – IP Multimedia System (IMS) interworking; Stage 2	R5	
23.864	Commonality and interoperability between IP Multimedia System (IMS) core networks	R6	
23.867	Internet Protocol (IP) based IP Multimedia Subsystem (IMS) emergency sessions	R6	
23.917	Dynamic policy control enhancements for end-to-end QoS, Feasibility study	R6	
23.979	3GPP enablers for Push-To-Talk over Cellular (PoC) services; Stage 2	R6	

Table B.2: 3GPP IMS-related specifications (part 2)

Spec#	Title	R#	Observations
23.981	Interworking aspects and migration scenarios for IPv4-based IP Multimedia Subsystem (IMS) Implementations	R6	
24.141	Presence service using the IP Multimedia (IM) Core Network (CN) subsystem; Stage 3	R6	
24.147	Conferencing using the IP Multimedia (IM) Core Network (CN) subsystem; Stage 3	R6	
24.228	Signalling flows for the IP multimedia call control based on Session Initiation Protocol (SIP) and Session Description Protocol (SDP); Stage 3	R5	
24.229	Internet Protocol (IP) multimedia call control protocol based on Session Initiation Protocol (SIP) and Session Description Protocol (SDP); Stage 3	R5	
24.247	Messaging using the IP Multimedia (IM) Core Network (CN) subsystem; Stage 3	R6	
26.235	Packet-switched conversational multimedia applications; Default codecs	R5	
26.236	Packet-switched conversational multimedia applications; Transport protocols	R5	
29.162	Interworking between the IM CN subsystem and IP networks	R6	
29.163	Interworking between the IP Multimedia (IM) Core Network (CN) subsystem and Circuit-Switched (CS) networks	R6	
29.207	Policy control over Go interface	R5	
29.208	End-to-end Quality of Service (QoS) signaling flows	R5	
29.209	Policy control over Gq interface	R6	
29.228	IP Multimedia (IM) Subsystem Cx and Dx Interfaces; Signalling flows and message contents	R5	
29.229	Cx and Dx interfaces based on the Diameter protocol; Protocol details	R5	

Table B.3: 3GPP IMS-related specifications (part 3)

Spec#	Title	R#	Observations
29.278	Customised Applications for Mobile network Enhanced Logic (CAMEL); CAMEL Application Part (CAP) specification for IP Multimedia Subsystems (IMS)	R5	
29.328	IP Multimedia Subsystem (IMS) Sh interface signaling flows and message content	R5	
29.329	Sh interface based on the Diameter protocol	R5	
29.962	Signalling interworking between the 3GPP profile of the Session Initiation Protocol (SIP) and non-3GPP SIP usage	R6	
31.103	Characteristics of the IP Multimedia Services Identity Module (ISIM) application	R5	
32.200	Telecommunication management; Charging management; Charging principles	R5	Only R5. Replaced by 32.240 in R6
32.225	Telecommunication management; Charging management; Charging data description for the IP Multimedia Subsystem (IMS)	R5	Only R5. Replaced by 32.260 in R6
32.240	Telecommunication management; Charging management; Charging Architecture and Principles	R6	
32.260	Telecommunication management; Charging management; IP Multimedia Subsystem (IMS) charging	R6	
32.421	Telecommunication management; Subscriber and equipment trace: Trace concepts and requirements	R6	
33.102	3G security; Security architecture	R5	
33.108	3G security; Handover interface for Lawful Interception (LI)	R5	
33.141	Presence service; Security	R6	
33.203	3G security; Access security for IP-based services	R5	
33.210	3G security; Network Domain Security (NDS); IP network layer security	R5	

Table B.4: 3GPP2 IMS-related specifications

Spec#	Title	3GPP #
X.S0013-000	Multi-Media Domain Overview	
X.S0013-002	IP Multimedia Subsystem (IMS); Stage 2	23.228
X.S0013-003	IP Multimedia (IM) session handling; IM call model	23.218
X.S0013-004	IP Multimedia Call Control Protocol based on SIP and SDP; Stage 3	24.229
X.S0013-005	IP Multimedia (IM) Subsystem Cx Interface; Signaling flows and message contents	29.228
X.S0013-006	Cx Interface based on the Diameter protocol; Protocol details	29.229
X.S0013-007	IP Multimedia Subsystem; Charging Architecture	32.200
X.S0013-008	IP Multimedia Subsystem; Accounting Information Flows and Protocol	32.225
X.S0013-010	IP Multimedia Subsystem (IMS) Sh Interface signaling flows and message contents	29.328
X.S0013-011	Sh interface based on the Diameter protocol	29.329
S.R0086-0	IMS Security Framework	33.203

Table B.5: ETSI NGN specifications (part 1)

Spec#	Title
TR 180 000	NGN Terminology
TR 180 001	NGN Release 1; Release definition
ES 282 001	NGN Functional Architecture Release 1
ES 282 002	NGN Release 1: Functional Architecture for PSTN/ISDN Emulation
ES 282 003	NGN Release 1: Functional Architecture; Resource and Admission Control Sub-system (RACS)
ES 282 004	NGN Functional Architecture; Network Attachment Sub-System (NASS)
—	NGN Architecture to support emergency communication from citizen to authority
—	NGN Requirements and Functional Architecture; Network Attachment using Fixed Wireless LAN (802.1x)
TS 182 005	NGN R1 Functional Architecture; Organization of user data
TS 182 006	NGN Release 1; NGN-IMS Stage 2 definition (endorsement of TS.23.228)
ES 282 007	NGN Release 1; Core IMS architecture
—	TISPAN NGN Architecture; IMS-based PSTN/ISDN Emulation Subsystem Architecture
TS 182 008	Presence Service; Architecture and functional description; (Endorsement of 3GPP TS 23.141)
EN 383 001	Interworking for SIP/SIP-T (BICC, ISUP) [ITU-T Recommendation Q.1912.5, modified]
ES 283 003	Endorsement of "IP Multimedia Call Control Protocol based on Session Initiation Protocol (SIP) and Session Description Protocol (SDP) Stage 3 (Release 6)" for NGN Release 1
TS 183 004	NGN Signalling Control Protocol; Communication Diversion (CDIV); PSTN/ISDN Simulation Services
TS 183 005	NGN Signalling Control Protocol; Conference (CONF); PSTN/ISDN Simulation services
TS 183 006	NGN Signalling Control Protocol; Message Waiting Indication (MWI); PSTN/ISDN Simulation Services
TS 183 007	NGN Signalling Control Protocol; Originating Identification Presentation (OIP) and Originating Identification Restriction (OIR); PSTN/ISDN Simulation Services
TS 183 008	NGN Signalling Control Protocol; Terminating Identification Presentation (TIP) and Terminating Identification Restriction (TIR); PSTN/ISDN Simulation Services
TS 183 009	NGN Signalling Control Protocol; Communication Waiting (CW); PSTN/ISDN Simulation Services

Table B.6: ETSI NGN specifications (part 2)

Spec#	Title
TS 183 010	NGN Signalling Control Protocol; Communication Hold (HOLD); PSTN/ISDN Simulation Services
TS 183 011	NGN Signalling Control Protocol; Anonymous Communication Rejection (ACR) and Communication Barring (CB); PSTN/ISDN Simulation Services
TS 183 012	NGN Signalling Control Protocol; Advice of Charge (AoC); PSTN/ISDN Simulation Services
TR 183 013	NGN Release 1 Analysis of relevant 3GPP IMS specifications for use in TISPAN NGN Release 1 specifications
TR 183 014	Verification of ETSI, IETF and ITU Specifications to prove the availability of Carrier class services for PSTN/ISDN Emulation
TS 183 015	NGN Signalling Control Protocol; Completion of Communications to Busy Subscriber (CCBS) and Completion of Communications on No Reply (CCNR); PSTN/ISDN Simulation Services
TS 183 016	NGN Signalling Control Protocol; Malicious Communication Identification (MCID); PSTN/ISDN Simulation Services
TS 183 017	NGN Release 1; Stage 3 description of the Gq' interface
TS 183 021	NGN Release 1; Endorsement of TS.29.162 Interworking between IM CN subsystem and IP networks
TS 183 023	Extensible Markup Language (XML) Configuration Access Protocol (XCAP) over the Ut interface for Manipulating NGN PSTN/ISDN Simulation Services
ES 283 027	Interworking SIP-ISUP for TISPAN-IMS
TS 183 028	NGN Signalling Control Protocol; Common Basic Communication procedures
TS 183 029	NGN Signalling Control Protocol; Explicit Communication Transfer (ECT); PSTN/ISDN Simulation Services
ES 283 030	Protocol support of presence capability in the NGN Release 1
TS 187 003	NGN Security (NGN EC); Security Architecture – NGN Release 1

References

[1] 3GPP. 3G security; Network Domain Security (NDS); IP network layer security. TS 33.210, 3rd Generation Partnership Project (3GPP), June 2004.

[2] 3GPP. Architectural requirements. TS 23.221, 3rd Generation Partnership Project (3GPP), June 2004.

[3] 3GPP. 3G security; Access security for IP-based services. TS 33.203, 3rd Generation Partnership Project (3GPP), June 2005.

[4] 3GPP. 3GPP enablers for Open Mobile Alliance (OMA) Push-to-talk over Cellular (PoC) services; Stage 2. TR 23.979, 3rd Generation Partnership Project (3GPP), June 2005.

[5] 3GPP. AMR speech Codec; General description. TS 26.071, 3rd Generation Partnership Project (3GPP), January 2005.

[6] 3GPP. AMR speech Codec; Transcoding Functions. TS 26.090, 3rd Generation Partnership Project (3GPP), January 2005.

[7] 3GPP. Characteristics of the IP Multimedia Services Identity Module (ISIM) application. TS 31.103, 3rd Generation Partnership Project (3GPP), June 2005.

[8] 3GPP. Characteristics of the USIM application. TS 31.102, 3rd Generation Partnership Project (3GPP), June 2005.

[9] 3GPP. Conferencing using the IP Multimedia (IM) Core Network (CN) subsystem; Stage 3. TS 24.147, 3rd Generation Partnership Project (3GPP), June 2005.

[10] 3GPP. Customized Applications for Mobile network Enhanced Logic (CAMEL); CAMEL Application Part (CAP) specification for IP Multimedia Subsystems (IMS). TS 29.278, 3rd Generation Partnership Project (3GPP), April 2005.

[11] 3GPP. Customized Applications for Mobile network Enhanced Logic (CAMEL) Phase 4; Stage 2; IM CN Interworking. TS 23.278, 3rd Generation Partnership Project (3GPP), June 2005.

[12] 3GPP. Cx and Dx interfaces based on the Diameter protocol; Protocol details. TS 29.229, 3rd Generation Partnership Project (3GPP), June 2005.

[13] 3GPP. End-to-end Quality of Service (QoS) concept and architecture. TS 23.207, 3rd Generation Partnership Project (3GPP), June 2005.

[14] 3GPP. End-to-end Quality of Service (QoS) signalling flows. TS 29.208, 3rd Generation Partnership Project (3GPP), June 2005.

[15] 3GPP. General Packet Radio Service (GPRS); Service description; Stage 2. TS 23.060, 3rd Generation Partnership Project (3GPP), June 2005.

[16] 3GPP. Internet Protocol (IP) multimedia call control protocol based on Session Initiation Protocol (SIP) and Session Description Protocol (SDP); Stage 3. TS 24.229, 3rd Generation Partnership Project (3GPP), June 2005.

[17] 3GPP. Interworking aspects and migration scenarios for IPv4-based IP Multimedia Subsystem (IMS) implementations. TR 23.981, 3rd Generation Partnership Project (3GPP), March 2005.

[18] 3GPP. Interworking between the IM CN subsystem and IP networks. TS 29.162, 3rd Generation Partnership Project (3GPP), June 2005.

[19] 3GPP. Interworking between the IP Multimedia (IM) Core Network (CN) subsystem and Circuit Switched (CS) networks. TS 29.163, 3rd Generation Partnership Project (3GPP), June 2005.

[20] 3GPP. IP Multimedia (IM) session handling; IM call model; Stage 2. TS 23.218, 3rd Generation Partnership Project (3GPP), April 2005.

[21] 3GPP. IP Multimedia (IM) Subsystem Cx and Dx Interfaces; Signalling flows and message contents. TS 29.228, 3rd Generation Partnership Project (3GPP), June 2005.

[22] 3GPP. IP Multimedia Subsystem (IMS) Sh interface; Signalling flows and message contents. TS 29.328, 3rd Generation Partnership Project (3GPP), June 2005.

[23] 3GPP. IP Multimedia Subsystem (IMS); Stage 2. TS 23.228, 3rd Generation Partnership Project (3GPP), June 2005.

[24] 3GPP. Media Gateway Control Function (MGCF) – IM Media Gateway (IM-MGW) Mn interface. TS 29.332, 3rd Generation Partnership Project (3GPP), June 2005.

[25] 3GPP. Messaging using the IP Multimedia (IM) Core Network (CN) subsystem; Stage 3. TS 24.247, 3rd Generation Partnership Project (3GPP), June 2005.

[26] 3GPP. Mobile Application Part (MAP) specification. TS 29.002, 3rd Generation Partnership Project (3GPP), June 2005.

[27] 3GPP. Mobile radio interface Layer 3 specification; Core network protocols; Stage 3. TS 24.008, 3rd Generation Partnership Project (3GPP), June 2005.

[28] 3GPP. Network architecture. TS 23.002, 3rd Generation Partnership Project (3GPP), June 2005.

[29] 3GPP. Open Service Access (OSA) Application Programming Interface (API); Part 1: Overview. TS 29.198-01, 3rd Generation Partnership Project (3GPP), January 2005.

[30] 3GPP. Packet switched conversational multimedia applications; Default codecs. TS 26.235, 3rd Generation Partnership Project (3GPP), April 2005.

[31] 3GPP. Policy control over Go interface. TS 29.207, 3rd Generation Partnership Project (3GPP), June 2005.

[32] 3GPP. Policy control over Gq interface. TS 29.209, 3rd Generation Partnership Project (3GPP), June 2005.

[33] 3GPP. Presence service; Architecture and functional description; Stage 2. TS 23.141, 3rd Generation Partnership Project (3GPP), June 2005.

[34] 3GPP. Service requirements for the Internet Protocol (IP) multimedia core network subsystem (IMS); Stage 1. TS 22.228, 3rd Generation Partnership Project (3GPP), March 2005.

[35] 3GPP. Sh interface based on the Diameter protocol; Protocol details. TS 29.329, 3rd Generation Partnership Project (3GPP), June 2005.

[36] 3GPP. Specification of the Subscriber Identity Module – Mobile Equipment (SIM-ME) interface. TS 51.011, 3rd Generation Partnership Project (3GPP), June 2005.

[37] 3GPP. Specification of the Subscriber Identity Module - Mobile Equipment (SIM-ME) Interface. TS 11.11, 3rd Generation Partnership Project (3GPP), June 2005.

[38] 3GPP. Speech codec speech processing functions; Adaptive Multi-Rate – Wideband (AMR-WB) speech codec; General description. TS 26.171, 3rd Generation Partnership Project (3GPP), January 2005.

[39] 3GPP. Speech codec speech processing functions; Adaptive Multi-Rate - Wideband (AMR-WB) speech codec; Transcoding functions. TS 26.190, 3rd Generation Partnership Project (3GPP), June 2005.

[40] 3GPP. Telecommunication management; Charging management; Charging architecture and principles. TS 32.240, 3rd Generation Partnership Project (3GPP), June 2005.

[41] 3GPP. Telecommunication management; Charging management; Charging data description for the IP Multimedia Subsystem (IMS). TS 32.225, 3rd Generation Partnership Project (3GPP), March 2005.

[42] 3GPP. Telecommunication management; Charging management; Charging principles. TS 32.200, 3rd Generation Partnership Project (3GPP), March 2005.

[43] 3GPP. Telecommunication management; Charging management; IP Multimedia Subsystem (IMS) charging. TS 32.260, 3rd Generation Partnership Project (3GPP), June 2005.

[44] 3GPP2. Multi-Media Domain Overview. TS X.S0013-000, 3rd Generation Partnership Project 2 (3GPP2).

[45] B. Aboba and M. Beadles. The Network Access Identifier. RFC 2486, Internet Engineering Task Force, January 1999.

[46] A. Allen. The P-Answer-State Header Extension to the Session Initiation Protocol (SIP) for the Open Mobile Alliance (OMA) Push to talk over Cellular (PoC). Internet-Draft draft-allen-sipping-poc-p-answer-state-header-00, Internet Engineering Task Force, June 2005. Work in progress.

[47] M. Arango, A. Dugan, I. Elliott, C. Huitema, and S. Pickett. Media Gateway Control Protocol (MGCP) Version 1.0. RFC 2705, Internet Engineering Task Force, October 1999.

[48] J. Arkko. Key Management Extensions for Session Description Protocol (SDP) and Real Time Streaming Protocol (RTSP). Internet-Draft draft-ietf-mmusic-kmgmt-ext-15, Internet Engineering Task Force, June 2005. Work in progress.

[49] J. Arkko, E. Carrara, F. Lindholm, M. Naslund, and K. Norrman. MIKEY: Multimedia Internet KEYing. RFC 3830, Internet Engineering Task Force, August 2004.

[50] J. Arkko, G. Kuijpers, H. Soliman, J. Loughney, and J. Wiljakka. Internet Protocol Version 6 (IPv6) for Some Second and Third Generation Cellular Hosts. RFC 3316, Internet Engineering Task Force, April 2003.

[51] J. Arkko, V. Torvinen, G. Camarillo, A. Niemi, and T. Haukka. Security Mechanism Agreement for the Session Initiation Protocol (SIP). RFC 3329, Internet Engineering Task Force, January 2003.

[52] M. Barnes. An Extension to the Session Initiation Protocol for Request History Information. Internet-Draft draft-ietf-sip-history-info-06, Internet Engineering Task Force, January 2005. Work in progress.

[53] M. Baugher, D. McGrew, M. Naslund, E. Carrara, and K. Norrman. The Secure Real-time Transport Protocol (SRTP). RFC 3711, Internet Engineering Task Force, March 2004.

[54] T. Berners-Lee, R. Fielding, and L. Masinter. Uniform Resource Identifiers (URI): Generic Syntax. RFC 2396, Internet Engineering Task Force, August 1998.

[55] S. Blake, D. Black, M. Carlson, E. Davies, Z. Wang, and W. Weiss. An Architecture for Differentiated Service. RFC 2475, Internet Engineering Task Force, December 1998.

[56] R. Braden, D. Clark, and S. Shenker. Integrated Services in the Internet Architecture: an Overview. RFC 1633, Internet Engineering Task Force, June 1994.

[57] R. Braden, L. Zhang, S. Berson, S. Herzog, and S. Jamin. Resource ReSerVation Protocol (RSVP) – Version 1 Functional Specification. RFC 2205, Internet Engineering Task Force, September 1997.

[58] S. Bradner, P. Calhoun, H. Cuschieri, S. Dennett, G. Flynn, M. Lipford, and M. McPheters. 3GPP2-IETF Standardization Collaboration. RFC 3131, Internet Engineering Task Force, June 2001.

[59] E. Burger. A Mechanism for Content Indirection in Session Initiation Protocol (SIP) Messages. Internet-Draft draft-ietf-sip-content-indirect-mech-05, Internet Engineering Task Force, October 2004. Work in progress.

[60] P. Calhoun, J. Loughney, E. Guttman, G. Zorn, and J. Arkko. Diameter Base Protocol. RFC 3588, Internet Engineering Task Force, September 2003.

[61] G. Camarillo. *SIP Demystified*. McGraw-Hill, 2001.

[62] G. Camarillo. Compressing the Session Initiation Protocol (SIP). RFC 3486, Internet Engineering Task Force, February 2003.

[63] G. Camarillo. Refering to Multiple Resources in the Session Initiation Protocol (SIP). Internet-Draft draft-ietf-sipping-multiple-refer-03, Internet Engineering Task Force, April 2005. Work in progress.

[64] G. Camarillo. The Binary Floor Control Protocol (BFCP). Internet-Draft draft-ietf-xcon-bfcp-04, Internet Engineering Task Force, May 2005. Work in progress.

[65] G. Camarillo, G. Eriksson, J. Holler, and H. Schulzrinne. Grouping of Media Lines in the Session Description Protocol (SDP). RFC 3388, Internet Engineering Task Force, December 2002.

[66] G. Camarillo and A. Johnston. Conference Establishment Using Request-Contained Lists in the Session Initiation Protocol (SIP). Internet-Draft draft-ietf-sipping-uri-list-conferencing-03, Internet Engineering Task Force, April 2005. Work in progress.

[67] G. Camarillo, W. Marshall, and J. Rosenberg. Integration of Resource Management and Session Initiation Protocol (SIP). RFC 3312, Internet Engineering Task Force, October 2002.

[68] G. Camarillo and A. Monrad. Mapping of Media Streams to Resource Reservation Flows. RFC 3524, Internet Engineering Task Force, April 2003.

[69] G. Camarillo and A. Roach. Requirements and Framework for Session Initiation Protocol (SIP) Uniform Resource Identifier (URI)-List Services. Internet-Draft draft-ietf-sipping-uri-services-03, Internet Engineering Task Force, April 2005. Work in progress.

[70] G. Camarillo and A. Roach. Subscriptions to Request-Contained Resource Lists in the Session Initiation Protocol (SIP). Internet-Draft draft-ietf-sipping-uri-list-subscribe-03, Internet Engineering Task Force, April 2005. Work in progress.

[71] G. Camarillo, A. B. Roach, J. Peterson, and L. Ong. Integrated Services Digital Network (ISDN) User Part (ISUP) to Session Initiation Protocol (SIP) Mapping. RFC 3398, Internet Engineering Task Force, December 2002.

[72] G. Camarillo, A. B. Roach, J. Peterson, and L. Ong. Mapping of Integrated Services Digital Network (ISDN) User Part (ISUP) Overlap Signalling to the Session Initiation Protocol (SIP). RFC 3578, Internet Engineering Task Force, August 2003.

[73] G. Camarillo and J. Rosenberg. The Alternative Network Address Types (ANAT) Semantics for the Session Description Protocol (SDP) Grouping Framework. RFC 4091, Internet Engineering Task Force, June 2005.

[74] G. Camarillo, H. Schulzrinne, and R. Kantola. Evaluation of Transport Protocols for the Session Initiation Protocol. *IEEE Network*, 17(5), 2003.

[75] B. Campbell. The Message Session Relay Protocol. Internet-Draft draft-ietf-simple-message-sessions-10, Internet Engineering Task Force, February 2005. Work in progress.

[76] B. Campbell, J. Rosenberg, H. Schulzrinne, C. Huitema, and D. Gurle. Session Initiation Protocol (SIP) Extension for Instant Messaging. RFC 3428, Internet Engineering Task Force, December 2002.

[77] K. Chan, R. Sahita, S. Hahn, and K. McCloghrie. Differentiated Services Quality of Service Policy Information Base. RFC 3317, Internet Engineering Task Force, March 2003.

[78] K. Chan, J. Seligson, D. Durham, S. Gai, K. McCloghrie, S. Herzog, F. Reichmeyer, R. Yavatkar, and A. Smith. COPS Usage for Policy Provisioning (COPS-PR). RFC 3084, Internet Engineering Task Force, March 2001.

[79] B. Davie, A. Charny, J.C.R. Bennet, K. Benson, J.Y. Le Boudec, W. Courtney, S. Davari, V. Firoiu, and D. Stiliadis. An Expedited Forwarding PHB (Per-Hop Behavior). RFC 3246, Internet Engineering Task Force, March 2002.

[80] F. Dawson and T. Howes. vCard MIME Directory Profile. RFC 2426, Internet Engineering Task Force, September 1998.

[81] S. Deering and R. Hinden. Internet Protocol, Version 6 (IPv6) Specification. RFC 2460, Internet Engineering Task Force, December 1998.

[82] T. Dierks and C. Allen. The TLS Protocol Version 1.0. RFC 2246, Internet Engineering Task Force, January 1999.

[83] R. Droms. Dynamic Host Configuration Protocol. RFC 2131, Internet Engineering Task Force, March 1997.

[84] R. Droms, J. Bound, B. Volz, T. Lemon, C. Perkins, and M. Carney. Dynamic Host Configuration Protocol for IPv6 (DHCPv6). RFC 3315, Internet Engineering Task Force, July 2003.

[85] D. Durham, J. Boyle, R. Cohen, S. Herzog, R. Rajan, and A. Sastry. The COPS (Common Open Policy Service) Protocol. RFC 2748, Internet Engineering Task Force, January 2000.

[86] D. Eastlake 3rd and P. Jones. US Secure Hash Algorithm 1 (SHA1). RFC 3174, Internet Engineering Task Force, September 2001.

[87] J. Elwell. SIP Reason header extension for indicating redirection reasons. Internet-Draft draft-elwell-sipping-redirection-reason-02, Internet Engineering Task Force, June 2005. Work in progress.

[88] ETSI. Digital cellular telecommunications system (Phase 2+); Full rate speech; Transcoding (GSM 06.10 version 5.1.1). ETS 300 961, European Telecommunications Standards Institute, May 1998.

[89] ETSI. Telecommunications and Internet Converged Services and Protocols for Advanced Networking (TISPAN); NGN Signalling Control Protocol; Advice of Charge (AoC); PSTN/ISDN Simulation Services. TS 183 012, European Telecommunications Standards Institute, July 2005.

[90] ETSI. Telecommunications and Internet Converged Services and Protocols for Advanced Networking (TISPAN); NGN Signalling Control Protocol; Anonymous Communication Rejection (ACR) and Communication Barring (CB); PSTN/ISDN Simulation Services. TS 183 011, European Telecommunications Standards Institute, July 2005.

[91] ETSI. Telecommunications and Internet Converged Services and Protocols for Advanced Networking (TISPAN); NGN Signalling Control Protocol; Communication Diversion (CDIV); PSTN/ISDN Simulation Services. TS 183 004, European Telecommunications Standards Institute, July 2005.

[92] ETSI. Telecommunications and Internet Converged Services and Protocols for Advanced Networking (TISPAN); NGN Signalling Control Protocol; Communication Hold (HOLD); PSTN/ISDN Simulation Services. TS 183 010, European Telecommunications Standards Institute, July 2005.

[93] ETSI. Telecommunications and Internet Converged Services and Protocols for Advanced Networking (TISPAN); NGN Signalling Control Protocol; Communication Waiting (CW); PSTN/ISDN Simulation Services. TS 183 009, European Telecommunications Standards Institute, July 2005.

[94] ETSI. Telecommunications and Internet Converged Services and Protocols for Advanced Networking (TISPAN); NGN Signalling Control Protocol; Completion of Communications to Busy Subscriber (CCBS); PSTN/ISDN Simulation Services; Completion of Communications by No Reply (CCNR); PSTN/ISDN Simulation Services. TS 183 015, European Telecommunications Standards Institute, July 2005.

[95] ETSI. Telecommunications and Internet Converged Services and Protocols for Advanced Networking (TISPAN); NGN Signalling Control Protocol; Conference (CONF); PSTN/ISDN Simulation Services. TS 183 005, European Telecommunications Standards Institute, July 2005.

[96] ETSI. Telecommunications and Internet Converged Services and Protocols for Advanced Networking (TISPAN); NGN Signalling Control Protocol; Malicious Call Identification (MCID); PSTN/ISDN simulation services. TS 183 016, European Telecommunications Standards Institute, July 2005.

[97] ETSI. Telecommunications and Internet Converged Services and Protocols for Advanced Networking (TISPAN); NGN Signalling Control Protocol; Message Waiting Indication (MWI); PSTN/ISDN Simulation Services. TS 183 006, European Telecommunications Standards Institute, July 2005.

[98] ETSI. Telecommunications and Internet Converged Services and Protocols for Advanced Networking (TISPAN); NGN Signalling Control Protocol; Originating Identification Presentation (OIP) and Originating Identification Restriction (OIR); PSTN/ISDN Simulation Services. TS 183 007, European Telecommunications Standards Institute, July 2005.

[99] ETSI. TISPAN NGN Release 1; PSTN/ISDN Simulation Services; Extensible Markup Language (XML) Configuration Access Protocol (XCAP) over the Ut interface for Manipulating NGN PSTN/ISDN Simulation Services. TS 183 023, European Telecommunications Standards Institute, July 2005.

[100] P. Faltstrom. E.164 number and DNS. RFC 2916, Internet Engineering Task Force, September 2000.

[101] R. Fielding, J. Gettys, J. Mogul, H. Frystyk, L. Masinter, P. Leach, and T. Berners-Lee. Hypertext Transfer Protocol – HTTP/1.1. RFC 2616, Internet Engineering Task Force, June 1999.

[102] J. Franks, P. Hallam-Baker, J. Hostetler, S. Lawrence, P. Leach, A. Luotonen, and L. Stewart. HTTP Authentication: Basic and Digest Access Authentication. RFC 2617, Internet Engineering Task Force, June 1999.

[103] N. Freed and N. Borenstein. Multipurpose Internet Mail Extensions (MIME) Part One: Format of Internet Message Bodies. RFC 2045, Internet Engineering Task Force, November 1996.

[104] J. Galvin, S. Murphy, S. Crocker, and N. Freed. Security Multiparts for MIME: Multipart/Signed and Multipart/Encrypted. RFC 1847, Internet Engineering Task Force, October 1995.

[105] M. García-Martín. A Session Initiation Protocol (SIP) Event Package and Data Format for Various Settings in Support for the Push-to-talk Over Cellular (PoC) Service. Internet-Draft draft-garcia-sipping-poc-isb-am-03, Internet Engineering Task Force, July 2005. Work in progress.

[106] M. García-Martín. Diameter Session Initiation Protocol (SIP) Application. Internet-Draft draft-ietf-aaa-diameter-sip-app-07, Internet Engineering Task Force, March 2005. Work in progress.

[107] M. García-Martín, C. Bormann, J. Ott, R. Price, and A. B. Roach. The Session Initiation Protocol (SIP) and Session Description Protocol (SDP) Static Dictionary for Signaling Compression (SigComp). RFC 3485, Internet Engineering Task Force, February 2003.

[108] M. García-Martín and G. Camarillo. Multiple-Recipient MESSAGE Requests in the Session Initiation Protocol (SIP). Internet-Draft draft-ietf-sipping-uri-list-message-03, Internet Engineering Task Force, April 2005. Work in progress.

[109] M. García-Martín, E. Henrikson, and D. Mills. Private Header (P-Header) Extensions to the Session Initiation Protocol (SIP) for the 3rd-Generation Partnership Project (3GPP). RFC 3455, Internet Engineering Task Force, January 2003.

[110] G. Good. The LDAP Data Interchange Format (LDIF) – Technical Specification. RFC 2849, Internet Engineering Task Force, June 2000.

[111] D. Grossman. New Terminology and Clarifications for Diffserv. RFC 3260, Internet Engineering Task Force, April 2002.

[112] L.-N. Hamer, B. Gage, B. Kosinski, and H. Shieh. Session Authorization Policy Element. RFC 3520, Internet Engineering Task Force, April 2003.

[113] L.-N. Hamer, B. Gage, and H. Shieh. Framework for Session Set-up with Media Authorization. RFC 3521, Internet Engineering Task Force, April 2003.

[114] M. Handley, S. Floyd, J. Padhye, and J. Widmer. TCP Friendly Rate Control (TFRC): Protocol Specification. RFC 3448, Internet Engineering Task Force, January 2003.

[115] M. Handley and V. Jacobson. SDP: Session Description Protocol. RFC 2327, Internet Engineering Task Force, April 1998.

[116] M. Handley, H. Schulzrinne, E. Schooler, and J. Rosenberg. SIP: Session Initiation Protocol. RFC 2543, Internet Engineering Task Force, March 1999.

[117] H. Hannu, J. Christoffersson, S. Forsgren, K.-C. Leung, Z. Liu, and R. Price. Signaling Compression (SigComp) – Extended Operations. RFC 3321, Internet Engineering Task Force, January 2003.

[118] D. Harkins and D. Carrel. The Internet Key Exchange (IKE). RFC 2409, Internet Engineering Task Force, November 1998.

[119] J. Heinanen, F. Baker, W. Weiss, and J. Wroclawski. Assured Forwarding PHB Group. RFC 2597, Internet Engineering Task Force, June 1999.

[120] S. Herzog, J. Boyle, R. Cohen, D. Durham, R. Rajan, and A. Sastry. COPS usage for RSVP. RFC 2749, Internet Engineering Task Force, January 2000.

[121] V. Hilt. A Delivery Mechanism for Session-Specific Session Initiation Protocol (SIP) Session Policies. Internet-Draft draft-hilt-sipping-session-spec-policy-03, Internet Engineering Task Force, July 2005. Work in progress.

[122] V. Hilt, G. Camarillo, and J. Rosenberg. Session Initiation Protocol (SIP) Session Policies – Document Format and Session-Independent Delivery Mechanism. Internet-Draft draft-ietf-sipping-session-indep-policy-03, Internet Engineering Task Force, July 2005. Work in progress.

[123] D. Hoffman, G. Fernando, V. Goyal, and M. Civanlar. RTP Payload Format for MPEG1/MPEG2 Video. RFC 2250, Internet Engineering Task Force, January 1998.

[124] R. Housley. Cryptographic Message Syntax (CMS). RFC 3369, Internet Engineering Task Force, August 2002.

[125] G. Huston and I. Leuca. OMA-IETF Standardization Collaboration. RFC 3975, Internet Engineering Task Force, January 2005.

[126] ISO. Coding of moving pictures and associated audio for digital storage media up to about 1,5 Mbits/s. Standard ISO/IEC 11172, International Organization for Standardization, November 1993.

[127] ISO. Generic coding of moving pictures and associated audio information. Standard ISO/IEC 13818, International Organization for Standardization, November 1994.

[128] ISO. Information technology – Coding of audio-visual objects – Part 1: Systems. Standard ISO/IEC 14496-1, International Organization for Standardization, June 2001.

[129] ISO. Information technology – Coding of audio-visual objects – Part 2: Visual. Standard ISO/IEC 14496-2, International Organization for Standardization, June 2001.

[130] ITU-T. Functional description of the Signalling System No. 7 Telephone User Part (TUP). Recommendation Q.721, International Telecommunication Union, November 1988.

[131] ITU-T. Pulse code modulation (PCM) of voice frequencies. Recommendation G.711, International Telecommunication Union, November 1988.

[132] ITU-T. 40, 32, 24, 16 kbit/s adaptive differential pulse code modulation (ADPCM). Recommendation G.726, International Telecommunication Union, December 1990.

[133] ITU-T. Functional description of the message transfer part (MTP) of Signalling System No. 7. Recommendation Q.701, International Telecommunication Union, March 1993.

[134] ITU-T. Introduction to CCITT Signalling System No. 7. Recommendation Q.700, International Telecommunication Union, March 1993.

[135] ITU-T. Video codec for audiovisual services at p x 64 kbit/s. Recommendation H.261, International Telecommunication Union, March 1993.

[136] ITU-T. Protocol for multimedia application text conversation. Recommendation T.140, International Telecommunication Union, February 1998.

[137] ITU-T. Video coding for low bit rate communication. Recommendation H.263, International Telecommunication Union, February 1998.

[138] ITU-T. Narrow-band visual telephone systems and terminal equipment. Recommendation H.320, International Telecommunication Union, May 1999.

[139] ITU-T. Signalling System No. 7 – ISDN User Part functional description. Recommendation Q.761, International Telecommunication Union, December 1999.

[140] ITU-T. Bearer Independent Call Control protocol. Recommendation Q.1901, International Telecommunication Union, June 2000.

[141] ITU-T. Information technology – Open Systems Interconnection – The Directory: Public-key and attribute certificate frameworks. Recommendation X.509, International Telecommunication Union, March 2000.

[142] ITU-T. H.263 Annex X: Profiles and levels definitions. Recommendation H.263 Annex X, International Telecommunication Union, April 2001.

[143] ITU-T. Gateway control protocol: Version 2. Recommendation H.248, International Telecommunication Union, May 2002.

[144] ITU-T. Terminal for low bit-rate multimedia communication. Recommendation H.324, International Telecommunication Union, March 2002.

[145] ITU-T. Packet-based multimedia communication systems. Recommendation H.323, International Telecommunication Union, July 2003.

[146] C. Jennings and R. Mahy. Relay Extensions for the Message Sessions Relay Protocol (MSRP). Internet-Draft draft-ietf-simple-msrp-relays-04, Internet Engineering Task Force, June 2005. Work in progress.

[147] C. Jennings, J. Peterson, and M. Watson. Private Extensions to the Session Initiation Protocol (SIP) for Asserted Identity within Trusted Networks. RFC 3325, Internet Engineering Task Force, November 2002.

[148] S. Josefsson. The Base16, Base32, and Base64 Data Encodings. RFC 3548, Internet Engineering Task Force, July 2003.

[149] S. Kent and R. Atkinson. IP Encapsulating Security Payload (ESP). RFC 2406, Internet Engineering Task Force, November 1998.

[150] S. Kent and R. Atkinson. Security Architecture for the Internet Protocol. RFC 2401, Internet Engineering Task Force, November 1998.

[151] J. Klensin. Simple Mail Transfer Protocol. RFC 2821, Internet Engineering Task Force, April 2001.

[152] G. Klyne and D. Atkins. Common Presence and Instant Messaging (CPIM): Message Format. RFC 3862, Internet Engineering Task Force, August 2004.

[153] E. Kohler. Datagram Congestion Control Protocol (DCCP). Internet-Draft draft-ietf-dccp-spec-11, Internet Engineering Task Force, March 2005. Work in progress.

[154] O. Levin. Suppression of Session Initiation Protocol REFER Method Implicit Subscription. Internet-Draft draft-ietf-sip-refer-with-norefersub-01, Internet Engineering Task Force, February 2005. Work in progress.

[155] M. Lonnfors and K. Kiss. User Agent Capability Extension to Presence Information Data Format (PIDF). Internet-Draft draft-ietf-simple-prescaps-ext-04, Internet Engineering Task Force, June 2005. Work in progress.

[156] J. Loughney. Diameter Command Codes for Third Generation Partnership Project (3GPP) Release 5. RFC 3589, Internet Engineering Task Force, September 2003.

[157] R. Ludwig and M. Meyer. The Eifel Detection Algorithm for TCP. RFC 3522, Internet Engineering Task Force, April 2003.

[158] R. Mahy. A Message Summary and Message Waiting Indication Event Package for the Session Initiation Protocol (SIP). RFC 3842, Internet Engineering Task Force, August 2004.

[159] R. Mahy, B. Biggs, and R. Dean. The Session Initiation Protocol (SIP) Replaces Header. RFC 3891, Internet Engineering Task Force, September 2004.

[160] A. Mankin, S. Bradner, R. Mahy, D. Willis, J. Ott, and B. Rosen. Change Process for the Session Initiation Protocol (SIP). RFC 3427, Internet Engineering Task Force, December 2002.

[161] W. Marshall. Private Session Initiation Protocol (SIP) Extensions for Media Authorization. RFC 3313, Internet Engineering Task Force, January 2003.

[162] L. Mattila, J. Koskinen, M. Stura, J. Loughney, and H. Hakala. Diameter Credit-control Application. Internet-Draft draft-ietf-aaa-diameter-cc-06, Internet Engineering Task Force, August 2004. Work in progress.

[163] D. Maughan, M. Schertler, M. Schneider, and J. Turner. Internet Security Association and Key Management Protocol (ISAKMP). RFC 2408, Internet Engineering Task Force, November 1998.

[164] K. McCloghrie, M. Fine, J. Seligson, K. Chan, S. Hahn, R. Sahita, A. Smith, and F. Reichmeyer. Structure of Policy Provisioning Information (SPPI). RFC 3159, Internet Engineering Task Force, August 2001.

[165] R. Moats. URN Syntax. RFC 2141, Internet Engineering Task Force, May 1997.

[166] P.V. Mockapetris. Domain names – concepts and facilities. RFC 1034, Internet Engineering Task Force, November 1987.

[167] A. Niemi. Session Initiation Protocol (SIP) Extension for Event State Publication. RFC 3903, Internet Engineering Task Force, October 2004.

[168] A. Niemi, J. Arkko, and V. Torvinen. Hypertext Transfer Protocol (HTTP) Digest Authentication Using Authentication and Key Agreement (AKA). RFC 3310, Internet Engineering Task Force, September 2002.

[169] Open Mobile Alliance. http://www.openmobilealliance.org.

[170] Open Mobile Alliance. OMA Provisioning Architecture Overview version 1.1. TS, Open Mobile Alliance, November 2002.

[171] Open Mobile Alliance. Enabler Release Definition for IMS in OMA Version 1.0. TS, Open Mobile Alliance, February 2005.

[172] Open Mobile Alliance. Enabler Release Definition for Push to Talk Over Cellular Version 1.0. TS, Open Mobile Alliance, April 2005.

[173] Open Mobile Alliance. IMS in OMA Version 1.0. Candidate Enabler Release, Open Mobile Alliance, February 2005.

[174] Open Mobile Alliance. OMA Device Management Protocol 1.2. TS, Open Mobile Alliance, March 2005.

[175] Open Mobile Alliance. Push to Talk Over Cellular Version 1.0. Candidate Enabler Release, Open Mobile Alliance, May 2005.

[176] Open Mobile Alliance. Push to Talk Over Cellular Version 1.0 – Architecture. TS, Open Mobile Alliance, April 2005.

[177] Open Mobile Alliance. Push to Talk Over Cellular Version 1.0 – Control Plane Specification. TS, Open Mobile Alliance, April 2005.

[178] Open Mobile Alliance. Push to Talk Over Cellular Version 1.0 – Requirements. TS, Open Mobile Alliance, March 2005.

[179] Open Mobile Alliance. Push to Talk Over Cellular Version 1.0 – User Plane Version. TS, Open Mobile Alliance, April 2005.

[180] Open Mobile Alliance. Push to Talk Over Cellular Version 1.0 – XDM Specification. TS, Open Mobile Alliance, April 2005.

[181] Open Mobile Alliance. Utilization of IMS Capabilities Version 1.0 – Architecture. TS, Open Mobile Alliance, February 2005.

[182] Open Mobile Alliance. Utilization of IMS Capabilities Version 1.0 – Requirements. TS, Open Mobile Alliance, February 2005.

[183] J. Peterson. A Privacy Mechanism for the Session Initiation Protocol (SIP). RFC 3323, Internet Engineering Task Force, November 2002.

[184] J. Peterson. Common Profile for Presence (CPP). RFC 3859, Internet Engineering Task Force, August 2004.

[185] J. Peterson. Session Initiation Protocol (SIP) Authenticated Identity Body (AIB) Format. RFC 3893, Internet Engineering Task Force, September 2004.

[186] J. Peterson and C. Jennings. Enhancements for Authenticated Identity Management in the Session Initiation Protocol (SIP). Internet-Draft draft-ietf-sip-identity-05, Internet Engineering Task Force, May 2005. Work in progress.

[187] D. Piper. The Internet IP Security Domain of Interpretation for ISAKMP. RFC 2407, Internet Engineering Task Force, November 1998.

[188] J. Postel. User Datagram Protocol. RFC 0768, Internet Engineering Task Force, August 1980.

[189] J. Postel. Internet Protocol. RFC 0791, Internet Engineering Task Force, September 1981.

[190] J. Postel. Transmission Control Protocol. RFC 0793, Internet Engineering Task Force, September 1981.

[191] R. Price, C. Bormann, J. Christoffersson, H. Hannu, Z. Liu, and J. Rosenberg. Signaling Compression (SigComp). RFC 3320, Internet Engineering Task Force, January 2003.

[192] B. Ramsdell. S/MIME Version 3 Message Specification. RFC 2633, Internet Engineering Task Force, June 1999.

[193] C. Rigney, A. Rubens, W. Simpson, and S. Willens. Remote Authentication Dial In User Service (RADIUS). RFC 2058, Internet Engineering Task Force, January 1997.

[194] C. Rigney, A. Rubens, W. Simpson, and S. Willens. Remote Authentication Dial In User Service (RADIUS). RFC 2138, Internet Engineering Task Force, April 1997.

[195] C. Rigney, S. Willens, A. Rubens, and W. Simpson. Remote Authentication Dial In User Service (RADIUS). RFC 2865, Internet Engineering Task Force, June 2000.

[196] R. Rivest. The MD5 Message-Digest Algorithm. RFC 1321, Internet Engineering Task Force, April 1992.

[197] A. Roach, J. Rosenberg, and B. Campbell. A Session Initiation Protocol (SIP) Event Notification Extension for Resource Lists. Internet-Draft draft-ietf-simple-event-list-07, Internet Engineering Task Force, January 2005. Work in progress.

[198] A.B. Roach. Session Initiation Protocol (SIP) – Specific Event Notification. RFC 3265, Internet Engineering Task Force, June 2002.

[199] J. Rosenberg. The Session Initiation Protocol (SIP) UPDATE Method. RFC 3311, Internet Engineering Task Force, October 2002.

[200] J. Rosenberg. A Presence Event Package for the Session Initiation Protocol (SIP). RFC 3856, Internet Engineering Task Force, August 2004.

[201] J. Rosenberg. A Session Initiation Protocol (SIP) Event Package for Registrations. RFC 3680, Internet Engineering Task Force, March 2004.

[202] J. Rosenberg. A Watcher Information Event Template-Package for the Session Initiation Protocol (SIP). RFC 3857, Internet Engineering Task Force, August 2004.

[203] J. Rosenberg. A Data Model for Presence. Internet-Draft draft-ietf-simple-presence-data-model-02, Internet Engineering Task Force, February 2005. Work in progress.

[204] J. Rosenberg. A Framework for Consent-Based Communications in the Session Initiation Protocol (SIP). Internet-Draft draft-ietf-sipping-consent-framework-01, Internet Engineering Task Force, February 2005. Work in progress.

[205] J. Rosenberg. A Session Initiation Protocol (SIP) Event Package for Conference State. Internet-Draft draft-ietf-sipping-conference-package-12, Internet Engineering Task Force, July 2005. Work in progress.

[206] J. Rosenberg. An INVITE Inititiated Dialog Event Package for the Session Initiation Protocol (SIP). Internet-Draft draft-ietf-sipping-dialog-package-06, Internet Engineering Task Force, April 2005. Work in progress.

[207] J. Rosenberg. Extensible Markup Language (XML) Formats for Representing Resource Lists. Internet-Draft draft-ietf-simple-xcap-list-usage-05, Internet Engineering Task Force, February 2005. Work in progress.

[208] J. Rosenberg. Interactive Connectivity Establishment (ICE): A Methodology for Network Address Translator (NAT) Traversal for Multimedia Session Establishment Protocols. Internet-Draft draft-ietf-mmusic-ice-04, Internet Engineering Task Force, February 2005. Work in progress.

[209] J. Rosenberg. Presence Authorization Rules. Internet-Draft draft-ietf-simple-presence-rules-02, Internet Engineering Task Force, February 2005. Work in progress.

[210] J. Rosenberg. The Extensible Markup Language (XML) Configuration Access Protocol (XCAP). Internet-Draft draft-ietf-simple-xcap-07, Internet Engineering Task Force, June 2005. Work in progress.

[211] J. Rosenberg. Traversal Using Relay NAT (TURN). Internet-Draft draft-rosenberg-midcom-turn-07, Internet Engineering Task Force, February 2005. Work in progress.

[212] J. Rosenberg and H. Schulzrinne. An Offer/Answer Model with Session Description Protocol (SDP). RFC 3264, Internet Engineering Task Force, June 2002.

[213] J. Rosenberg and H. Schulzrinne. Reliability of Provisional Responses in Session Initiation Protocol (SIP). RFC 3262, Internet Engineering Task Force, June 2002.

[214] J. Rosenberg and H. Schulzrinne. Session Initiation Protocol (SIP): Locating SIP Servers. RFC 3263, Internet Engineering Task Force, June 2002.

[215] J. Rosenberg, H. Schulzrinne, G. Camarillo, A. Johnston, J. Peterson, R. Sparks, M. Handley, and E. Schooler. SIP: Session Initiation Protocol. RFC 3261, Internet Engineering Task Force, June 2002.

[216] J. Rosenberg, H. Schulzrinne, and P. Kyzivat. Caller Preferences for the Session Initiation Protocol (SIP). RFC 3841, Internet Engineering Task Force, August 2004.

[217] J. Rosenberg, H. Schulzrinne, and P. Kyzivat. Indicating User Agent Capabilities in the Session Initiation Protocol (SIP). RFC 3840, Internet Engineering Task Force, August 2004.

[218] J. Rosenberg, J. Weinberger, C. Huitema, and R. Mahy. STUN – Simple Traversal of User Datagram Protocol (UDP) Through Network Address Translators (NATs). RFC 3489, Internet Engineering Task Force, March 2003.

[219] R. Sahita, S. Hahn, K. Chan, and K. McCloghrie. Framework Policy Information Base. RFC 3318, Internet Engineering Task Force, March 2003.

[220] H. Schulzrinne. The tel URI for Telephone Numbers. RFC 3966, Internet Engineering Task Force, December 2004.

[221] H. Schulzrinne. CIPID: Contact Information in Presence Information Data Format. Internet-Draft draft-ietf-simple-cipid-05, Internet Engineering Task Force, June 2005. Work in progress.

[222] H. Schulzrinne. RPID: Rich Presence Extensions to the Presence Information Data Format (PIDF). Internet-Draft draft-ietf-simple-rpid-07, Internet Engineering Task Force, June 2005. Work in progress.

[223] H. Schulzrinne. Timed Presence Extensions to the Presence Information Data Format (PIDF) to Indicate Status Information for Past and Future Time Intervals. Internet-Draft draft-ietf-simple-future-04, Internet Engineering Task Force, June 2005. Work in progress.

[224] H. Schulzrinne and S. Casner. RTP Profile for Audio and Video Conferences with Minimal Control. RFC 3551, Internet Engineering Task Force, July 2003.

[225] H. Schulzrinne, S. Casner, R. Frederick, and V. Jacobson. RTP: A Transport Protocol for Real-Time Applications. RFC 3550, Internet Engineering Task Force, July 2003.

[226] H. Schulzrinne, D. Oran, and G. Camarillo. The Reason Header Field for the Session Initiation Protocol (SIP). RFC 3326, Internet Engineering Task Force, December 2002.

[227] H. Schulzrinne and S. Petrack. RTP Payload for DTMF Digits, Telephony Tones and Telephony Signals. RFC 2833, Internet Engineering Task Force, May 2000.

[228] H. Schulzrinne and B. Volz. Dynamic Host Configuration Protocol (DHCPv6) Options for Session Initiation Protocol (SIP) Servers. RFC 3319, Internet Engineering Task Force, July 2003.

[229] R. Sparks. The Session Initiation Protocol (SIP) Refer Method. RFC 3515, Internet Engineering Task Force, April 2003.

[230] R. Stewart, Q. Xie, K. Morneault, C. Sharp, H. Schwarzbauer, T. Taylor, I. Rytina, M. Kalla, L. Zhang, and V. Paxson. Stream Control Transmission Protocol. RFC 2960, Internet Engineering Task Force, October 2000.

[231] H. Sugano, S. Fujimoto, G. Klyne, A. Bateman, W. Carr, and J. Peterson. Presence Information Data Format (PIDF). RFC 3863, Internet Engineering Task Force, August 2004.

[232] A. Vemuri and J. Peterson. Session Initiation Protocol for Telephones (SIP-T): Context and Architectures. RFC 3372, Internet Engineering Task Force, September 2002.

[233] WAP Forum. WAP architecture. Recommendation WAP architecture, Wireless Application Protocol Forum, July 2001.

[234] D. Willis and A. Allen. Requesting Answering and Alerting Modes for the Session Initiation Protocol (SIP). Internet-Draft draft-willis-sip-answeralert-00, Internet Engineering Task Force, June 2005. Work in progress.

[235] D. Willis and B. Hoeneisen. Session Initiation Protocol (SIP) Extension Header Field for Registering Non-Adjacent Contacts. RFC 3327, Internet Engineering Task Force, December 2002.

[236] D. Willis and B. Hoeneisen. Session Initiation Protocol (SIP) Extension Header Field for Service Route Discovery During Registration. RFC 3608, Internet Engineering Task Force, October 2003.

[237] T. Ylonen and C. Lonvick. SSH Protocol Architecture. Internet-Draft draft-ietf-secsh-architecture-22, Internet Engineering Task Force, March 2005. Work in progress.

Index